IET ENERGY ENGINEERING 171

Utility-scale Wind Turbines and Wind Farms

Other volumes in this series:

Volume 1	**Power Circuit Breaker Theory and Design** C.H. Flurscheim (Editor)
Volume 4	**Industrial Microwave Heating** A.C. Metaxas and R.J. Meredith
Volume 7	**Insulators for High Voltages** J.S.T. Looms
Volume 8	**Variable Frequency AC Motor Drive Systems** D. Finney
Volume 10	**SF_6 Switchgear** H.M. Ryan and G.R. Jones
Volume 11	**Conduction and Induction Heating** E.J. Davies
Volume 13	**Statistical Techniques for High Voltage Engineering** W. Hauschild and W. Mosch
Volume 14	**Uninterruptible Power Supplies** J. Platts and J.D. St Aubyn (Editors)
Volume 15	**Digital Protection for Power Systems** A.T. Johns and S.K. Salman
Volume 16	**Electricity Economics and Planning** T.W. Berrie
Volume 18	**Vacuum Switchgear** A. Greenwood
Volume 19	**Electrical Safety: a guide to causes and prevention of hazards** J. Maxwell Adams
Volume 21	**Electricity Distribution Network Design, 2nd Edition** E. Lakervi and E.J. Holmes
Volume 22	**Artificial Intelligence Techniques in Power Systems** K. Warwick, A.O. Ekwue and R. Aggarwal (Editors)
Volume 24	**Power System Commissioning and Maintenance Practice** K. Harker
Volume 25	**Engineers' Handbook of Industrial Microwave Heating** R.J. Meredith
Volume 26	**Small Electric Motors** H. Moczala *et al.*
Volume 27	**AC-DC Power System Analysis** J. Arrillaga and B.C. Smith
Volume 29	**High Voltage Direct Current Transmission, 2nd Edition** J. Arrillaga
Volume 30	**Flexible AC Transmission Systems (FACTS)** Y-H. Song (Editor)
Volume 31	**Embedded generation** N. Jenkins *et al.*
Volume 32	**High Voltage Engineering and Testing, 2nd Edition** H.M. Ryan (Editor)
Volume 33	**Overvoltage Protection of Low-Voltage Systems, Revised Edition** P. Hasse
Volume 36	**Voltage Quality in Electrical Power Systems** J. Schlabbach *et al.*
Volume 37	**Electrical Steels for Rotating Machines** P. Beckley
Volume 38	**The Electric Car: Development and future of battery, hybrid and fuel-cell cars** M. Westbrook
Volume 39	**Power Systems Electromagnetic Transients Simulation** J. Arrillaga and N. Watson
Volume 40	**Advances in High Voltage Engineering** M. Haddad and D. Warne
Volume 41	**Electrical Operation of Electrostatic Precipitators** K. Parker
Volume 43	**Thermal Power Plant Simulation and Control** D. Flynn
Volume 44	**Economic Evaluation of Projects in the Electricity Supply Industry** H. Khatib
Volume 45	**Propulsion Systems for Hybrid Vehicles** J. Miller
Volume 46	**Distribution Switchgear** S. Stewart
Volume 47	**Protection of Electricity Distribution Networks, 2nd Edition** J. Gers and E. Holmes
Volume 48	**Wood Pole Overhead Lines** B. Wareing
Volume 49	**Electric Fuses, 3rd Edition** A. Wright and G. Newbery
Volume 50	**Wind Power Integration: Connection and system operational aspects** B. Fox *et al.*
Volume 51	**Short Circuit Currents** J. Schlabbach
Volume 52	**Nuclear Power** J. Wood
Volume 53	**Condition Assessment of High Voltage Insulation in Power System Equipment** R.E. James and Q. Su
Volume 55	**Local Energy: Distributed generation of heat and power** J. Wood
Volume 56	**Condition Monitoring of Rotating Electrical Machines** P. Tavner, L. Ran, J. Penman and H. Sedding
Volume 57	**The Control Techniques Drives and Controls Handbook, 2nd Edition** B. Drury
Volume 58	**Lightning Protection** V. Cooray (Editor)
Volume 59	**Ultracapacitor Applications** J.M. Miller
Volume 62	**Lightning Electromagnetics** V. Cooray
Volume 63	**Energy Storage for Power Systems, 2nd Edition** A. Ter-Gazarian
Volume 65	**Protection of Electricity Distribution Networks, 3rd Edition** J. Gers
Volume 66	**High Voltage Engineering Testing, 3rd Edition** H. Ryan (Editor)
Volume 67	**Multicore Simulation of Power System Transients** F.M. Uriate
Volume 68	**Distribution System Analysis and Automation** J. Gers
Volume 69	**The Lightening Flash, 2nd Edition** V. Cooray (Editor)
Volume 70	**Economic Evaluation of Projects in the Electricity Supply Industry, 3rd Edition** H. Khatib
Volume 72	**Control Circuits in Power Electronics: Practical issues in design and implementation** M. Castilla (Editor)
Volume 73	**Wide Area Monitoring, Protection and Control Systems: The enabler for Smarter Grids** A. Vaccaro and A. Zobaa (Editors)
Volume 74	**Power Electronic Converters and Systems: Frontiers and applications** A. M. Trzynadlowski (Editor)
Volume 75	**Power Distribution Automation** B. Das (Editor)
Volume 76	**Power System Stability: Modelling, analysis and control** B. Om P. Malik
Volume 78	**Numerical Analysis of Power System Transients and Dynamics** A. Ametani (Editor)
Volume 79	**Vehicle-to-Grid: Linking electric vehicles to the smart grid** J. Lu and J. Hossain (Editors)
Volume 81	**Cyber-Physical-Social Systems and Constructs in Electric Power Engineering** S. Suryanarayanan, R. Roche and T.M. Hansen (Editors)

Volume 82 **Periodic Control of Power Electronic Converters** F. Blaabjerg, K.Zhou, D. Wang and Y. Yang
Volume 86 **Advances in Power System Modelling, Control and Stability Analysis** F. Milano (Editor)
Volume 87 **Cogeneration: Technologies, Optimisation and Implentation** C. A. Frangopoulos (Editor)
Volume 88 **Smarter Energy: from Smart Metering to the Smart Grid** H. Sun, N. Hatziargyriou, H. V. Poor, L. Carpanini and M. A. Sánchez Fornié (Editors)
Volume 89 **Hydrogen Production, Separation and Purification for Energy** A. Basile, F. Dalena, J. Tong and T.N.Veziroğlu (Editors)
Volume 90 **Clean Energy Microgrids** S. Obara and J. Morel (Editors)
Volume 91 **Fuzzy Logic Control in Energy Systems with Design Applications in Matlab/Simulink®** İ.H. Altaş
Volume 92 **Power Quality in Future Electrical Power Systems** A. F. Zobaa and S. H. E. A. Aleem (Editors)
Volume 93 **Cogeneration and District Energy Systems: Modelling, Analysis and Optimization** M. A. Rosen and S. Koohi-Fayegh
Volume 94 **Introduction to the Smart Grid: Concepts, technologies and evolution** S.K. Salman
Volume 95 **Communication, Control and Security Challenges for the Smart Grid** S.M. Muyeen and S. Rahman (Editors)
Volume 96 **Industrial Power Systems with Distributed and Embedded Generation** R Belu
Volume 97 **Synchronized Phasor Measurements for Smart Grids** M.J.B. Reddy and D.K. Mohanta (Editors)
Volume 98 **Large Scale Grid Integration of Renewable Energy Sources** A. Moreno-Munoz (Editor)
Volume 100 **Modeling and Dynamic Behaviour of Hydropower Plants** N. Kishor and J. Fraile-Ardanuy (Editors)
Volume 101 **Methane and Hydrogen for Energy Storage** R. Carriveau and D. S-K. Ting
Volume 104 **Power Transformer Condition Monitoring and Diagnosis** A. Abu-Siada (Editor)
Volume 106 **Surface Passivation of Industrial Crystalline Silicon Solar Cells** J. John (Editor)
Volume 107 **Bifacial Photovoltaics: Technology, applications and economics** J. Libal and R. Kopecek (Editors)
Volume 108 **Fault Diagnosis of Induction Motors** J. Faiz, V. Ghorbanian and G. Joksimović
Volume 110 **High Voltage Power Network Construction** K. Harker
Volume 111 **Energy Storage at Different Voltage Levels: Technology, integration, and market aspects** A.F. Zobaa, P.F. Ribeiro, S.H.A. Aleem and S.N. Afifi (Editors)
Volume 112 **Wireless Power Transfer: Theory, Technology and Application** N.Shinohara
Volume 114 **Lightning-Induced Effects in Electrical and Telecommunication Systems** Y. Baba and V. A. Rakov
Volume 115 **DC Distribution Systems and Microgrids** T. Dragičević, F.Blaabjerg and P. Wheeler
Volume 116 **Modelling and Simulation of HVDC Transmission** M. Han (Editor)
Volume 117 **Structural Control and Fault Detection of Wind Turbine Systems** H.R. Karimi
Volume 119 **Thermal Power Plant Control and Instrumentation: The control of boilers and HRSGs, 2nd Edition** D. Lindsley, J. Grist and D. Parker
Volume 120 **Fault Diagnosis for Robust Inverter Power Drives** A. Ginart (Editor)
Volume 121 Monitoring and Control using Synchrophasors in Power Systems with Renewables I. Kamwa and C. Lu (Editors)
Volume 123 **Power Systems Electromagnetic Transients Simulation, 2nd Edition** N. Watson and J. Arrillaga
Volume 124 **Power Market Transformation** B. Murray
Volume 125 **Wind Energy Modeling and Simulation Volume 1: Atmosphere and plant** P.Veers (Editor)
Volume 126 **Diagnosis and Fault Tolerance of Electrical Machines, Power Electronics and Drives** A.J. M. Cardoso
Volume 128 **Characterization of Wide Bandgap Power Semiconductor Devices** F. Wang, Z. Zhang and E.A. Jones
Volume 129 **Renewable Energy from the Oceans: From wave, tidal and gradient systems to offshore wind and solar** D. Coiro and T. Sant (Editors)
Volume 130 **Wind and Solar Based Energy Systems for Communities** R. Carriveau and D. S-K. Ting (Editors)
Volume 131 **Metaheuristic Optimization in Power Engineering** J. Radosavljević
Volume 132 **Power Line Communication Systems for Smart Grids** I.R.S Casella and A. Anpalagan
Volume 139 **Variability, Scalability and Stability of Microgrids** S. M. Muyeen, S. M. Islam and F. Blaabjerg (Editors)
Volume 145 **Condition Monitoring of Rotating Electrical Machines** P. Tavner, L. Ran, C. Crabtree
Volume 146 **Energy Storage for Power Systems, 3rd Edition** A.G. Ter-Gazarian
Volume 147 **Distribution Systems Analysis and Automation 2nd Edition** J. Gers
Volume 152 **Power Electronic Devices: Applications, failure mechanisms and reliability** F Iannuzzo (Editor)
Volume 153 **Signal Processing for Fault Detection and Diagnosis in Electric Machines and Systems** Mohamed Benbouzid (Editor)
Volume 155 **Energy Generation and Efficiency Technologies for Green Residential Buildings** D. Ting and R. Carriveau (Editors)
Volume 157 **Electrical Steels, 2 Volumes** A. Moses, K. Jenkins, Philip Anderson and H. Stanbury
Volume 158 **Advanced Dielectric Materials for Electrostatic Capacitors** Q Li (Editor)
Volume 159 **Transforming the Grid Towards Fully Renewable Energy** O. Probst, S. Castellanos and R. Palacios (Editors)
Volume 160 **Microgrids for Rural Areas: Research and case studies** R.K. Chauhan, K. Chauhan and S.N. Singh (Editors)
Volume 166 **Advanced Characterization of Thin Film Solar Cells** N Haegel and M Al-Jassim (Editors)
Volume 167 **Power Grids with Renewable Energy** Storage, integration and digitalization A. S. Sallam and O. P. Malik
Volume 172 **Lighting interaction with Power Systems, 2 volumes** A. Piantini (Editor)
Volume 193 **Overhead Electric Power Lines: Theory and practice** S. Chattopadhyay and A. Das
Volume 905 **Power system protection, 4 volumes**

Utility-scale Wind Turbines and Wind Farms

Edited by
Ahmad Vasel-Be-Hagh and David S.-K. Ting

The Institution of Engineering and Technology

Published by The Institution of Engineering and Technology, London, United Kingdom

The Institution of Engineering and Technology is registered as a Charity in England & Wales (no. 211014) and Scotland (no. SC038698).

First published 2021

The Institution of Engineering and Technology
Michael Faraday House
Six Hills Way, Stevenage
Herts, SG1 2AY, United Kingdom

www.theiet.org

British Library Cataloguing in Publication Data
A catalogue record for this product is available from the British Library

ISBN 978-1-83953-099-9 (Hardback)
ISBN 978-1-83953-100-2 (PDF)

Typeset in India by Exeter Premedia Services Private Limited

Contents

List of figures

List of tables

About the editors

Ahmad Vasel-Be-Hagh is an assistant professor of Mechanical Engineering and the founder of the Fluid Mechanics Research Laboratory at Tennessee Technological University, USA. His expertise is in thermal-fluid sciences and their applications to a wide range of engineering areas, including wind turbines and wind farms. He has published his research findings in more than twenty-five journal articles and chapters. He has edited eight books, proceedings, and special issues.

David S-K. Ting is a professor in Mechanical, Automotive and Materials Engineering and the founder of the Turbulence & Energy Laboratory (www.turbulenceandenergylab.com) at University of Windsor, Canada. To date, he has co/supervised over eighty graduate students primarily in the energy and turbulence areas and co-authored more than one hundred forty related journal papers. He has also published four textbooks and co-edited ten volumes.

Preface

Many recent advances have been achieved in multiple aspects of utility-scale wind power. These aspects fall into five major categories: aerodynamics, control, environmental concerns, intermittency issues, and offshore wind. The editors believed there was a need for another research book that reviews recent progress in all these critical aspects of utility wind power while providing some directions for more effective future research to achieve more efficient wind power industry. Hence, a diverse team of wind energy researchers was assembled to cover essential topics associated with utility-scale wind turbines and wind farms, focusing on horizontal-axis wind turbines.

We open up the book by briefing the readers on the history of wind power and its current status. Then, we turn our focus on the aerodynamics of wind turbines and wind farms by devoting four chapters to this critical aspect of wind power conversion technology. Chapters 2 and 3 address the aerodynamics of wind turbines, and Chapter 4 deals with more general aspects of the aerodynamics of wind farms. The book further dives into the aerodynamics concept by discussing wind farm design optimization in Chapter 5.

We then start shifting our attention toward the control of wind turbines and wind farms by providing a chapter on analyzing data obtained via wind farm supervisory control and data acquisition (Chapter 6). The next two chapters are devoted to the innovative control strategies for wind turbines (Chapter 7) and recent advances in vibration control for wind turbines under multiple hazards (Chapter 8).

We also address two major environmental concerns of wind turbines: noise propagation (Chapter 9) and bird and bat collisions (Chapter 10). Strategies to minimize these concerns are presented.

Another critical issue of wind-based electricity is its intermittency. Three viable approaches to address this issue are developing accurate short-term and long-term wind power forecast models, integrating storage technologies, and hybridizing wind farms with other renewable sources. These three strategies are discussed in detail in Chapters 11, 12, and 13, respectively.

In Chapter 14, we close the book by reflecting on offshore wind, which is accelerating to dominate the wind power industry and leapfrogging toward the front line of renewables.

We hope *Utility-Scale Wind Turbines and Wind Farms* will provide the readers with a convenient source for learning the latest wind power development.

Ahmad Vasel-Be-Hagh and David S.-K. Ting
January 2021

Acknowledgment

We acknowledge IET, Christoph von Friedeburg, and Olivia Wilkins for making this book possible. Tennessee Technological University and the University of Windsor have been instrumental as well in our research and professional endeavors, including this book project. Finally, we are indebted to our colleagues who sacrificed their precious time in this extraordinary year to furnish the quality chapters and for reviewing and providing invaluable comments to enhance the quality of their peers' work. We could not have completed this understanding without you. Thank you.

Vasel-Be-Hagh and Ting

Chapter 1

The current status of wind power

Michael R. Hackler[1], Ahmad Vasel-Be-Hagh[1], and David S.-K. Ting[2]

At the beginning of 2020, wind power capacity worldwide exceeded approximately 650 GW, covering less than 5% of the global electricity demand. This current global wind power capacity is enough to power more than 400 million average houses. The International Renewable Energy Agency projects that wind will generate approximately 35% of the total required electricity by 2050. Technological developments in towers, foundations, rotors, and drivetrains will enable this accelerated expansion of the wind industry. After presenting a brief on the current state of these significant wind energy technology pillars, the chapter lays out the various topics that the present book covers.

1.1 Introduction

Wind power is expected to address one-third of the global electricity demand by 2050. The industry is far away from such a high production capacity. The research community, including research universities, national laboratories, and industrial R&D teams, is leading the way into more efficient turbines and farms that can deliver this promise. Hence, smarter blades [1], rotors [2], towers [3], foundations [4], drivetrains, farm layouts [5, 6], and control strategies [7, 8] are introduced rapidly. In this opening chapter, we present a high-level view of the current status of these major aspects of the wind power industry. The following chapters will take a deep dive into many of them.

1.2 History of wind power harvesting

From very early on, civilizations have experienced wind power and asked themselves how they could use this limitless resource to make their lives easier. From

[1]Fluid Mechanics Research Laboratory, Tennessee Tech University, Cookeville, TN38505, USA (mrhackler42@tntech.edu)
[2]Turbulence and Energy Laboratory, University of Windsor, Windsor, ON, Canada

sailing on the open ocean and pumping water to new heights, to grinding anything from cloth to corn mill, wind power has been greatly utilized throughout history.

The invention of the windmill, as it was documented, was around 644 AD in the Middle East as the area experienced a large amount of strong winds [9]. A place called Sistan, located in present-day Iran, was an entire province that had implemented wind power into their everyday society. Windmills were used as pumps for irrigation in the farms, grinders for raw materials, and they even used them to level out massive sand dunes that towered over their towns. Many people also consider Sistan to be the birthplace of the vertical-axis windmill in which the direction of the wind was perpendicular to the "blades" that were usually made from cloth or woven fibers. These technologies soon found their way into different regions through trade and other means, and thus wind power was born.

Although vertical windmills proved to be extremely useful for places with high wind, the windmill's design usually had one or two blades at a time supporting the movement of the entire structure, causing a considerable loss of efficiency. While this type of windmill was being used in and around the Middle East, a new type of windmill was being introduced. The horizontal-axis windmill was invented in Europe and had its blades oriented with their rotating axis parallel to the wind direction. This allowed for a more uniform use of all blades, leading to higher efficiencies than vertical windmills. Even though both types of windmills were used for the same sort of applications, some historians believed they were developed separately with horizontal windmills emerging, instead, from adaptations of the Roman waterwheel [10].

In Holland, for many centuries, many designs improving the windmill emerged. The first iteration yielded the post design in which the entire mill itself rotated around a central post connected to grindstones. It was made primarily of wood, which proved to be dangerous as the mill and grain friction could cause fires. The second design was the tower design. It consisted of a tower made of brick with a separate top cap that rotated freely from the base so that it was easier to adjust for changing wind patterns. The final design that surfaced in Holland was the smock design, with the only difference being that it was made with a stone base and weatherboarding on the entire height of the mill.

As turbines continue to evolve every year with advances in construction materials, efficiency, and power generation, wind power looks to have a larger footprint in the future energy industry. Wind turbines have become larger and larger to generate more power (see Figure 1.1). For instance, Enercon's wind turbines have advanced from the E-15 model with rotor diameter and rated capacity of 15 m and 50 kW to the E-126 model with rotor diameter and rated power of 126 m and 7.5 MW over the last four decades [11]. The two most powerful wind turbines currently dominating the industry are the Vestas V164, having a rotor diameter of 164 m and rated power output of 10 MW, and Siemens-Gamesa 8.0–167 DD, having a rotor diameter of 167 m and rated power output of 8 MW. General Electric (GE) is developing the world's first 12 MW offshore wind turbine, Haliade-X. The goal is to manufacture the most powerful and efficient offshore wind turbine in the world with a capacity factor of 63%. This wind turbine is expected to have a rotor diameter of 220 m. These modern

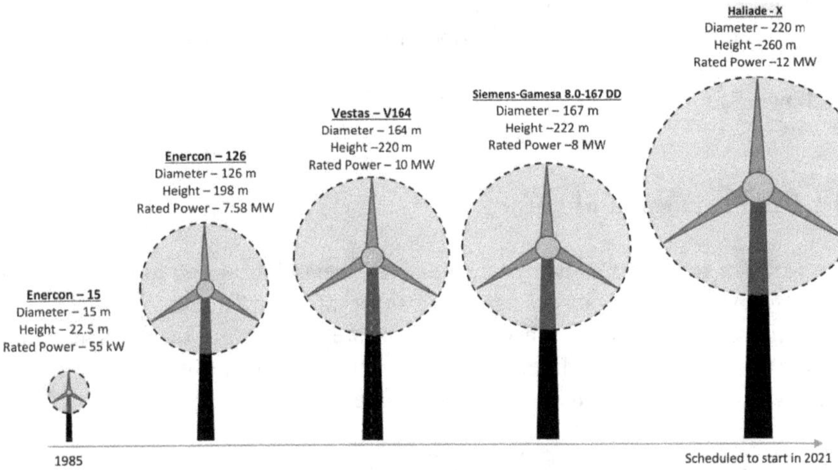

Figure 1.1 Evolution of commercial horizontal-axis wind turbines

turbines are being installed at higher hub heights. With every 10 m increase in the windmill's height, the wind speed increases by approximately 20% [12], leading to a significant increase in the available mechanical power. More wind speed and larger rotor diameters would translate to more power output, explaining why commercial wind turbines continue to grow more massive.

1.3 Current installed wind capacity

Countries have devoted time and money to power the future through clean and renewable wind power throughout the world. As more places move away from the traditional model of using fossil fuels and natural gas for most power generation, wind power will have an enormous responsibility to pick up the slack to power an ever-growing world population with increased energy demand.

Wind power provides a cost-effective way for countries that lack traditional ways of producing power to generate what they need for their population. In 2019, the global installed wind capacity expanded by 19% and hit approximately 645 GW of power. China and the USA are leading the world with approximately 236 GW and 105.6 GW installed. The global installed wind capacity is projected to more than triple to 2110 GW by the year 2030 [13]. This projection comes from renewable energy sources becoming much cheaper due to advances in materials, design, and manufacturing.

Another major thrust of the wind energy sector is the implementation of offshore wind farms. Due to advances in foundation design coupled with improved turbine construction, countries have begun to utilize the vast surface area of oceans to generate useful power for their population. Large-scale operations are already underway in Europe, for example, with the collective of the European Union having already installed 25.3 GW of wind capacity offshore. That number is expected to triple by 2030 to 76 GW as the utilization of offshore wind power takes off in full force [14]. Being out on the open water also allows turbines to grow even more massive allowing for more energy production and efficiency.

1.4 Current status of rotors

In the world of rotor design, many techniques are being introduced to increase efficiency of power production. Changing the shape of rotors, number of rotor sets, and addition of new complimentary components are just a few of the ways in which wind turbines are being modified for a more efficient future. Technology is also being developed that can detect if rotor blades are in need of repair.

1.4.1 Rotor shape modification through bio-mimicry

In many ways, the rotor's shape is critical, as each environment constitutes its own type of rotor configuration. Researchers have even begun to look at nature for inspiration for the design. One of the latest configurations mimics that of a maple seed in which researchers noticed the auto-rotation of the seeds when they fall from trees and believed that this attribute would translate well into the application to wind turbines [15]. They determined that this is yet another practical way to design wind turbine rotors. This is not the only bio-inspired rotor. A single-bladed helical-shaped rotor inspired by seed pods of two Indian trees was proposed and numerically examined by Venkataraman and Manabendra [16]. Several other bio-inspired rotors can be found in the literature [17, 18].

1.4.2 Dual rotor Design

The concept of adding rotor sets and experimenting with their rotational orientation in relation to each other has interested many researchers as of late. In one of the latest tries, researchers of the Lodz University of Technology in Poland and Arts et Métiers ParisTech in France [19] have developed a fully resolved rotor model to investigate a diffuser-augmented double rotor design with both sets rotating in the same and opposite directions. In their experiment, both of their rotor diameters were of equal size. Other researchers have focused on how the generator of a dual-rotor design needs to be constructed [20].

The impact of using dual rotors on the wake of wind turbines and aerodynamics of wind farms has also been investigated. In a piece titled, "Wind Farms With Counter-Rotating Wind Turbines," Vasel-Be-Hagh and Archer reported their large-eddy simulations' results on a wind farm with dual-rotor wind turbines in which each turbine has two rotors, with the first rotor rotating counter-clockwise and the second rotor rotating clockwise [21]. The power production of each rotor was calculated using the actuator line model. The dual-rotor wind farm was found to produce 22.6% more power than the control wind farm.

Many different aspects of dual-rotor wind turbines have been investigated over the last few years. This includes evaluating their power efficiency [22], studying the aerodynamic wake of these turbines numerically and experimentally [23–25], blade [2] and design optimization [26–28], dynamic modeling and developing efficient dynamic mechanisms [29, 30], developing control strategies [31, 32], and studying the effect of the axial distance between the rotors [33].

1.4.3 Addition of complimentary components to wind turbines

Addition of different complimentary components into rotor design has added a different set of assets to improve efficiency in wind turbines. Researchers at the Poznan University of Technology [34] implemented and tested adding two different configurations of a convex bowl centered at the horizontal axis of a wind turbine. Promising conclusions were made in which the installed convex bowl increased wind speeds to the turbine blades. This increase in speed would yield better power production and efficiency for turbines thus making the topic of complimentary components a viable research topic that could help the wind industry take another step forward. Two comprehensive reviews on such diffuser-augmented wind turbines have been published recently [35, 36].

1.4.4 Rotor blade life and post-life management

As time moves forward and wind turbine technology continues to advance, the question of what to do with older, less-efficient turbines presents itself. To keep up with larger scale turbines, Fraunhofer Institute for Wind Energy Systems [37] in Bremerhaven, Germany, aimed to battle the decommissioning of smaller turbines by introducing blade extensions. This would help to produce more power and keep the old turbines online. Less downtime from a retrofit instead of complete decommissioning would prove to be much more economical with the minimal extra strain on the old turbines themselves.

Keeping turbines online with minimal downtime can be controlled with well-timed maintenance. The ability to predict when a turbine rotor is beginning to fail could prove advantageous to any company as they could fix the problem before it even surfaces. This would increase the safety and profitability of the unit. Researchers in Hannover, Germany [38], developed a way to predict turbine failure based on the frequency with which the rotors operate. The algorithm could predict if something was not right and notify maintenance teams to correct the error before it turned catastrophic.

No matter how much a current system is modified and monitored, it will eventually need to be decommissioned. The problem with removing old turbines is the recycling process. It is increasingly harder to recycle wind turbines as they become more complex with time. In a recent article titled "Gone with the Wind: The Life and Death of a Wind Turbine Rotor Blade," Eva Karatairi [39] discusses processes that are being tested to help break down turbine materials into their elementary fibers. This would allow turbines to be recycled more readily into new systems, thus keeping the green image of wind turbines at the forefront of their construction. Maintenance and recycling of turbine components are just as important as the design of the turbine itself in powering communities efficiently.

1.5 Current status of towers

The tower of a horizontal-axis wind turbine holds the nacelle and the rotor. The type of tower is chosen based on aerodynamics, structural, and financial considerations. Towers come in five general types: lattice towers, Guyed towers, tubular steel towers, concrete towers, and hybrid towers.

While lattice and Guyed towers have several unique advantages, such as less erection time, reduced wind resistance, and cost-effective foundations [40], developers do not use them on the utility-scale anymore. Their height typically ranges between 20 and 40 m [41] and can only carry tiny, low-capacity turbines (a few kilowatts). Szafraski and Maachowski [42], for instance, developed a numerical model of a Guyed tower to minimize manufacturing and installation costs of only a 3.5-kW wind turbine. Axisa *et al.* [43] designed a lattice tower to carry a turbine with a rotor diameter of only 3.4 m.

Tubular steel towers are the most popular turbine towers. With the diameter increasing toward the base, these conical towers come in sections that are either flanged or bolted. The use of stiffeners, such as horizontal rings and vertical strips, to enhance the strength of these towers has been recommended [44]. The dynamic behavior of these towers has been extensively investigated from different points of view. The article by Koulatsou [45] focusing on how efforts to avoid resonance of the soil-foundation-tower system affects tower optimization is among the most recent studies on the dynamic behavior of Tubular steel towers.

The possibility of building taller towers via concrete to harvest more wind power has been investigated. A comprehensive study funded by the US Department of Energy conducted at the Iowa State University concluded that concrete could be used to form towers as high as 140 m to carry utility-scale nacelles and towers [46]. The use of concrete towers is also recommended in earthquake areas. The behavior of concrete supported towers under earthquake action has been investigated by Cao *et al.* [47]. Concrete towers are typically fabricated in segmented slices, as it makes the transportation more comfortable and, hence, reduces the overall costs. Yue *et al.* [48] investigated the effect of the size, shape, and interconnection mechanism of these slices on the stiffness and reliability of the structure. The research community is actively searching for ways to further improve the durability performance of concrete towers. Among all, incorporating protective skins made from engineered cementitious composites (ECC) seems very promising [49, 50].

Hybrid approaches are recommended for addressing specific issues that each of the above-described individual towers might suffer from. For instance, Chen *et al.* [51] proposed a steel-concrete hybrid wind turbine tower where the lower part of the steel tower is replaced with a concrete segment. This improves the lateral stiffness of the tower and addresses the excessive vibration of the steel tower. See the article by Kim *et al.* [52] for more on hybrid steel-concrete towers. Life-cycle assessments of steel-concrete towers have been conducted by Oebels and Pacca [53], Gkantou *et al.* [54], and Gervásio *et al* [55].

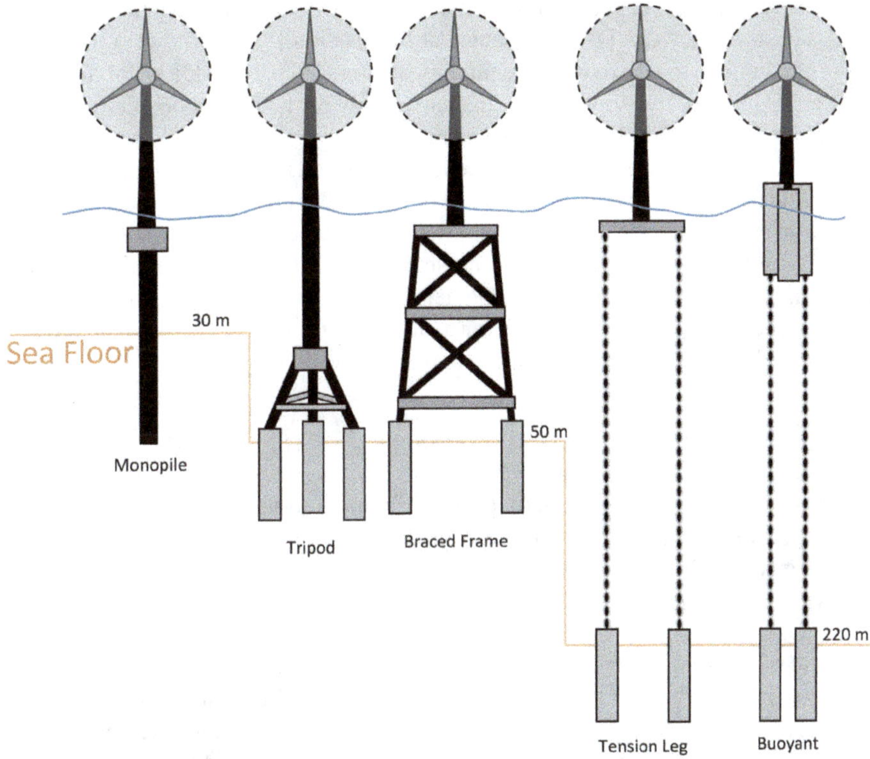

Figure 1.2 Major types of wind turbine foundations

1.6 Current status of offshore foundations

Turbines continue to grow in size and stature each year to capture more wind energy for the ever-increasing demand for power. While the size of the rotor blades and the height of the tower are essential for harnessing wind energy, a turbine must support its growing size. Improvements in foundation technology paired with reductions in installation costs have allowed for these increases to take place.

There are many different types of applicable foundations for offshore wind turbines. The main types are buoyant, tension leg, monopile, tripod, and braced frame. Application for each type depends on the depth of the body of water in which the turbines are being placed (see Figure 1.2). Monopile foundations are mainly implemented when the water depth is 30 m or less. This type of foundation can reach sizes of 5 m in diameter with a weight of over 600 tons and must be lifted out to the site and driven into the seabed for structural integrity [56]. Tripod and braced frame foundation designs are implemented at medium depths of 20–50 m and must also be driven into the seabed by large machinery, which adds to installation costs and requires more time. Floating foundations, however, only have to be secured to the seabed by anchors, which are much easier to install. This foundation type is the most up-and-coming technology, and they provide a promising future for the wind industry. They can be implemented at depths up to 220 m allowing a broader range of countries to utilize wind technology offshore [57]. The addition of the semi-submersible hull design yields even greater stability for floating foundations, with a part of the structure being under the water [58].

Enhanced strength and performance testing have also been introduced recently to help the design and installation of wind turbines offshore. Sabaliauskas and Ibsen [59] developed a testing strategy that allows researchers to monitor the stiffness of the foundation of the turbine as the seabed is disturbed from cyclic loading-unloading from wind patterns. This can help to design more stable foundations that allow for flexibility, which would, in turn, allow the useful life of the foundation to be extended, saving money on repair and maintenance costs.

1.7 Current status of drive trains

The drive train converts the kinetic energy of wind into usable mechanical energy for the generators inside wind turbines. Currently, the gearbox is most widely used, but designs like direct-drive and hydrostatic drive trains have become more competitive with advances in their assembly aspects.

The gearbox is widely used in wind turbines today. It converts the high-torque low-rpm of spinning blades to a high-rpm low-torque drive that can be used in generators to produce energy [60]. The gears inside the gearboxes are more expensive to manufacture, and they add complexity to the system as a whole. Gearboxes also have much stress on them as varying wind speeds can cause bearings to wear out and components to break [61]. Research is being done to refine the manufacturing process in gearboxes with affordability without loss of reliability in mind.

The direct-drive drivetrain does not use any gears at all, and the driving shaft in the generator has the same low rotational speed of the turbine blades. Although one can replace one of the most complex parts of the turbine by removing the gearbox and adding a direct-drive mechanism, direct drives are usually heavier and more massive.

Hydrostatic drivetrains have been around for a long time, but here recently, there has been a renewed interest in their applied technology. In contrast to gearboxes, which wear out quickly when different magnitudes of load are applied, the hydrostatic drivetrain absorbs fluctuations in wind patterns allowing for greater efficiency and less maintenance [62]. Hydrostatic drivetrains also allow the generator to be oriented at the bottom of the turbine tower resulting in a decrease in the nacelle's weight at the top. This would yield cheaper construction costs both in the tower and foundation. On average, this configuration weighs three to four times less than the traditional gearbox setup [63].

1.8 Advances in utility-scale wind turbines and wind farms

The current book, *Utility-Scale Wind Turbines and Wind Farms*, covers some of the most critical advances recently made in utility-scale wind power. This includes advances made in (i) technology of horizontal-axis wind turbines and (ii) design, control, and operation of large-scale wind farms. The book aims to provide readers with a convenient source for learning the latest developments in the following specific areas.

1.8.1 Aerodynamics of wind turbines and wind farms

Chapter 2 presents advances in the aerodynamics of horizontal-axis wind turbines. This includes aerodynamics of blades as well as wind lenses. The chapter critiques the possibility of developing new aerodynamic designs to exceed Betz's limit. Some of the recent approaches adopted to delay stall are covered. Some of the existing and proposed concepts to reduce the aerodynamic loading, particularly in the gusty atmosphere, are presented. The possibility of using control surfaces, such as flaps, to enhance turbines' aerodynamic performance is also discussed. Potential impacts of the number of blades and their rotational direction are covered too. This is followed by Chapter 3 on the scaling utility-scale wind turbines for laboratory experiments.

Chapter 4 focuses on recent advances in the aerodynamics of wind farms. This includes progress made in simulating flow through large-scale wind farms using computational fluid dynamics, the interaction of wind turbines via their aerodynamic wakes, analytical models to predict wind speed deficit within wind farms, analytical approaches to overlap the impact of multiple upstream wakes, in-farm wind speed and direction variability, the impact of complex terrains on the aerodynamics of wind farms, the impact of surface roughness on the aerodynamics of wind farms, and existing and potential concepts to minimize the aerodynamic interaction of wind turbines within wind farms. Wind farm wake control, both passive and adaptive, will be covered in this chapter.

Optimizing turbine type, number of turbines, their position, and their hub heights to maximize overall power production, annual energy production, or annual revenue, or to minimize levelized cost of electricity is one of the most critical tasks for developing an efficient design for a wind farm. Chapter 5 reports on the most modern layout optimization algorithms with single or multiple objective functions.

1.8.2 Field data collection and control of wind turbines and wind farms

Field observations play a significant role in gaining physical insights into the underlying physics of wind farms. Collecting quality data and then analyzing data clouds obtained via supervisory control and data acquisition (SCADA) and laser measuring instrumentations such as light detection and ranging (LIDAR) is a challenge, which could lead to misleading conclusions if the users do not consider a series of important operational facts. For instance, rotors are never perfectly normal to the wind directions. One needs to make corrections to filter out the effect of this unintentional yaw. Chapter 6 aims at introducing different strategies and methods of collecting, analyzing, and then making use of such data for different requirements.

Wind turbine control mechanisms and strategies play an essential role in reducing aerodynamic loads and adjusting power production. Wind turbines can be controlled via several mechanisms including pitch, yaw, tilt, and cone. Researchers actively develop new control strategies for both inland and offshore wind turbines, including installation of LIDARs on each turbine to provide some look-ahead capacity for more efficient functionality. All these modern strategies will be introduced in Chapters 7 and 8.

1.8.3 Addressing the environmental issues

US Department of Energy has set a target of 20% for wind power to contribute to the national end-use electricity demand. A considerable percentage of this number will be generated by land-based wind farms. This requires a special attention to the wind turbine noise propagation issue. Accurate prediction of wind farm noise propagation over short and long distances, their impact on nearby communities and wildlife, and feasible strategies to reduce such noises either during the design and development process or using aftermarket technologies have been the subject of several investigations. Chapter 9 would present recent progress made by researchers of this essential aspect of wind power technology.

Chapter 10 further investigates the impact of wind farms on wildlife by looking into the issue of birds colliding with wind turbines. Most recent advances in developing collision rate models as well as technologies to reduce such collisions will be presented.

1.8.4 Addressing the intermittency issue

Wind, like any other renewable energy, is intermittent and unpredictable, reducing its penetration into the grid. One way to mitigate this issue without using backup fossil-fuel generators or costly storage plants is developing wind power forecasting approaches. The most modern wind power forecasting techniques will be discussed in detail in Chapter 11.

Another viable solution to the intermittence and unpredictability issues of wind power is developing hybrid wind-storage plants to store surplus energy during off-peak hours to ensure the plant can provide a uniform outlet regardless of instantaneous meteorological conditions or the gird demand. Recent advances in developing hybrid storage-wind plants will be covered in Chapter 12. This chapter will also touch on the concept of ancillary grid services, including frequency control, spinning reserves, and operating reserves. Such operations aim at maintaining grid stability and security.

Hybridizing renewable energy resources, such as wind-solar power plants, are another viable strategy to overcome intermittence issue. Although recent developments in wind power forecasting (discussed in Chapter 11) and integrated storage plants (discussed in Chapter 12) can substantially address this issue, the problem has not been eliminated yet. The hybridizing concept can significantly contribute to mitigating this issue without wasting the generated power due to roundtrip losses associated with storage plants. Researchers have proposed multiple hybridization concepts and strategies. Such advances will be covered in Chapter 13.

1.8.5 Offshore wind

Chapter 14 will deal with the site selection problem. There has been a significant surge in wind power investment in recent years. The push for developing more utility-scale wind farms demands much more accurate and reliable methods to identify optimal sites for both inland and offshore wind farms. Site selection is complicated by several factors, including available wind resources, the feasibility of installation, safety, and interaction with the surrounding infrastructures. This chapter will report on the most recent progress made in developing more efficient site-selection strategies.

References

[1] Chen B., Hua X., Zhang Z., *et al.* 'Active flutter control of the wind turbines using double-pitched blades'. *Renewable Energy*. 2021;**163**:2081–97.

[2] Moghadassian B., Sharma A. 'Designing wind turbine rotor blades to enhance energy capture in turbine arrays'. *Renewable Energy*. 2020;**148**:651–64.

[3] Hernandez-Estrada E., Lastres-Danguillecourt O., Robles-Ocampo J.B., *et al.* 'Considerations for the structural analysis and design of wind turbine towers: a review'. *Renewable and Sustainable Energy Reviews*. 2021;**137**(8):110447.

[4] Wang X., Zeng X., Li J., *et al.* 'A review on recent advancements of substructures for offshore wind turbines'. *Energy Conversion and Management*. 2018;**158**:103–19.

[5] Archer C., Vasel-Be-Hagh A., Yan C., *et al.* 'Review and evaluation of wake loss models for wind energy applications'. *Applied Energy*. 2018;**226**:1187–207.

[6] Azlan F., Kurnia J.C., Tan B.T., Ismadi M.-Z. 'Review on optimisation methods of wind farm array under three classical wind condition problems'. *Renewable and Sustainable Energy Reviews*. 2021;**135**:110047.

[7] Sierra-Garca J., Santos M. 'Wind turbine pitch control with an RBF neural network'. *Advances in Intelligent Systems and Computing 1268 AISC*. 2021:397–406.

[8] Archer C., Vasel-Be-Hagh A. 'Wake steering via yaw control in multi-turbine wind farms: Recommendations based on large-eddy simulation'. *Sustainable Energy Technologies and Assessments*. 2019;**33**:34–43.

[9] Ragheb M. 'History of harnessing wind power' in Letcher T.M. (ed.). *Wind Energy Engineering*. Academic Press; 2017. pp. 127–43.

[10] Shepherd D.G.Sat. 'Historical development of the windmill'. United States; 1990. Available from https://www.osti.gov/servlets/purl/6342767.

[11] Renewables First. Enercon wind turbines. 2020. Available from https://www.renewablesfirst.co.uk/windpower/wind-turbines/enercon-wind-turbines/ [Accessed 21 Nov 2020].

[12] Tawfiq K., Mansour A., Ramadan H., *et al.* 'Wind energy conversion system topologies and converters: Comparative review'. *Energy Procedia.* 2019;**162**:38–47.
[13] Darwish A., Al-Dabbagh R. 'Wind energy state of the art: present and future technology advancements'. *Renewable Energy Environmental Sustainability.* 2020;**5**:8.
[14] Soares-Ramos E.P., de Oliveira-Assis L., Sarrias-Mena R. 'Current status and future trends of offshore wind power in Europe'. *Energy.* 2020;**202**(2015):117787.
[15] Gaitan-Aroca J., Sierra F., Castellanos Contreras J.U. 'Bio-inspired rotor design characterization of a horizontal axis wind turbine'. *Energies.* 2020;**13**(14):3515.
[16] Venkataraman P., Manabendra M.D. 'Numerical investigation of standstill characteristics of a bio-inspired vertical axis wind turbine rotor'. *IOP Conference Series: Materials Science and Engineering.* 2018;**377**:012014.
[17] Seidel C., Jayaram S., Kunkel L.,*et al.* 'Structural analysis of biologically inspired small wind turbine blades'. *International Journal of Mechanical and Materials Engineering.* 2017;**12**(19).
[18] Coe M. *Design and Analysis of Bio-Inspired Nacelle for Current Energy Turbine [Master's thesis].* San José State University; 2017.
[19] Lipian M., Dobrev I., Massouh F., *et al.* 'Small wind turbine augmentation: Numerical investigations of shrouded- and twin-rotor wind turbines'. *Energy.* 2020;**201**:117588.
[20] Kutt F., Blecharz K., Karkosiński D. 'Axial-flux permanent-magnet dual-rotor generator for a counter-rotating wind turbine'. *Energies.* 2020;**13**(11):2833.
[21] Vasel-Be-Hagh A., Archer C.L. 'Wind farms with counter-rotating wind turbines'. *Sustainable Energy Technologies and Assessments.* 2017;**24**:19–30.
[22] Mohamed A., El-Baz A., Abd-Elaziz N., *et al.* Computational investigation of ducted dual rotor wind turbine. Novel Intelligent and Leading Emerging Sciences Conference; 2019. pp. 29–33.
[23] Hollands E., He C., Gan L. 'A particle image velocimetry study of dual-rotor counter-rotating wind turbine near wake'. *Journal of Visualization.* 2020;**23**(3):425–35.
[24] Asfar K., Mahasneh A. Wake patterns in a dual-rotor wind turbine. American Society of Mechanical Engineers, Fluids Engineering Division (Publication) FEDSM 3; 2020.
[25] Wang Z., Ozbay A., Tian W., *et al.* 'An experimental study on the aerodynamic performances and wake characteristics of an innovative dual-rotor wind turbine'. *Energy.* 2018;**147**:94–109.
[26] Thelen A., Leifsson L., Sharma A., *et al.* 'Variable-fidelity shape optimization of dual-rotor wind turbines'. *Engineering Computations.* 2018;**35**(7):2514–42.
[27] Moghadassian B., Sharma A. 'Inverse design of single- and multi-rotor horizontal axis wind turbine blades using computational fluid dynamics'. *Journal of Solar Energy Engineering.* 2018;**140**(2):021003.

[28] Thelen A., Leifsson L., Sharma A., *et al.* 'RANs-based design optimization of dual-rotor wind turbines'. *Engineering Computations.* 2018;**35**(1):35–52.

[29] Neagoe M., Jaliu C., Saulescu R., *et al.* 'Steady-state response of a dual-rotor wind turbine with counter-rotating electric generator and planetary gear increaser'. *Mechanisms and Machine Science.* 2020;**83**:106–15.

[30] Filsoof O., Hansen M., Yde A., *et al.* 'Dynamic modeling and stability analysis of a dual-rotor wind turbine'. *Proceedings of the ASME Design Engineering Technical Conference.* 2018;**6**:7.

[31] Yahdou A., Boudjema Z., Taleb R., *et al.* Backstepping sliding mode control of a dual rotor wind turbine system. Proceedings of 2018 3rd International Conference on Electrical Sciences and Technologies in Maghreb. CISTEM 2018; 2019.

[32] Kahal H., Taleb R., Boudjema Z., *et al.* 'Control of a dual rotor wind turbine system'. *Lecture Notes in Networks and Systems.* 2019;**62**:263–9.

[33] Nugroho S., Diana L., Ariyanti D. 'The effect of axial distance on dual rotor wind turbine's performance'. *Journal of Physics: Conference Series.* 2019;**1367**(1):012031.

[34] Bugała A., Roszyk O. 'Investigation of an innovative rotor modification for a small-scale horizontal axis wind turbine'. *Energies.* 2020;**13**(10):2649.

[35] Bontempo R., Manna M. 'Diffuser augmented wind turbines: review and assessment of theoretical models'. *Applied Energy.* 2020;**280**(3):115867.

[36] Nunes M.M., Brasil Junior A.C.P., Oliveira T.F. 'Systematic review of diffuser-augmented horizontal-axis turbines'. *Renewable and Sustainable Energy Reviews.* 2020;**133**(6):110075.

[37] Rosemeier M., Saathoff M. 'Assessment of a rotor blade extension retrofit as a supplement to the lifetime extension of wind turbines'. *Wind Energy Science.* 2020;**5**(3):897–909.

[38] Krause T., Ostermann J. 'Damage detection for wind turbine rotor blades using airborne sound'. *Structural Control and Health Monitoring.* 2020;**27**(5):e2520.

[39] Karatairi E., Bischler R. 'Gone with the wind: The life and death of a wind turbine rotor blade'. *MRS Bulletin.* 2020;**45**(3):178–9.

[40] Jespersen M., Strup-Andersen U. Guyed wind turbine towers: Developments and outlook. The 14th Nordic Steel Construction Conference; 2019. pp. 779–84.

[41] Agbayani N., Vega R. The rapid evolution of wind turbine tower structural systems: A historical and technical overview. Proceedings of the 2012 Structures Congress; 2012. pp. 1201–12.

[42] Szafraski T., Maachowski J. 'Numerical analysis of guyed mast for small wind turbine at normal operating conditions'. *Lecture Notes in Mechanical Engineering.* 2017:551–9.

[43] Axisa R., Muscat M., Sant T., *et al.* 'Structural assessment of a lattice tower for a small, multi-bladed wind turbine'. *International Journal of Energy and Environmental Engineering.* 2017;**8**(4):343–58.

[44] Hu Y., Yang J., Baniotopoulos C., *et al.* 'A comparison of structural performance enhancement of horizontally and vertically stiffened tubular steel wind turbine towers'. *Structural Engineering and Mechanics*. 2020;**73**(5):487–500.

[45] Koulatsou K., Kazakis G., Gantes C., *et al.* 'Resonance investigation and its effects on weight optimization of tubular steel wind turbine towers'. *Procedia Manufacturing*. 2020;**44**:4–11.

[46] Sritharan S. 'Hexcrete Tower for Harvesting Wind Energy at Taller Hub Heights - Budget Period 2'. *U.S. Department of Energy, Office of Scientific and Technical Information*. 2017

[47] Cao Y., He M., Ma R., *et al.* 'Beam-column modeling and seismic fragility analysis of a prestressed segmental concrete tower for wind turbines'. *Advances in Structural Engineering*. 2020;**23**(8):1715–27.

[48] Yue Y., Tian J., Mu Q., *et al.* 'Feasibility of segmented concrete in wind turbine tower: numerical studies on its mechanical performance'. *International Journal of Damage Mechanics*. 2021;**30**(4):518–36.

[49] Jin Q., Li V.C. 'Structural and durability assessment of ECC/concrete dual-layer system for tall wind turbine towers'. *Engineering Structures*. 2019;**196**:109338.

[50] Jin Q., Li V. 'Development of lightweight engineered cementitious composite for durability enhancement of tall concrete wind towers'. *Cement and Concrete Composites*. 2019;**96**:87–94.

[51] Chen J., Li J., He X. 'Design optimization of steel–concrete hybrid wind turbine tower based on improved genetic algorithm'. *The Structural Design of Tall and Special Buildings*. 2020;**29**(10):e1741.

[52] Kim M., Kim T., Lee D., *et al.* 'Experimental investigation of the steel-concrete joint in a hybrid tower for a wind turbine under fatigue loading'. *KSCE Journal of Civil Engineering*. 2019;**23**(7):2971–82.

[53] Oebels K.B., Pacca S. 'Life cycle assessment of an onshore wind farm located at the northeastern coast of brazil'. *Renewable Energy*. 2013;**53**:60–70.

[54] Gkantou M., Rebelo C., Baniotopoulos C. 'Life cycle assessment of tall Onshore hybrid steel wind turbine towers'. *Energies*. 2020;**13**(15):3950.

[55] Gervásio H., Rebelo C., Moura A., Veljkovic M., Simões da Silva L. 'Comparative life cycle assessment of tubular wind towers and foundations – Part 2: life cycle analysis'. *Engineering Structures*. 2014;**74**(2012):292–9.

[56] Manzano-Agugliaro F., Sánchez-Calero M., Alcayde A., San-Antonio-Gómez C., Perea-Moreno A.-J., Salmeron-Manzano E. 'Wind turbines offshore foundations and connections to grid'. *Inventions*. 2020;**5**(1):8.

[57] Sánchez S., López-Gutiérrez J.-S., Negro V., Esteban M.D. 'Foundations in offshore wind farms: evolution, characteristics and range of use. Analysis of main dimensional parameters in monopile foundations'. *Journal of Marine Science and Engineering*. 2019;**7**(12):441.

[58] Odijie A., Wang F., Ye J. 'A review of floating semisubmersible hull systems: Column stabilized unit'. *Ocean Engineering*. 2017;**144**:191–202.

[59] Sabaliauskas T., Ibsen L. 'The new scope of frictionless triaxial apparatus—disturbed sand testing'. *Geotechnical Testing Journal*. 2018;**41**(6):1117–30.

[60] Taherian-Fard E., Sahebi R., Niknam T., *et al.* 'Wind turbine drivetrain technologies'. *IEEE Transactions on Industry Applications*. 2020;**56**(2):1729–41.
[61] Rezamand M., Kordestani M., Orchard M.E. 'Improved remaining useful life estimation of wind turbine drivetrain bearings under varying operating conditions (VOC)'. *IEEE Transactions on Industrial Informatics*. 2021;**17**(3):1742–52.
[62] Schulte H. 'Control-oriented description of large scale wind turbines with hydrostatic transmission using takagi-sugeno models'. IEEE Conference on Control Applications (CCA); 2014. pp. 664–8.
[63] Rajabhandharaks D., Hsu P. 'Optimal aerodynamic energy capture strategies for hydrostatic transmission wind turbine'. IEEE Conference on Technologies for Sustainability (SusTech); 2014. pp. 140–7.

Part I

Aerodynamics

Chapter 2

Advances in the aerodynamics of horizontal axis wind turbines

Oluseyi O. Ajayi[1,2], *Logan Unser*[2], *and Joseph O. Ojo*[3]

Wind turbine technology enhancements have, in the last two to three decades, been through developmental progression. Several efforts tailored toward improvement in power generation at reduced cost have dominated global research activities. These activities include wind turbine design improvements, control efficiency enhancements, blade shape design and optimization, flow separation and control, system stability measure, and condition monitoring. Among all of these research efforts, the contributions of the aerodynamics of rotor blades vis-à-vis the complexities of the flow structure around the rotor-blade system have dominated recent research focus. The efforts aim to have an operable system that is more efficient and environmentally friendlier. Thus, this study focuses on the advances in the aerodynamics of horizontal axis wind turbines. The chapter reviews some of the research efforts that have been carried out on this topic and the developments that have emanated from such efforts. Considerations are given to blade geometry modifications for effective power production, flow regimes and control, application of wind-lenses, and flow stream modification through the implementation of various flaps, airfoil concavity, shrouds, and flanges. Some of the researchers' documented efforts to augment the power coefficients beyond Betz's limit are also surveyed and criticized.

2.1 Introduction

Wind as a veritable resource for energy production has come through many years of developmental progression. Its earliest use is from prehistoric times and includes its use for boats driving, water pumping, and powering windmills for grain grinding. The first wind power machine, invented by Charles F. Brush in 1888 in Cleveland, Ohio, was a 50 feet (17 m) diameter and 144 rotor blades wind turbine. It generated

[1]Department of Mechanical Engineering, Covenant University, Ota, Ogun, Nigeria
[2]Fluid Mechanics Research Laboratory, Tennessee Technological University, Cookeville, TN, USA
[3]Electrical and Computer Engineering Department, Tennessee Technological University, Cookeville, TN, USA

about 12 kilowatts (kW) of electricity [1]. Ever since then, the technology of wind turbines for electricity generation has witnessed progressive growth that ranges from design improvement, size reduction, generation capacity increase, control, efficiency improvement, decrease in the number of blades to blade shape optimization, among others [2]. These developmental improvements have, over the years, brought about a reduction in cost per kW electricity production and an increase in the magnitude of energy harvestable from an installed wind turbine. Thus, driving the adoption and encouraging new annual installed capacities across various continents of the world. Presently, wind energy is one of the fast-growing electricity-production resources globally. It is a renewable, environmentally- and human-friendly source of sustainable electricity. It is easy to apply with comparable minimal operation and maintenance costs, operable as grid-connected, and off-grid installation. It is recognized, alongside solar power, as the leading renewable power source in the energy market.

Based on this, global installed wind power capacity at the end of the year 2018 was about 597 gigawatts (GW) with installed capacity addition of approximately 50.1 GW between 2017 and 2018 [3]. Currently, China leads the world with a cumulative installed capacity of about 217 GW, as of 2018, followed by the USA with about 96 GW. Some other countries from Europe, Asia, South, and North America have adopted wind energy systems operation for power generation. Some of the top countries, based on 2018 statistics, include Germany (59 GW), India (35 GW), Spain (23 GW), United Kingdom (20 GW), France (15 GW), Brazil (15 GW), and Canada (13 GW) [2, 3]. China has become the first nation to reach an accumulated installed capacity of 200 GW or more, while the USA is just about the 100 GW capacity mark. Figure 2.1 displays the trend of global total installed wind power capacity from the year 2009 to 2018. It shows a progressive increase in the adoption

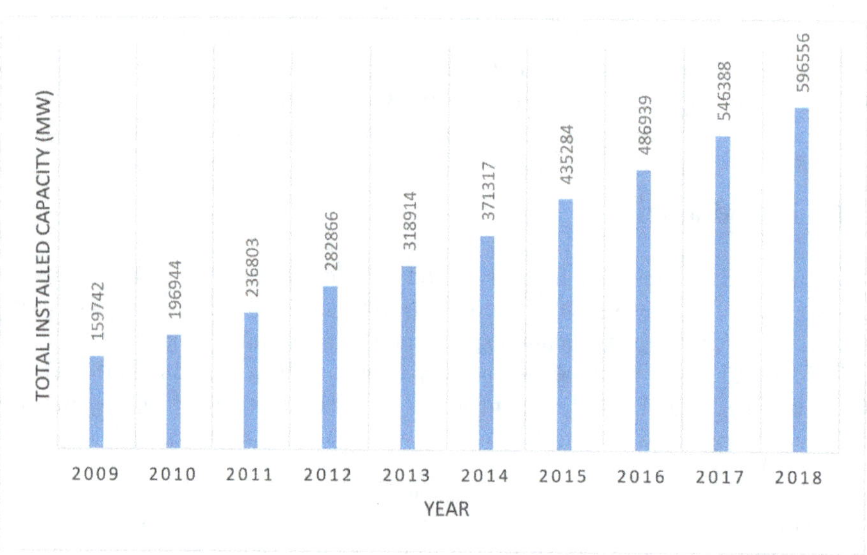

Figure 2.1 Global cumulative installed wind power capacity (MW) [2, 3]

and deployment of wind power generation. This growth in the acceptance and deployment of wind power is predicated on the positive results of concerted efforts to improve the turbine system's performance and the consistent reduction in electricity production cost per kW. All of the wind turbine system's main components have received significant research attention over the years. Besides those mentioned earlier, these research efforts include consideration of the hub heights for optimum generation, startup capacity rating, siting, environmental influence, and resource assessment and characterization. Others include reducing noise pollution, condition monitoring, design optimization, advanced control technology for effective performance, and aerodynamics of the rotor blades.

Wind energy research projects focus on seeking ways to make the system more efficient and environmentally friendlier. One main component at the center of research efforts is the aerodynamics of the turbine blade, as the blade of a wind turbine plays a crucial role in its operation and performance. A careful and proper blade design is required for the system to function at maximum efficiency [4]. Thus, research efforts are still ongoing to seek ways to improve the aerodynamics of blades in such a way as to improve the glide ratio over a range of incidence angles. This chapter's interest is to review some of the researches on the aerodynamics of wind turbines with a particular focus on horizontal axis wind turbine (HAWT).

2.2 *Wind turbine's blade technology development*

The shape of the blades is one of the most critical factors that affect turbine performance. Blade design affects the stiffness and stability of the rotor as well as the efficiency of the turbine [5, 6]. The blade is the part that is directly in charge of extracting the kinetic energy of wind, and its aerodynamics is critical to the system's performance. Hence, attempts to improve the aerodynamic efficiency of the blade profile will eventually boost the performance of the turbine system [7]. For instance, Kwon *et al.* [8] performed a sectional aerodynamic design optimization of a HAWT to improve the aerodynamic performance of the turbine blades. They employed computational fluid dynamics to analyze and optimize the design by studying nine spanwise blade sections. The results show that the aerodynamic optimization brought about an 11% enhancement of torque and better turbine performance. Also, Yu *et al.* [9] employed a computational fluid dynamics solver to optimize the shape of a HAWT blade to enhance the glide ratio. The outcome indicated that the optimization and reconfiguration of the blade section increased the lift-to-drag ratio and consequently improved the aerodynamic efficiency at both the designed and off-designed operating conditions. Lee and Kwon [10] also carried out the aerodynamic optimization and aeroelastic deformation study of a HAWT. The purpose was to demonstrate the influence of shape optimization on the cost of energy. The study employed the coordinates of the blade section contours normal to the chord as the design variables. It used an artificial neural network technique to optimize these variables to maximize the blades' aerodynamic performance, thus minimizing energy production costs. It is important to note that optimizing to maximize energy production versus optimizing to minimize electricity costs would lead to different designs. Benini and Toffolo [11] tried finding the optimal trade-off between maximizing annual energy production (AEP) per square meter of a wind farm and minimizing energy costs. They found that the minimization of electricity costs requires large-sized HAWTs having high AEP but low blade loads and low blade weights. On the other hand, maximizing AEP density requires small HAWTs with relatively low AEP but high blade loads, leading to overly expensive design solutions.

In another research, Zhang *et al.* [12] demonstrated the impact of operating parameters on the aerodynamics and aeroacoustics of a HAWT. They discussed how the tip speed ratio, the inlet flow speed, and the turbulent intensity could affect the aerodynamic performance of a wind turbine's rotor blade. More so, the investigation reported by Bani-Hani *et al.* [13] stated categorically that the shape of the rotor blades has effects on the turbine performance and reported explicitly that a lack of an airfoil shape has the capacity to reduce the turbine performance by 5%. The aerodynamics of a given turbine design is a function of three non-dimensional parameters: Reynolds number, Mach number, and the tip speed ratio [14]. Mach number implies the ratio of the rotor speed to the speed of sound at room temperature and describes the compressibility effect; the tip speed ratio demonstrates the relative importance of the speed induced by the blade tip rotation compared to the upstream wind speed; and finally, the Reynolds number compares the significance of the inertial to the viscous forces and describes the laminarity and turbulence of fluid flow around a rotor boundary. One needs to ensure dynamic similarity between simulations or experiments and the field scenarios to ensure the blade optimization is realistic. This requires a match between all three dimensionless numbers. Among all, it was found that the Reynolds number can significantly alter turbine performance even at relatively large values. According to Miller *et al.* [14], there is a strong Reynolds number dependence in the turbine's power coefficient, even at very high Reynolds numbers, up to 10 million.

2.3 Progress in the aerodynamics of HAWT rotor blades

Without a rigorous airfoil design, a blade may rely on the wind's push to drive its rotation. This will require that the wind force be high enough to overcome its inertia. Consequently, the moment it begins to move, the drag force will actively act on it with such a tendency as to halt the motion. In essence, the blade acts as if it is initially insensitive to the wind until the wind force becomes large enough, leading to high cut-in wind speeds.

Due to the differential airflow between the rounded and flat surfaces of the airfoil, there is a low-pressure region around the downward curved surface, and high pressure occurs around the flat surface. The low-pressure region sucks the blade downward, cursing a lift, and the high-pressure region (which is at the upwind face of the airfoil) pushes on the blade, thereby slowing it down, a phenomenon described as drag. For an explanation of the two forces, consider Figure 2.2. An important design criterion, therefore, is for the lift-to-drag ratio (called glide ratio) to be as high as possible at the appropriate incidence angle (called angle of attack). The idea is to maximize the glide ratio across a range of angles of attack than at a single

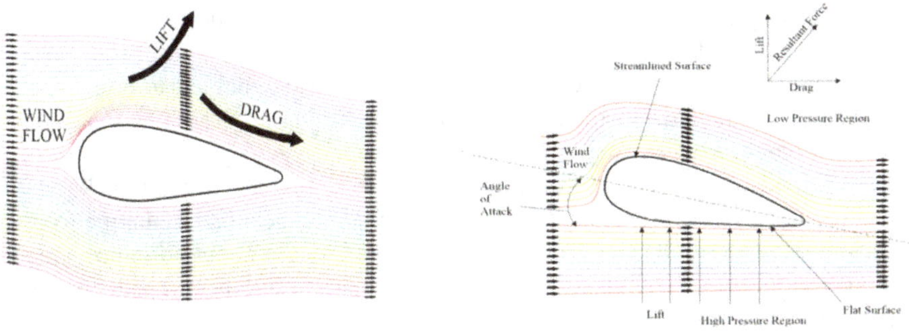

Figure 2.2 Diagrammatic representation of the blade airfoil showing the aerodynamic forces and flowstream

peak point. To demonstrate the dynamics of the glide ratio over a range of angles of attack, Ajayi *et al.* [15] designed and analyzed the aerodynamic performance of a new airfoil applied in a small HAWT. They interpolated the coordinates of three standard airfoils (S1210, E214, and NACA7409). They used the attached flow stage airfoil behavior with XFLR5 and the blade element momentum theory to develop and analyze the new airfoil. The performance results were compared with those of the parent airfoils. The outcome showed that the new airfoil had a preferable glide ratio over a longer range of angles of attack, chord distribution, and blade twist. Overall, the performance of the new airfoil was better than that of its parents. One beneficial conclusion of Ajayi *et al.*'s [15] study was that combining the existing airfoils by finding the average of the coordinate of the points, making up their profile, could lead to more efficient airfoils.

Several other studies have been carried out to determine the effect of modifying an airfoil's aerodynamic parameters on the performance of a turbine blade. For instance, Chattot [16] investigated the effects of blade tip modifications on a wind turbine rotor with a focus on the forward (−10%) or backward (+10%) sweeps, dihedral, or winglet curves. The idea was to improve the aerodynamic performance of the blades. It used the vortex model based on the Goldstein approach and treated each blade as a lifting line, which is given sweep in the plane of rotation or the plane containing the blade and the rotor axis. Each of the modifications increased power generation. The overall conclusion was that the aerodynamic performance could be enhanced with tip modifications. Based on this, there has been a growing interest in studying rotor tip geometry modification toward enhancing the aerodynamic efficiency of the turbine blades and thereby making them less sensitive to gustiness. Kaya *et al.* [7] studied the aerodynamic performance of a HAWT with different forward and backward swept blades using computational fluid dynamics. Prior to the study, investigation on HAWT with swept blades was mostly directed toward loads on the blades. However, this study focused on the effect of the blade sweep direction, the location of the sweep startup, and blade tip offset on the aerodynamic performance of the system. The forward-swept blades improved the performance while the backward-swept blades should be preferred when the desire is to decrease the thrust coefficient.

Several studies have also focused on the application of winglets to improve the aerodynamic characteristics of turbine blades. Khaled *et al.* [17] studied the effect of winglet length and cant angle on the performance and aerodynamic efficiency of a HAWT blade. They employed the ANSYS-Fluent, a CFD software package, to simulate a steady-state flow and used the artificial neural network technique for the optimization. The results showed that the winglets brought about an improvement in power and thrust coefficients. Various other studies have been conducted that highlighted the advantages of adding winglets to the blade profile. The advantages include its ability to reduce downwash and drag coefficient, improve thrust and load distribution, and diffuse the wing-tip vortex away from the rotor plane, among many other benefits [18–21].

Göçmen and Özerdem [22] studied the geometric modifications around the trailing edge and pressure side of airfoil profiles to enhance the aerodynamic performance and reduce the aerodynamic noise. The modifications include the use of a thinner edge and engraved surface in the airfoils' lower part. The idea was predicated on the fact that most of the noise emanates from the side of the trailing edge, and the pressure side is aerodynamically critical. The outcome demonstrated that the geometrical optimization induced an improved glide ratio and reduced noise level up to 5 dB. Meng *et al.* [23] experimentally investigated the performance of a novel downwind forward-folding rotor operated on a HAWT. The turbine design aligned the thrust, gravitational, and centripetal forces along the blade path. It considered different pitch angles and their resulting coefficient of power. They found that the fold angle has a good influence on the turbine's efficiency. At a fold angle of 25°, the maximum coefficient of power increased with increasing freestream speed. The study also reveals the implication of the blade pitch angle on the rotor power output. The purpose of the downwind turbine configuration is to eliminate the structural problems associated with large blade loads due to size, weight, stiffness, shear stress, and span of the rotor blades. Adapting the blade/rotor geometry to cope with the condition and relax the stiffness constraints is a step in the right direction [24–27].

Although the aerodynamic efficiency of turbines is largely dependent on the blade geometry, the performance may differ depending on its two or three dimensionalities. Based on this, Shen *et al.* [28] investigated the aerodynamic shape optimization of non-straight small wind turbine blades. The methodology was focused on giving sweep to the blades with the three-dimensional (3D) stacking line in the plane of rotation and dihedral in the plane containing the blade and rotor axis. It employed a lifting surface method with a free wake model to represent the aerodynamic model. The evaluation was based on the blade element method, and the optimization of the non-straight, aerodynamically shaped blades was carried out with a micro-genetic algorithm. The basic idea was to study the effects of the optimization of chord and twist angle distribution and the 3D stacking line. The results demonstrated that blades with a properly designed 3D stacking line could be used in wind turbines to produce more energy with better starting behavior than the 2D-optimized blade geometry. Strikingly, the study noted that the shorter chord length throughout the blade length gives the blade a low second moment of inertia. In contrast, the larger twist angle gives high torque at a low speed ratio, leading to a better startup. Also, the 3D geometry gives additional driving force at the blade tip, thereby producing more torque to the main shaft, but does not enhance the startup behavior of small turbines at low speeds. It is worth noting that the 3D stacking line can be used to reduce the turbine's operation noise, improve aerodynamic performance, and reduce blade loads [29]. More so, the lifting line and lifting surface methods (a part of the vortex methods) present the turbine blades as a series of bound and trailing vortices and provide the opportunity to analyze 3D aerodynamic shapes more appropriately. In the same vein, Melo *et al.* [30] examined the flow around the rotor blade boundaries of a HAWT by using the lifting line theory. It is noted that the flow around a wind turbine is complex and challenging to analyze and, as such, may require simplification. The lifting line theory provides the opportunity to model the aerodynamics of the turbine rotor and demonstrate the radially bound vortices. One advantage of the lifting line theory when coupled with a non-linear vortex wake alignment scheme is that it becomes simpler with fewer computation tasks than the viscous flow model such as the Reynolds-Averaged-Navier Stokes model.

In other studies, Singh and Ahmed [31] investigated the performance of specially designed low Reynolds number airfoils that permit lower cut-in wind speeds while increasing the startup torque. The airfoils operated on a two-bladed rotor. The idea was to eliminate the low wind speed effect that small wind turbines face due to laminar flow separation and laminar separation bubbles caused by low wind speeds and small rotor size. Laminar flow separation occurs when the separated laminar boundary layer does not reattach to the downstream surface as a result of an unfavorable pressure gradient, thereby creating a reduction in aerodynamic performance. The design incorporated taper and twist to the airfoil and varied the pitch between 15° and 20° with a speed range of 3–6 m/s. The outcome demonstrated that the two-bladed designed airfoil rotor produced more electrical power with a better coefficient of power than a conventional three-bladed turbine when operated under similar conditions.

Similarly, Refan and Hangan [32] investigated the three-dimensionality effect on a small HAWT and found that the three-dimensionality effects on the turbine rotor blade can be due to flow separation delay at the upwind section, tip vortex, radial flow, and downwash effects. Also, Li *et al.* [33] affirmed that the characteristics of the flow field near the boundary layer of a rotating blade are very complex due to the 3D unstable separated flow. Therefore, to have an accurate assessment of the blade's aerodynamic characteristics in the spanwise direction, it is worthwhile to consider the effect of the 3D stall delay in the boundary layer. Based on this, Li *et al.* [33] investigated the surface flow characteristics of a HAWT rotor. They found among other things that the airfoil surface flow at low tip speed ratio displayed a stronger three-dimensionality effect than at optimum tip speed ratio. Further to this, Li *et al.* [34] considered the 3D effect, in the spanwise direction, of the surface flow of an airfoil to study the performance characteristics and near wake effect of a three-bladed upwind HAWT. They measured the pressure distribution and the velocity characteristics around the rotor surface and found that the pressure coefficient is stable between chord positions (x/c) of 0.3 and 0.5 for $r/R = 0.4$ and 0.5, respectively. They also found that the flow around the blade surface accelerated and decelerated around the leading and trailing edges, respectively.

Airfoil concavity is a method that has been used to prevent or postpone flow separation around the wings of aircraft. With the blade of wind turbines mostly shaped to mimic aircraft wings, it is worthwhile to investigate the effectiveness of applying airfoil concavity on the aerodynamic performance of wind turbine blades. This is because, according to Ma *et al.* [35], the flow field around the blade of a wind turbine is more complicated than that of an aircraft wing. Hence, Ma *et al.* [35] experimentally studied the effect of airfoil concavity near the trailing edge of the blade suction surface on a small HAWT. The results indicate that the airfoil concavity reduced aerodynamic noise, improved blade stiffness, and increased aerodynamic efficiency of the turbine by 3–15%. The ideology behind airfoil concavity is its capability to control flow separation on the surface of the airfoil. The point of flow separation is regarded as the notable position of concavity. Aerodynamic performance can be improved via flow control. Flow control is a technique that may be used to improve the fluid dynamics around the airfoils. Flow around the airfoil can be controlled actively or passively [36]. It involves introducing a device around the rotor system to delay the boundary layer separation so that turbulence is suppressed. In line with this, Wang *et al.* [37] reported that using an off-surface small structure (referred to simply as micro-cylinder) in front of the blade's leading edge has the potential of preventing the flow separation without affecting the turbine stability, thus enhancing aerodynamic performance by improving the lift, reducing the drag and flow-induced noise. Menon *et al.* [38] earlier investigated the effect of flaps attached to the trailing edge of a turbine's blade. It demonstrated the impact of active control, unlike Wang *et al.* [37]. The study demonstrated that appropriately positioned flaps have the advantage to revitalize the boundary layer on the upper surface of the blade section by reattaching the separated boundary layer. The phenomenon, therefore, delayed the onset of flow separation and enhanced aerodynamic efficiency. Amini *et al.* [39] studied the effect of an airfoil attachment on the lift-to-drag ratio of a blade. They attached a short flat plate to the trailing edge of the airfoil at the pressure side. The attachment is called a Gurney flap (see Figure 2.3). The purpose of the Gurney

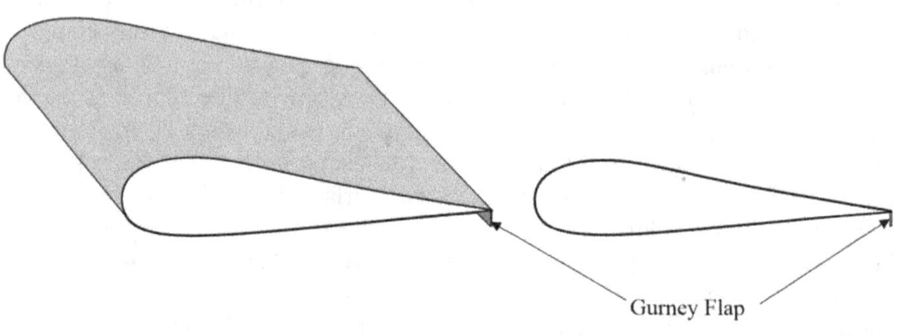

Figure 2.3 Schematic representation of the Gurney flap attached to an airfoil

flap is to increase the lift of the airfoil considerably. However, while doing that, it also increases the drag on the pressure side, as a result of the flow separation, downstream of the trailing edge. To decrease the drag coefficient, the study employed a continuous adjoint formulation to optimize the shape of the airfoil. The results show that the method improves the aerodynamic efficiency of the airfoil.

Amini *et al.* [40] also studied the effects of the angle of attack and Gurney flap height on the aerodynamics lift-to-drag characteristics. Equally, Zhang *et al.* [41] studied the effect of implementing two configurations of the Gurney flaps at the inboard part of the blade, located at 85% of the chord on the pressure surface. They used the Reynolds-Averaged-Navier Stokes with multiple reference frames to model the turbulence flow over the rotor. The study focused on the fact that although the root section of the blade contributes little to the magnitude of generated power compared with the blade tip section, it accounts for the majority of aerodynamic losses. This is because flow separation and dynamic stall are predominant around the blade root section. Hence, unlike Amini *et al.* [39], Zhang *et al.* [41] installed the Gurney flap at the root section, and the results indicate an improvement in power coefficient up to about 21%. The study showed that deploying Gurney flaps creates a 3D effect on spanwise thrust and torque distribution and increases the power coefficients. Worthy of note is that, according to Zhang *et al.* [41], the Gurney flap application is a passive flow control technique that can be used to improve the design lift coefficients of airfoils. Wang *et al.* [42], on the other hand, studied the impact of leading-edge slat on the aerodynamic performance of an S809 airfoil and phase VI blade. The leading-edge slat was employed as a flow control device to control the flow separation around the boundary layer. The study demonstrated that the positional arrangement of the slat influenced the lift and pressure drag coefficient of the airfoil with the ability to delay flow separation and improve the aerodynamic efficiency of the blade shape.

In addition, Abdelwaly *et al.* [43] numerically investigated the influence of wind turbine shroud on the flow and its corresponding impact on turbine performance. Different shroud geometries were considered on the flow field velocity at the rotor. Wind turbine shrouds are used to increase the rotor flow speed, and the shroud parameters have a great impact on the aerodynamic performance of the turbine system [44]. The outcome shows that, when the shrouded turbine performance was compared with the unshrouded, more torque and power were produced by the turbine with the shroud. Sedighi *et al.* [45] numerically investigated the effect of dimples on blades on the aerodynamic performance of HAWT. The influence of location, quantity, and radius of dimples on the flow separation, power generation, torque, and thrust was considered. The results indicate that appropriately designed dimples on blades can increase torque and power generation. In light of this, several other studies investigated the impact of the aerodynamics of blades vis-à-vis different blade profile modifications and its corresponding complex flow stream to determine the performance of the airfoil shapes on power generation [46–53].

However, knowing that an important aspect of the aerodynamic design and optimization of a wind turbine is the distribution of the chord and twist angle along the blade, modifications of these parameters have effects on the performance and output of a wind turbine [54, 55], especially very small wind turbine. Hence, Pourrajabian *et al.* [56] studied the aerodynamic design and optimization of the blades of three micro-HAWTs operating in low wind speed regions. It considered geometric modifications of chord and twist angle distribution while at the same time seeks for a better startup behavior and power output. The blade element momentum theory was employed together with a genetic algorithm optimization technique. The outcome demonstrates that to enhance the startup performance, larger chord and twist values at the root part of the blade are essential for low wind speed operation. The results further indicate that a higher number of blades, up to 5, improves the torque of micro-HAWTs and, therefore, enhances the startup time and power output. This is because tiny wind turbines operating in low-speed regions have torque requirements, especially the resisting torque that must be overcome. Thus, the increase in the number of blades increases the aerodynamic torque, leading to better power coefficient and improved startup time. Tang *et al.* [55] employed a direct mathematical and optimization approach to design a small wind turbine. The method focused on evaluating the optimal blade chord and twist angle distribution while considering the effects of the Reynolds number, hub, drag, and blade tip, respectively. The results show that the tip-hub loss brought about a visible effect on the chord and twist angle distributions at the tip and hub sections.

Further to this, Mouhsine *et al.* [57] reported that the orientation of the twist angle and the chord distribution affect the lift-to-drag characteristics and, consequently, the aerodynamic efficiency. Pourrajabian *et al.* [58] showed that an increase in the twist and chord distribution enables the reduction in the rotational inertia and consequently reduces the startup time. Kaya and Elfarra [59] demonstrated that the chord lengths at the tip, mid-span, and root affect the magnitudes of the thrust and torque. It noted that when the tip chord length is about 25% of the mid-span and root chord lengths, the torque increases, while it decreases when the root chord is longer than both the tip and mid-span chords. However, minimum thrust and maximum torque are obtained when the tip chord is as small as possible. Tahani *et al.* [60] further applied a novel linearization technique on the chord and twist distributions to determine the optimum points along with the chord and twist distributions where higher power coefficients can be derived. The outcome demonstrated that the optimum locations for higher power coefficients are from 30% to 37% along the blade length for twist distribution and from 60% to 67% along the blade length for the chord distribution, respectively. In a previous study, Hoogedoom *et al.* [61] used a 2D computational analysis to investigate the aeroelastic behavior of a flexible HAWT blade. The effects of parameters such as pitch angle, cambering, airfoil thickness, and wind speeds on the flexibility of the aerodynamically shaped blade were considered. The study was premised on the plausibility of replacing rigid blades with elastic blade material that allows for the adaptation of the blade shape to the direction and speed of the wind, causing an improvement in blade characteristics and performance efficiency. The study concluded by stating that flexible cambered airfoils can improve the glide ratio and also increase the turbine's performance efficiency.

Similarly, Keshavarzzadeh *et al.* [62] investigated the impact of design quality and response on the shape optimization of the rotor blades of a HAWT. They redesigned the blades by optimizing the twist angle along the blade span, focusing on maximizing power generation as a function of the blade stiffness. The wind loads on the blades were evaluated via an aerodynamic analysis using the blade element method. They demonstrated the effects of time-dependent radial and tangential gravitational forces and time-independent centrifugal forces alongside the aerodynamic loads. Kavari *et al.* [63] studied the impact of wind shear on the aerodynamic coefficients along the length of a HAWT blade. It used the blade element momentum theory with a vertically changing wind speed profile along the blade length. The results indicate that the aerodynamic coefficient is marginally affected by wind shear. However, for a wind turbine designed for constant inflow, the existence of a wind shear profile causes a reduction in the harvestable energy and aerodynamic performance. As a result of the advantage of direct grid-connection using induction generators, fixed-pitch fixed-speed HAWTs have gained prominence over the counterpart fixed-pitch variable-speed wind turbines that use permanent magnet synchronous generators. It is simple and costs less, therefore widely deployed. However, the chord and twist angle distribution of fixed-pitch fixed-speed HAWTs are influenced by the designed rotational speed of the rotor blades. Liu *et al.* [64] employed optimum rotor and blade element momentum theories with both Prandtl tip loss correction and wake to design the blade profile and determine the power performance of the blade. They employed linear radial profiles of blade chord and twist angle with fixed and floating values at the blade tip and root, respectively. The outcome indicates that the new profile produced a better wind power harvest at low material and manufacturing costs. Moreover, contrary to the assertion of fixed-speed over variable-speed, Sedaghat *et al.* [65] carried out a study on the aerodynamic performance of variable speed HAWT with optimized blades. They used a compact blade element momentum technique to optimize the blades and introduced a generalized quadratic equation on the angular induction factor while variable rotor speed was assumed. The outcome shows that the variable-speed wind turbine outperformed the conventional constant speed turbines, especially at higher wind speeds. This suggests that the aerodynamic designs of the wind turbines have great impacts on their performance irrespective of the speed types.

2.4 Augmenting horizontal axis wind turbines with diffusers

When a homogenous freestream moves through the rotor of a wind turbine, it suffers retardation in speed due to the conversion of the wind's kinetic energy into electrical power. In line with this analogy, Betz employed the vortex theory to achieve the ideal efficiency of a wind turbine. He showed that the maximum theoretically extractable energy from an ideal wind turbine is limited to 59.3% of the available mechanical power in the wind [66–68]. Based on this, several turbine systems have been designed with the idea of Betz's limit [69–72]. Some studies have recently considered designs that led to generation at higher coefficients of power above the Betz limit. However, it is worth arguing against that these designs may actually not exceed Betz's limit. We believe these novel designs provide more input power to the turbine; hence, its output increases. A critical point that users need to consider is that if a diffuser is added to the turbine to guide more wind into the rotor, the setup's total area, that is, the diffuser's largest cross-sectional area, must be used to calculate the power coefficient, defined as $c_p = \frac{P/1}{2}\rho A V^3$. Of course, if one uses the rotor area to calculate the power coefficient, a c_p larger than Betz's limit might be found; however, this is not a fair comparison. For instance, Abe *et al.* [73] installed a flanged diffuser and demonstrated its impact on the flow field in the wake of a small HAWT. They found that the turbine with the flanged diffuser generated a power coefficient that was higher in magnitude than Betz's limit and four times higher than that without the diffuser. As was mentioned, the augmented turbine does not exceed Betz's limit if one used the diffuser's maximum area to calculate the power coefficient. Also, the power coefficient is four times larger than the small original turbine. Some critics believe it would be fairer to compare the diffuser augmented turbine's power production with a turbine that is as large as the diffuser, not the small original turbine. Abe *et al.* [73] also elucidated the flow mechanism in the wake of the turbines with and without the diffuser and found a considerable difference in the vortex structure. Rapid destruction of the vortex structure occurred in the diffuser-shrouded turbine, while the vortex structure downwind of the turbine without the flanged diffuser recovered. Worthy of note is that the flanged diffuser serves the purpose to accelerate the approaching wind and, by so doing, improves the power generation.

Ohya *et al.* [74] investigated the influence of the flanged diffuser parameters on power generation. They employed a diffuser shroud with a broad-ring (large) flange at the exit periphery. The large flange is a distinguishing feature of the study's diffuser shroud. The strategy adopted was to position the flange at the diffuser shroud's exit to form a vortex, which leads to a low-pressure zone around the exit region of the diffuser. The low pressure creates suction at the inlet, leading to much more airflow toward the turbine inside the diffuser. Aside from the large flange, the optimum diffuser parameters include the inlet shroud shape, hub ratio, center body length, flange height, and the diffuser open angle. The result indicated that the flanged diffuser, which also acted as yawing motion control, led to approximately four to five times more power production and a significant reduction in the aerodynamic noise. Wang *et al.* [75] carried out an experimental investigation into the influence of the flanged diffuser on the dynamics of the blade of a 3 kW wind turbine shroud with a flanged diffuser. The flanged diffuser component complicates the wind load acting on the turbine system and creates a complex interaction between the blade rotation and the flow stream. The results revealed that the turbine with the flanged diffuser rotates faster, and the dynamic strain is higher than that without the flanged diffuser.

Khamlaj and Rumpfkeil [76] considered optimizing the turbine and wind-lens (flanged diffuser shroud assembly) parameters to maximize the power generation and simultaneously decrease the thrust and drag forces. They employed a multi-objective genetic algorithm as the optimization technique while the focus was given to the blade's chord, twist distributions, and flange height, among others. The blade element momentum theory and the Reynolds-Averaged- Navier Stokes equation with the k-ε turbulence model were engaged for evaluating the aerodynamic performance and flow phenomenon. The outcome demonstrated that simultaneously optimizing the blade shape and wind-lens parameters lead to improved shrouded turbine performance. Siavash *et al.* [77] developed a nozzle-diffuser duct made of a fixed ring and two-piece diffusers capable of rotating over each other so that the diffuser wall can be opened up to 180°. The diffuser assembly works on a micro-HAWT and serves to control speed-up ratio and drag forces. The outcome showed that the ducted turbine improved the power generation. The study further found that the presence of rings in the ducts amplifies the power and rotor speed. They also found that in higher ranges of wind speed, the half-opened duct exceeds the conventional shrouded turbine in power generation and rotational speed.

Vaz and Wood [78] also studied the effectiveness of a diffuser surrounding a rotor and claimed it could improve the power coefficient of a wind turbine above the Betz's limit, which again, can be argued against if one used the diffuser's maximum area to calculate the power coefficient—it does improve the power production of the turbine though. They employed the blade element theory extended to include the diffuser and mathematically solve for the associated diffuser losses. The numerical results were compared with those of the experiment. The outcome demonstrated the dependency of the power and thrust coefficients on the diffuser efficiency. In another study, Leloudas *et al.* [79] developed a modular scheme to optimize the aerodynamic shapes of diffuser-augmented HAWT shrouds. Their study employed both numerical and experimental approaches and found that the diffuser-augmented turbine shrouds were able to: provide high-velocity acceleration and to reduce flow separation around the exit region, which itself increases the inlet speed, causes a reduction in drag coefficient, and prevents recirculation. Agha *et al.* [80] also performed an experimental study on a full-scale diffuser-augmented HAWT to evaluate the response behavior and the augmentation ratio of the augmented turbine. Response behavior refers to the time lapse between free stream wind speeds experienced at the rotor and the turbine's output power. The augmentation ratio was defined as $r = \left(C_D/C_p\right)\left(V_t/V_\infty\right)^3$, where C_D is the rotor's drag coefficient, C_p is the pressure recovery, V_t is the velocity of wind at the turbine, and V_∞ is the freestream velocity. The results showed that the diffuser shroud accelerated the wind speed by about 136%, the augmentation ratio was about 8.32, and the turbine exhibited improved behavior compared to the unshrouded turbine.

2.5 Dual-rotor turbines

Moreover, it is reported that employing a dual-rotor system (see Figure 2.4), with one in the wake of the other, has the potential of increasing the power coefficient compared to when a single-rotor system is used. Bani-Hani *et al.* [13] reported that

Figure 2.4 Double-rotor-horizontal axis wind turbine

adding a second rotor system to a HAWT can increase the Betz limit and further boost the energy harvest. Also, it was noted that making the second rotor rotate in a counter direction to the first greatly improved the efficiency/performance and, consequently, the energy harvest [81]. In light of this, as a result of the complexities arising from the aerodynamic interaction between the two rotor systems, Lee *et al.* [82] investigated the effects of design parameters, such as rotation speed, radii of the rotors, and the pitch angles, on the aerodynamic efficiency of a counter-rotating HAWT. They employed the blade element momentum theory for the analysis. The outcome indicated that the aerodynamic efficiency could be improved by variations in the rotation speed ratio, radii of the two rotors, and the combination of pitch angles. Kumar and Saha [83] theoretically analyzed the performance of a shrouded-twin-rotor HAWT using the actuator disk theory to determine the power production efficiency of the system. The study focused on assessing the optimum performance and the effects of the shroud parameters, flow efficiency, viscous loss, and back efficiency, among others. Kumar and Saha's [83] design has no flange. The outcome demonstrated that the first rotor area, nozzle efficiency, and back pressure coefficient have a significant impact on the magnitude of power production and showed that the system's efficiency exceeds the Betz limit. In addition, it was found that the shroud increased the air mass flow rate, and the second rotor accounted for additional power production.

2.6 Conclusion

The shape of wind turbine's rotor blades is one of the critical factors that influence turbine performance, and the impact of its aerodynamics on flow dynamics and operation characteristics has, over the years, been the focus of various research efforts. The focus of the research efforts is majorly geared toward improving aerodynamic performance and reducing electricity production costs while making the system more environmentally friendly. Various researches centered on the improvements of the aerodynamics of HAWTs' rotor blades were surveyed in this chapter. Several influential concepts, including flow separation and flow control techniques, and shape modification, were considered, among others. Also, the contributions of twin-rotor systems and shrouded turbine systems to improving the turbine's performance were reviewed.

References

[1] Third Planet Windpower (TPW). *History: early wind power [online]*. 2020. Available from www.thirdplanetwind.com/energy/history.aspx [Accessed 31 Jan 2020].

[2] Ajayi O.O., Ojo O., Vasel A. 'On the need for the development of low wind speed turbine generator system'. *IOP Conference Series: Earth and Environmental Science*. 2019;**331**:012062–12.

[3] World Wind Energy Association (WWEA). *Wind power capacity worldwide reaches 597 GW, 50.1 GW added in 2018 [online]*. 2019. Available from https://wwindea.org/blog/2019/02/25/wind-power-capacity-worldwide-reaches-600-gw-539-gw-added-in-2018/ [Accessed 31 Jan 2020].

[4] Chaudhary M.K., Roy A. 'Design & optimization of a small wind turbine blade for operation at low wind speed'. *World Journal of Engineering*. 2015;**12**(1):83–94.

[5] Jureczko M., Pawlak M., Mężyk A. 'Optimisation of wind turbine blades'. *Journal of Materials Processing Technology*. 2005;**167**(2-3):463–71.

[6] Tenguria N., Mittal N.D., Ahmed S. 'Review on horizontal axis wind turbine rotor design and optimization'. *Trends in Applied Sciences Research*. 2011;**6**(4):309–44.

[7] Kaya M.N., Kose F., Ingham D., Ma L., Pourkashanian M. 'Aerodynamic performance of a horizontal axis wind turbine with forward and backward swept blades'. *Journal of Wind Engineering and Industrial Aerodynamics*. 2018;**176**(2):166–73.

[8] Kwon H.I., You J.Y., Kwon O.J. 'Enhancement of wind turbine aerodynamic performance by a numerical optimization technique'. *Journal of Mechanical Science and Technology*. 2012;**26**(2):455–62.

[9] Yu D.O., Lee H.M., Kwon O.J. 'Aerodynamic shape optimization of wind turbine rotor blades considering aeroelastic deformation effect'. *Journal of Mechanical Science and Technology*. 2016;**30**(2):705–18.

[10] Lee H.M., Kwon O.J. 'Performance improvement of horizontal axis wind turbines by aerodynamic shape optimization including aeroealstic deformation'. *Renewable Energy*. 2020;**147**:2128–40.

[11] Benini E., Toffolo A. 'Optimal design of horizontal-axis wind turbines using blade-element theory and evolutionary computation'. *Journal of Solar Energy Engineering*. 2002;**124**(4):357–63.

[12] Zhang S., Luo K., Yuan R., *et al.* 'Influences of operating parameters on the aerodynamics and aeroacoustics of a horizontal-axis wind turbine'. *Energy*. 2018;**160**(1):597–611.

[13] Bani-Hani E., Sedaghat A., Saleh A., *et al.* 'Evaluating performance of horizontal axis double rotor wind Turbines'. *Energy Engineering*. 2019;**116**(1):26–40.

[14] Miller M.A., Kiefer J., Westergaard C., Hansen M.O.L., Hultmark M. 'Horizontal axis wind turbine testing at high Reynolds numbers'. *Physical Review Fluids*. 2019;**4**(110504):1–22.

[15] Ajayi O.O., Okeowo O., Aasa S.A., Aboyade A.O., Willoughby A. 'Novel airfoil design for small horizontal axis wind turbine: a preliminary result'. *Proceedings of the Third Southern African Solar Energy Conference, 11–13 May, Kruger National Park, South Africa*; 2015. pp. 181–6.

[16] Chattot J.-J. 'Effects of blade tip modifications on wind turbine performance using vortex model'. *Computers & Fluids*. 2009;**38**(7):1405–10.

[17] Khaled M., Ibrahim M.M., Abdel Hamed H.E., AbdelGwad A.F. 'Investigation of a small horizontal–axis wind turbine performance with and without winglet'. *Energy*. 2019;**187**(2):115921.

[18] Devit I. 'Design and analysis of wind turbine rotor blade and winglet study'. *Journal of Mechanical and Aeronautical Engineering Research*. 2016;**1**(1):17–21.

[19] Farhan A., Hassanpour A., Burns A., Motlagh Y.G. 'Numerical study of effect of winglet planform and airfoil on a horizontal axis wind turbine performance'. *Renewable Energy*. 2019;**131**(1):1255–73.

[20] Hansen T.H., Mühle F. 'Winglet optimization for a model-scale wind turbine'. *Wind Energy*. 2018;**21**(8):634–49.

[21] Tobin N., Hamed A., Chamorro L. 'An experimental study on the effects of winglets on the wake and performance of a model wind turbine'. *Energies*. 2015;**8**(10):11955–72.

[22] Göçmen T., Özerdem B. 'Airfoil optimization for noise emission problem and aerodynamic performance criterion on small scale wind turbines'. *Energy*. 2012;**46**(1):62–71.

[23] Meng H., Ma Z., Dou B., Zeng P., Lei L. 'Investigation on the performance of a novel forward-folding rotor used in a downwind horizontal-axis turbine'. *Energy*. 2020;**190**(4):116384.

[24] Jamieson P.M., Jaffrey A. 'Advanced wind turbine design'. *Journal of Solar Energy Engineering*. 1997;**119**(4):315–20.

[25] Kress C., Chokani N., Abhari R.S. 'Downwind wind turbine yaw stability and performance'. *Renewable Energy*. 2015;**83**(A):1157–65.

[26] Netzband S., Schulz C.W., Abdel-Maksoud M. 'Self-aligning behaviour of a passively yawing floating offshore wind turbine'. *Ship Technology Research.* 2020;**67**(1):15–25.

[27] Picot N., Verelst D.R., Larsen T.J. 'Free yawing stall-controlled downwind wind turbine with swept blades and coned rotor'. *Proceedings of the European Wind Energy Association (EWEA).* Brissels; 2011.

[28] Shen X., Yang H., Chen J., Zhu X., Du Z. 'Aerodynamic shape optimization of non-straight small wind turbine blades'. *Energy Conversion and Management.* 2016;**119**(6):266–78.

[29] Elfarra M.A., Sezer-Uzol N., Akmandor I.S. 'NREL VI rotor blade: numerical investigation and winglet design and optimization using CFD'. *Wind Energy.* 2014;**17**(4):605–26.

[30] Melo D.B., Baltazar J., Falcão de Campos J.A.C. 'A numerical wake alignment method for horizontal axis wind turbines with the lifting line theory'. *Journal of Wind Engineering and Industrial Aerodynamics.* 2018;**174**(4):382–90.

[31] Singh R.K., Ahmed M.R. 'Blade design and performance testing of a small wind turbine rotor for low wind speed applications'. *Renewable Energy.* 2013;**50**:812–9.

[32] Refan M., Hangan H. 'Aerodynamic performance of a small horizontal axis wind turbine'. *Journal of Solar Energy Engineering.* 2012;**134**(2):1–7.

[33] Li Qing'an., Xu J., Maeda T., *et al.* 'Laser Doppler velocimetry (LDV) measurements of airfoil surface flow on a horizontal axis wind turbine in boundary layer'. *Energy.* 2019;**183**(5):341–57.

[34] Li Qing'an., Xu J., Kamada Y., *et al.* 'Experimental investigations of airfoil surface flow of a horizontal axis wind turbine with LDV measurements'. *Energy.* 2020;**191**(3):116558.

[35] Ma J., Duan Y., Zhao M., *et al.* 'Effect of airfoil concavity on wind turbine blade performances'. *Shock and Vibration.* 2019;**2019**(4):1–11.

[36] Aramendia I., Fernandez-Gamiz U., Ramos-Hernanz J.A. 'Flow control devices for wind turbines' in Bizon N., Mahdavi Tabatabaei N., Blaabjerg F., Kurt E. (eds.). *Energy Harvesting and Energy Efficiency. Lecture Notes in Energy.* Cham: Springer; 2017.

[37] Wang Y., Li G., Shen S., Huang D., Zheng Z. 'Influence of an off-surface small structure on the flow control effect on horizontal axis wind turbine at different relative inflow angles'. *Energy.* 2018;**160**(3):101–21.

[38] Menon M., Ponta F., Sun X., Dai Q. 'Aerodynamic analysis of flow-control devices for wind turbine applications based on the trailing-edge slotted-flap concept'. *Journal of Aerospace Engineering.* 2016;**29**(5):04016037–12.

[39] Amini Y., Emdad H., Farid M. 'Adjoint shape optimization of airfoils with attached Gurney flap'. *Aerospace Science and Technology.* 2015;**41**:216–28.

[40] Amini Y., Liravi M., Izadpanah E. 'The effects of Gurney flap on the aerodynamic performance of NACA 0012 airfoil in the rarefied gas flow'. *Computers & Fluids.* 2018;**170**:93–105.

[41] Zhang Y., Ramdoss V., Saleem Z., Wang X., Schepers G., Ferreira C. 'Effects of root Gurney flaps on the aerodynamic performance of a horizontal axis wind turbine'. *Energy*. 2019;**187**(1):115955.

[42] Wang H., Jiang X., Chao Y., *et al.* 'Effects of leading edge SLAT on flow separation and aerodynamic performance of wind turbine'. *Energy*. 2019;**182**(99):988–98.

[43] Abdelwaly M., El-Batsh H., Bassily Hanna M., Hanna M.B. 'Numerical study for the flow field and power augmentation in a horizontal axis wind turbine'. *Sustainable Energy Technologies and Assessments*. 2019;**31**(12):245–53.

[44] Matsushima T., Takagi S., Muroyama S. 'Characteristics of a highly efficient propeller type small wind turbine with a diffuser'. *Renewable Energy*. 2006;**31**(9):1343–54.

[45] Sedighi H., Akbarzadeh P., Salavatipour A. 'Aerodynamic performance enhancement of horizontal axis wind turbines by dimples on blades: numerical investigation'. *Energy*. 2020;**195**(1):117056.

[46] Hao W., Ding Q., Li C. 'Optimal performance of adaptive flap on flow separation control'. *Computers & Fluids*. 2019;**179**(9):437–48.

[47] Khalafallah M.G., Ahmed A.M., Emam M.K. 'The effect of using winglets to enhance the performance of swept blades of a horizontal axis wind turbine'. *Advances in Mechanical Engineering*. 2019;**11**(9):1–10.

[48] Pasquale L., Durante D., Broglia R. 'Flow separation prevention around a NACA0012 profile through multivariable feedback controlled plasma actuators'. *Computers & Fluids*. 2019;**182**(3):85–107.

[49] Prasad C.S., Dimitriadis G. 'Application of a 3D unsteady surface panel method with flow separation model to horizontal axis wind turbines'. *Journal of Wind Engineering and Industrial Aerodynamics*. 2017;**166**(5):74–89.

[50] Saenz-Aguirre A., Fernandez-Gamiz U., Zulueta E., Aramendia I., Teso-Fz-Betono D. 'Flow control based 5Â MW wind turbine enhanced energy production for hydrogen generation cost reduction'. *International Journal of Hydrogen Energy*. 2020;**32**(12).

[51] Sedighi H., Akbarzadeh P., Salavatipour A., Sessarego M., Ramos-García N., Shen W.Z. 'Analysis of winglets and sweep on wind turbine blades using a lifting line vortex particle method in complex inflow conditions'. *Journal of Physics. Conference Series*. 2018;**1037**:1–12.

[52] Shivaramaiah S., Nagpurwala Q.H., Varpe M.K., Narahari H.K. 'Effect of rotor tip winglet on the performance and stability of a transonic axial compressor'. *Proceedings of the ASME 2017 Gas Turbine India Conference; Bangalore, India, 7-8 Dec. 2017, 1–12*; 2017.

[53] Zhou Y., Hou L., Huang D. 'The effects of Mach number on the flow separation control of airfoil with a small plate near the leading edge'. *Computers & Fluids*. 2017;**156**(1):274–82.

[54] Lee S. 'Inverse design of horizontal axis wind turbine blades using a vortex line method'. *Wind Energy*. 2015;**18**(2):253–66.

[55] Tang X., Huang X., Peng R., Liu X. 'A direct approach of design optimization for small horizontal axis wind turbine blades'. *Procedia CIRP*. 2015;**36**(5):12–16.

[56] Pourrajabian A., Ebrahimi R., Mirzaei M. 'Applying micro scales of horizontal axis wind turbines for operation in low wind speed regions'. *Energy Conversion and Management*. 2014;**87**(3):119–27.

[57] Mouhsine S.E., Oukassou K., Ichenial M.M., Kharbouch B., Hajraoui A. 'Aerodynamics and structural analysis of wind turbine blade'. *Procedia Manufacturing*. 2018;**22**:747–56.

[58] Pourrajabian A., Nazmi Afshar P.A., Ahmadizadeh M., Wood D. 'Aero-structural design and optimization of a small wind turbine blade'. *Renewable Energy*. 2016;**87**(3):837–48.

[59] Kaya M., Elfarra M. 'Effect of taper modification on the performance of NREL VI wind turbine blade for low and mid wind speeds'. *Wind Engineering*. 2019;**43**(4):392–403.

[60] Tahani M., Kavari G., Masdari M., Mirhosseini M. 'Aerodynamic design of horizontal axis wind turbine with innovative local linearization of chord and twist distributions'. *Energy*. 2017;**131**(4):78–91.

[61] Hoogedoorn E., Jacobs G.B., Beyene A. 'Aero-elastic behavior of a flexible blade for wind turbine application: a 2D computational study'. *Energy*. 2010;**35**(2):778–85.

[62] Keshavarzzadeh V., Ghanem R.G., Tortorelli D.A. 'Shape optimization under uncertainty for rotor blades of horizontal axis wind turbines'. *Computer Methods in Applied Mechanics and Engineering*. 2019;**354**(115):271–306.

[63] Kavari G., Tahani M., Mirhosseini M. 'Wind shear effect on aerodynamic performance and energy production of horizontal axis wind turbines with developing blade element momentum theory'. *Journal of Cleaner Production*. 2019;**219**:368–76.

[64] Liu X., Wang L., Tang X. 'Optimized linearization of chord and twist angle profiles for fixed-pitch fixed-speed wind turbine blades'. *Renewable Energy*. 2013;**57**:111–9.

[65] Sedaghat A., El Haj Assad M., Gaith M. 'Aerodynamics performance of continuously variable speed horizontal axis wind turbine with optimal blades'. *Energy*. 2014;**77**(6):752–9.

[66] Huleihil M., Mazor G. 'Wind turbine power: the Betz limit and beyond.' Advances in Wind Power, Rupp Carriveau, IntechOpen. 2012. Available from https://www.intechopen.com/books/advances-in-wind-power/wind-turbine-power-the-betz-limit-and-beyond.

[67] Okulov V.L., Sørensen J.N. 'An ideal wind turbine with a finite number of blades'. *Doklady Physics*. 2008;**53**(6):337–42.

[68] Stiebler M. *Wind Energy Systems for Electric Power Generation, Green Energy and Technology*. Berlin: Springer; 2008.

[69] Jiang H., Li Y., Cheng Z. 'Performances of ideal wind turbine'. *Renewable Energy*. 2015;**83**(3):658–62.

[70] Schubel P.J., Crossley R.J. 'Wind turbine blade design'. *Energies*. 2012;**5**(9):3425–49.
[71] Vaz J.R.P., Wood D.H. 'Performance analysis of wind turbines at low tip-speed ratio using the Betz-Goldstein model'. *Energy Conversion and Management*. 2016;**126**:662–72.
[72] De Lellis M., Reginatto R., Saraiva R., Trofino A. 'The Betz limit applied to airborne wind energy'. *Renewable Energy*. 2018;**127**(D12):32–40.
[73] Abe K., Nishida M., Sakurai A., *et al*. 'Experimental and numerical investigations of flow fields behind a small wind turbine with a flanged diffuser'. *Journal of Wind Engineering and Industrial Aerodynamics*. 2005;**93**(12):951–70.
[74] Ohya Y., Karasudani T., Sakurai A., Abe K.-ichi., Inoue M. 'Development of a shrouded wind turbine with a flanged diffuser'. *Journal of Wind Engineering and Industrial Aerodynamics*. 2008;**96**(5):524–39.
[75] Wang W.-X., Matsubara T., Hu J., *et al*. 'Experimental investigation into the influence of the flanged diffuser on the dynamic behavior of CFRP blade of a shrouded wind turbine'. *Renewable Energy*. 2015;**78**(1):386–97.
[76] Khamlaj T.A., Rumpfkeil M.P. 'Analysis and optimization of ducted wind turbines'. *Energy*. 2018;**162**(12):1234–52.
[77] Keramat Siavash N., Najafi G., Tavakkoli Hashjin T., Ghobadian B., Mahmoodi E. 'An innovative variable shroud for micro wind turbines'. *Renewable Energy*. 2020;**145**(1):1061–72.
[78] Vaz J.R.P., Wood D.H. 'Effect of the diffuser efficiency on wind turbine performance'. *Renewable Energy*. 2018;**126**:969–77.
[79] Leloudas S.N., Lygidakis G.N., Eskantar A.I., Nikolos I.K. 'A robust methodology for the design optimization of diffuser augmented wind turbine shrouds'. *Renewable Energy*. 2020;**150**:722–42.
[80] Agha A., Chaudhry H.N., Wang F. 'Determining the augmentation ratio and response behaviour of a diffuser augmented wind turbine (DAWT)'. *Sustainable Energy Technologies and Assessments*. 2020;**37**(4):100610.
[81] Vasel-Be-Hagh A., Archer C.L. 'Wind farms with counter-rotating wind turbines'. *Sustainable Energy Technologies and Assessments*. 2017;**24**(1):19–30.
[82] Lee S., Kim H., Son E., Lee S. 'Effects of design parameters on aerodynamic performance of a counter-rotating wind turbine'. *Renewable Energy*. 2012;**42**(2):140–4.
[83] Kumar V., Saha S. 'Theoretical performance estimation of shrouded-twin-rotor wind turbines using the actuator disk theory'. *Renewable Energy*. 2019;**134**(3):961–9.

Chapter 3

Scaling utility-scale wind turbines

Logan Unser[1] and Ahmad Vasel-Be-Hagh[1]

This chapter provides guidelines for designing utility-scale wind turbines for wind tunnel testing. While one might think scaling down all dimensions of a utility-scale wind turbine with the same scale factor suffices, the aerodynamic results obtained for such a scale model significantly differ from those of the full-scale turbine. It is essential to scale a turbine so that its power coefficient and tip speed ratio remain unchanged.

3.1 Why test scale turbines?

Wind tunnel testing has been a boon since the very earliest forms of aerodynamic testing. Along with the increase in computational and theoretical capabilities, wind tunnel testing of wind turbines still represents an economical and efficient means of exploring novel control concepts [1], understanding wake interactions within farms [2, 3], and confirming, experimentally, aeroelastic numeric methods [4]. Although full-scale experimental testing and observational data collected may provide a more accurate picture of the aeroelastic interaction of turbine and atmosphere, scale turbines still have a place synergistic with observational and computational studies. The wind tunnel allows for control over the boundary conditions of the system, experimental farm layouts, and for the simple testing of new aeroelastic features and modern control theories. All too often, when considering in-field studies, limited access to proprietary data, the costs associated with instrumenting a full-scale turbine, and the unavoidable external effects, such as a change in surface properties or the surrounding buildings' wakes, can be prohibitive. A scale model as an alternative can help create a more controlled testing apparatus with better-defined errors and conditions. While computational tools are quickly becoming the preferred research method in wind energy, testing scale turbines at wind tunnels can validate these models and further the development of larger, more efficient power generation.

[1]Fluid Mechanics Research Laboratory, Mechanical Engineering Department, Tennessee Technological University, Cookeville, TN, USA

Scale models of wind turbines must overcome many challenges to be an effective research tool. To accurately develop utility-scale multi-megawatt turbines, the control hardware and the turbine blades themselves must be miniaturized. Understanding the aerodynamic performance of the system as well as the behavior of the airfoils at inherently lower Reynolds numbers is necessary to ensure a representative model. Furthermore, the instrumentation and control of the model pose its own set of unique challenges. This chapter will briefly discuss various applications of model turbines and then dive deeper into the scaling considerations and instrumentation of model turbines.

3.2 How to properly scale wind turbines for wind tunnel testing?

As researchers approach scaling a turbine down, it is often useful to look at the nondimensional groups that influence the performance of a turbine. One can see the influence of the Froude number, Reynolds number, tip speed ratio (TSR), the Lock number, and the Mach number, among others [5]. The Froude number, a ratio of inertia of the flow to the gravitational effect, is crucial in studying giant turbines and offshore floating turbines to describe the waves' effects. The similarity of the Reynolds number, the ratio of inertial forces to viscous forces, is sacrificed to maintain the similitude of the TSR while keeping the inlet velocity at a controllable level. The TSR is the ratio of the tip speed of the turbine to the free stream velocity. The Lock number is a dimensionless quantity that represents the ratio of aerodynamic lifting capability and the effect of the weight of the blade. The Mach number is the ratio of local wind speed to the speed of sound. Often researchers neglect Mach number scaling, as the compressibility effects are considered negligible for both scale and prototype turbines. However, recent research at the University of Delaware may call this into question for larger turbines [6].

Other important nondimensional parameters include the coefficient of thrust and the coefficient of performance, as these quantify the structural and energy harvesting characteristics of the turbine. The coefficient of performance is the ratio of power extracted from the wind to the power available, while the coefficient of thrust is the ratio of total axial force acting upon the turbine to the dynamic thrust of the swept area. Both of these parameters are highly reliant on the TSR, and therefore scale turbines must accurately match the TSR of the larger turbine. When assessing a turbine as an aeroelastic system, it is essential to consider the natural frequency of the blades. It would be helpful here to talk specifically about an example scale turbine.

The Lifes50+ scale turbine was developed as part of a project furthering research about offshore wind energy. The scale model was used in conjunction with ocean basin experiments to experimentally confirm numeric simulations done with FAST. The model was designed so that the coefficient of thrust versus TSR curve and the coefficient of power versus TSR curve closely resembled the DTU 10MW reference turbine that it was based on. Additionally, the first flap-wise natural frequency was matched, and the weights of the blades were scaled via appropriate scaling laws [7].

Matching the C_p and C_t of the prototype turbine is a nontrivial task. A simple geometric scaling will result in a turbine with comparatively low performance due to the poor lift/drag characteristics of the airfoils at lower Reynolds numbers. Instead, researchers at Politecnico di Milano opted for a different airfoil entirely. The airfoil SD7032 was chosen from the Selig low Reynolds airfoil database. The lift and drag characteristics were found experimentally for the Reynolds number ranges that they were expecting to operate at.

Once this initial quantification of airfoil performance was completed, they began the aeroelastic scaling of the turbine. The coefficient of thrust and the first flap-wise natural frequency were the design parameters of interest. These two were chosen because the coefficient of thrust has a large effect on the overall dynamics of the floater of an offshore turbine, and the natural frequency similitude is necessary to effectively compare the model to numeric aeroelastic simulations. To match the C_t and the natural frequency, a modified Blade Element Momentum Theory (BEMT) approach was implemented. BEMT, classically, iteratively alters the pitch, thickness, and chord length of turbine blade sections to optimize rotor performance. Instead, the researchers sought to match the coefficient of thrust of the DTU 10MW reference turbine. First, a blade geometry was generated using the Selig airfoil and the DTU 10MW blades' chord length thickness-to-chord ratio (*tc*), and twist. The lift and drag coefficient of each blade section is then estimated via a linear interpolation between the well-defined bounds of the Selig airfoil ($tc = 0.10367$) and a cylinder ($tc = 1$) (see Figure 3.1).

The coefficient of thrust would then be calculated via the modified BEMT, and the thickness distribution (*tc*) would be modeled as a cantilever beam with varying thickness. The natural frequency would be calculated using finite element methods, and the thickness (*tc*) of each blade element would be altered toward the prototype natural frequency. The modified BEMT would recalculate the C_t and move toward the prototype C_t. This process would be repeated until the C_t and natural frequency are within reasonable levels (5%) compared to those of the full-scale blade. This design process differs from many other researchers as it takes into account the aeroelastic properties of the turbine. A more simplified approach develops the optimal rotor given their airfoil using BEMT methods while disregarding the natural frequency [9]. While others still use XFOIL to generate airfoil performance data and a program called Propid to develop a turbine that operates well at scale [10]. It is the author's opinion that the scaling method used for the Lifes50+ is superior to the alternate methods as the resulting turbine performed closer to the prototype. In summary, the steps to scale a turbine are presented below:

1. Choose a well-characterized prototype turbine with available geometry, C_p versus TSR curve, C_t versus TSR curve, natural frequency, and weight.

   Commonly researchers opt for the National Renewable Energy Laboratory (NREL) 5MW [11] or the DTU 10MW reference turbines [12]. The NREL 5MW reference turbine is commonly used in research as it is the default turbine for most FAST simulations. The DTU 10MW is a rational upscale of the NREL turbine that came four years later and is very well documented with easy access

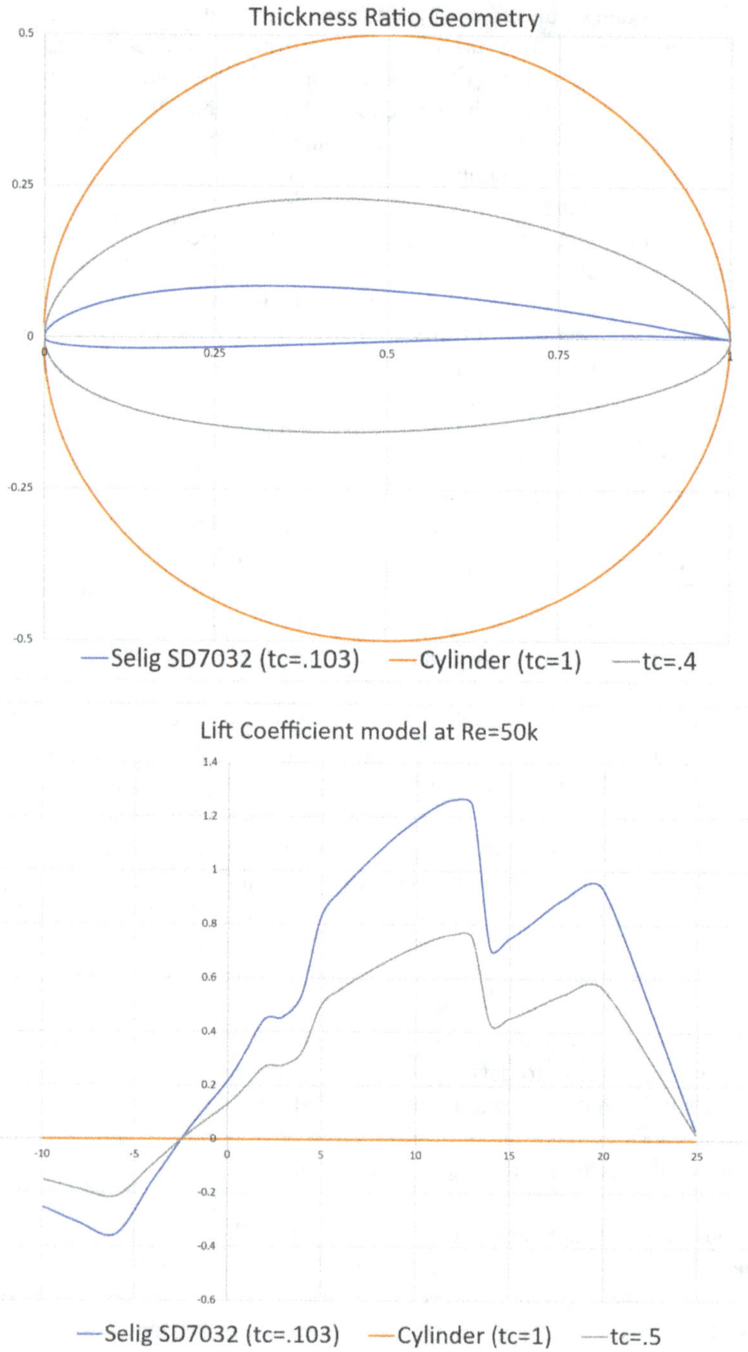

Figure 3.1 *Thickness over chord model for SD7032 airfoil based on data from Selig's summary of low-speed airfoil data [8]*

to structural and performance characteristics through the DTU website. Both of these turbines are optimal as a prototype because of their open-source nature.

2. Choose aspects of performance that are important during testing. This will impact the scaling technique that will be implemented.

   In most cases, the TSR, coefficient of power, and coefficient of thrust are chosen as the nondimensional terms that must be matched during scaling of rigid models.

3. Choose appropriate airfoils that are either well documented or perform experimental evaluations of lift and drag coefficients at Reynolds numbers the turbine is expected to operate at.

   Researchers often opt to use XFOIL as an alternative, but it is well known that XFOIL can become inaccurate at low Reynolds numbers. Researchers who chose well-characterized airfoils or did their own wind tunnel campaigns often achieve better similarity between numerical performance coefficients and experimental results.

4. Scale the turbine through iterative techniques that result in the match of desired performance characteristics.

These techniques are usually implementations of modified Blade Element Theory that alter the chord, thickness, and twist of each blade element to achieve the correct performance characteristics. Many researchers choose to match the twist of the blade elements defined by the prototype and vary the chord length to match the coefficient of thrust or power versus TSR curve. In theory, both the coefficient of thrust and the coefficient of power could be matched by changing both the angle of twist and chord length, but opening up this third degree of freedom for the design adds unneeded complexity to the algorithms necessary to reach an optimal design. The thickness is also altered to maintain the same first flap-wise natural frequency of the blade. Again, if aeroelastic concerns are not in consideration, then the thickness-to-chord ratio could be altered to achieve matching power and thrust coefficient curves.

3.3 Practical steps for implementation in Python

1. Using a geometrically scaled turbine, set initial conditions for the optimization process [11, 12]. Since a utility-scale turbine does not use the thickness-to-chord ratio in design, it is necessary to develop a guideline for how to set initial thickness-to-chord ratios. An exponential decay from 1 to the thickness-to-chord ratio of the airfoil has the desired output of a strong, circular root and a thin airfoil at the tip. This could be of the form:

   $$tc(r) = e^{-ar} + tc' \tag{3.1}$$

   where tc is the thickness-to-chord ratio, tc' is the thickness-to-chord ratio of the airfoil, r is the radius from the hub, and a is a tuning parameter used to alter how quickly the blade matches the shape of the airfoil.

2. Estimate coefficient of lift and drag using linear interpolation from selected airfoil and a cylinder using the thickness-to-chord ratio. Although these are Reynolds-dependent ($C_l(tc,\mathrm{Re})$ and $C_d(tc,\mathrm{Re})$), the solver for BEMT that is built into Wind-Plant Integrated System Design and Engineering Model (WISDEM) does not solve directly for local velocity [13]. The small variance in expected Reynolds number as well as the subtle change in lift and drag due changes in Reynolds means we can confidently model the lift and drag coefficient as only dependent on the thickness-to-chord ratio. Either iteratively coupling this step with the BEMT solution or using an alternative solver could create a model that has Reynolds dependence.
3. Evaluate BEMT equations for given geometry at multiple TSRs, solving for distributed loads, C_p, and C_t. This is implemented in WISDEM as CCBlade. Details and useful examples can be found in the documentation of WISDEM and on the GitHub [14].
4. Evaluate for the first natural frequency using the distributed loads and a simple FEA code. This is implemented in WISDEM as pBEAM [14].
5. Evaluate a cost function (or cost functions, depending on optimization approach) that emphasizes coefficient of thrust and power at the operating TSR. This could be of the form:

$$cost = a_1\left(C_p - C_p'\right) + a_2\left(C_t - C_t'\right) + a_3\left(f_n - f_n'\right) \tag{3.2}$$

   where the arrays a_1, a_2, and a_3 act as tuning parameters for C_p, C_t, and f_n at each TSR, and prime denotes the prototype turbine characteristics.
6. Use an optimization technique to drive the cost function down. Although the exact function is left to the reader to decide, the developers of WISDEM suggest that OpenMDAO is well suited for optimization problems using their framework [15]. In this step, you will set bounds for the chord and *tc* search space.
7. Smooth resulting blade to have realistic chord distribution. This technique often results in a blade that has unrealistic jumps in chord and *tc*.

3.4 Blade element theory when applied to wind turbine scaling

BEMT is a standard method of modeling a turbine by breaking the blade down into many small elements along the span and deriving the thrust and performance coefficients through a series of iterative computations. BEMT's central assumption is that those small elements do not interact with their neighbors. The flow across these sections is assumed to be two-dimensional, from which the aerodynamic forces are computed from the coefficients of lift and drag versus angle of attack tables. The tables are developed either using XFOIL, CFD, or sectional test in a wind tunnel and an extension of that data to ±180 degrees usually via Viterna technique [16]. The aerodynamic forces are then summed for each blade element to compute the axial force and the moment on the turbine. The second half of BEM, momentum theory, relies on the fact that there is pressure loss due to the work done as the wind goes through the rotor plane. Momentum theory can be used to calculate the induced velocity in the axial and tangential directions, affecting the local inflow angle and, therefore, the aerodynamic forces. The coupling of these two concepts is the definition of BEMT and can be used to set up an iterative solution to a 2 degrees of freedom system of nonlinear equations. Recently, Ning at NREL developed a parameterization of BEMT to only the local inflow angle. This dodges the difficulty of solving a second-degree nonlinear system, although it does sacrifice the precise inflow velocity for Reynolds number reliance. An approximation of the local velocity can be calculated, but there will be a small variance between that number and reality. Ning's model is used for the CCBlade module of the WISDEM farm analysis tool developed at NREL [13].

BEMT is beautiful in its simplicity but fails to accurately predict the performance of most turbines. Some of these problems with the model are overcome by including correction factors that compensate for the altered flow at the hub and tip of the blade. At the hub, the induced velocity is effected by the hub itself, whereas the near to the tip is largely effected by vortex shedding. If vortex shedding and hub model are used, the induced velocity changes drastically. Additionally, the model breaks down at high axial induction factors. Fortunately, empirical methods were developed to compensate for this fact. Hopefully, by this point, the readers have gained a healthy respect for the model.

3.5 Blockage corrections for model turbines

Commonly, model turbine designers opt for a low blockage ratio (<10%), but it is important to examine the effects of the wind tunnel walls on turbine's performance. When measured in a wind tunnel as opposed to a freestream flow, the thrust and power coefficients increase [17]. To better understand the results, a blockage correction calculation must take place. Some of the very first blockage corrections for wind turbines were derived from Glauerts corrections for rotors [18]. This blockage correction method employs axial momentum theory and a set of five equations to obtain a closed-loop form solution of the axial momentum theory equation. Bahaja et al. used this model to create an iterative procedure that will produce a blockage correction equation through numeric methods [19]. Both of these turbine-specific techniques provide blockage correction equations that can be applied during post-processing of data to provide an accurate sense of the coefficient of performance and thrust that a turbine is imposing. These corrections are necessary for a deep understanding of the model turbine. However, in the end, the boundary layer produced by the wind tunnel may provide an important tool to understand full-scale turbines.

3.6 Summary

This chapter has provided a basic overview of "why wind tunnel testing of wind turbines is necessary?" and the important nondimensional terms that affect the performance of a turbine. The basic steps to accurately scale a turbine were outlined.

Furthermore, practical steps were developed to facilitate further scale turbine research. A very basic overview of the BEMT equations was introduced, and finally the concept of blockage corrections for turbines is presented.

References

[1] Verelst D.R., Van Wingerden J.-W. *Numerical and experimental results of a passive free yawing downwind wind turbine. DTU wind energy*. DTU Wind Energy PhD No. 0023(EN); 2013.

[2] Iungo G.V., Viola F., Camarri S., Porté-Agel F., Gallaire F. 'Linear stability analysis of wind turbine wakes performed on wind tunnel measurements'. *Journal of Fluid Mechanics*. 2013;**737**:499–526.

[3] Campagnolo F., Petrović V., Schreiber J., Nanos E.M., Croce A., Bottasso C.L. 'Wind tunnel testing of a closed-loop wake deflection controller for wind farm power maximization'. *Journal of Physics: Conference Series*. 2016;**753**:032006.

[4] Bayati I., Belloli M., Bernini L., Zasso A. 'Wind tunnel validation of AeroDyn within LIFES50+ project: imposed surge and pitch tests'. *Journal of Physics: Conference Series*. 2016;**753**:092001.

[5] Bayati I., Belloli M., Bernini L., Zasso A. 'Aerodynamic design methodology for wind tunnel tests of wind turbine rotors'. *Journal of Wind Engineering and Industrial Aerodynamics*. 2017;**167**(6):217–27.

[6] Yan C., Archer C.L., Yan A. 'Assessing compressibility effects on the performance of large horizontal-axis wind turbines'. *Applied Energy*. 2018;**212**:33–45.

[7] Bayati I., Belloli M., Bernini L., Zasso A., Mikkelsen R. 'On the aero-elastic design of the DTU 10MW wind turbine blade for the LIFES50+ wind tunnel scale model'. *Journal of Physics: Conference Series*. 2016;**753**:022028.

[8] Selig M. *Summary of Low-Speed Airfoil Data, Virginia Beach*. Virginia Beach (VI): SoarTech Publications; 1995.

[9] Bastankhah M., Porté-Agel F. 'A new Miniature wind turbine for wind tunnel experiments. Part I: design and performance'. *Energies*. 2017;**10**(7).

[10] de Ridder E.J. 'Development of a scaled-down floating wind turbine for offshore basin testing'. *Proceedings of the ASME 2014 33rd International Conference on Ocean, Offshore and Arctic Engineering June 8-13, 2014*; San Francisco, California, USA; 2014.

[11] Jonkman J., Butterfield S., Musial W., Scott G. 'Definition of a 5-MW reference wind turbine for offshore system development'. *NREL, Golden*. 2009.

[12] Bak C., Zahle F., Bitsche R., *et al*. 'Description of the DTU 10 MW reference wind turbine'. *DTU Wind Energy Report*. 2013.

[13] Ning A. 'A simple solution method for the blade element momentum equations with guaranteed convergence'. *Wind Energy*. 2014;**17**(9):1327–45.

[14] Available from https://github.com/WISDEM/WISDEM.

[15] Gray J.S., Hwang J.T., Martins J.R.R.A., Moore K.T., Naylor B.A. 'OpenMDAO: an open-source framework for multidisciplinary design, analysis, and optimization'. *Structural and Multidisciplinary Optimization*. 2019;**59**(4):1075–104.

[16] Viterna L., Janetzke D. *Theoretical and Experimental Power from Large Horizontal-Axis Wind*. Cleveland, OH: NASA; 1982.

[17] Barlow J.B., Rae W.H., Pope A. *Low-Speed Wind Tunnel Testing*. Third Edition. USA and Canada: John Wiley & Sons; 1999.

[18] Glauert H. 'Airplane propellers' in Durand W.F. (ed.). *Aerodynamic Theory, Vol. IV, Division L*. Springer, New York; 1963. pp. 169–360.

[19] Bahaja A.S., Mollandb A.F., Chaplina J.R., Battena W.M.J. 'Power and thrust measurements of marine current turbines under various hydrodynamic flow conditions'. *Renewable Energy*. 2007;**32**(3):407–26.

Chapter 4

Advances in aerodynamics of wind farms

Tanmoy Chatterjee[1] *and Yulia T. Peet*[2]

4.1 Introduction

Wind farms comprising massive arrays of wind turbines (mostly horizontal axis wind turbines or HAWTs) have become popular among the alternative source of energy/ renewable energy since the last couple of decades as a popular and viable method of extracting wind energy from atmospheric boundary layer (ABL) and converting it into electric power. The United States is the home to one of the fastest growing wind energy markets in the world, with a total of roughly 105 GW of installed wind plant capacity in 2019 and contributing to about 7.29% of the total electricity generation in the country. The major motivation of wind turbine manufacturers and industrial wind farm owners is to design the wind farm layout so as to maximize the annual energy production (AEP, in megawatt-hours) or minimize the levelized cost of energy (LCOE—dollar/megawatt-hours) of farms. Wind farms are extremely complex machines comprising typically tens to hundreds of turbines and interacting with the multiscalar turbulent ABL as well as each other. The AEP, depends on many factors, e.g., the variability of wind speed and wind direction, atmospheric stability condition (neutral, stably stratified, or convectively unstable), aerodynamic surface roughness, presence of complex terrain, diameter and hub-height of wind turbine's installed, and the inter-turbine distances (layout). To understand the complex interaction of wind turbines with the atmosphere, it is imperative that we have a proper understanding of the physics via high-fidelity computational/experimental tools. The three-dimensional turbulent flow involving the interaction of wind farms and ABLs are exceedingly complex and inherently multiscalar in nature. At such high Reynolds number flows ($Re \sim 10^6 - 10^{10}$, based on bulk velocity), researchers have medium-fidelity computational tools such as Reynolds averaged Navier–Stokes (RANS) equation (CPU hours $\sim O(10^1 - 10^2)$) secs.) or high-fidelity computational tools like large eddy simulation (LES) (CPU hours $\sim O(10^3 - 10^5$ secs.) to capture the effects of wind turbine atmospheric flow interaction. This involves understanding the physics of various flow regions, such as atmospheric

[1]GE Global Research (General Electric Company), New York, NY, USA
[2]Arizona State University, Tempe, AZ, USA

shear flow/boundary layer modulated by the wind turbines, the induction region that experiences deceleration of the wind flow prior to wind turbine interception (and flow acceleration sideways), and the velocity-deficit or "wake" region immediate downstream of the wind turbines [1]. Understanding these flow phenomenon plays a critical role toward creating reduced order models of wind farms to optimize their AEP. In this chapter, we will talk about the progress made toward understanding the physics of wind turbine–atmospheric interaction, the recent advances in using high-fidelity computational tools such as LES, which will pave the way for better analytical modeling of wind farms for AEP estimation and developing better strategies for wind farm controls for maximizing the energy harvest and reducing the aerodynamics-induced structural loads in the wind turbines (maximizing turbine life).

In particular, this chapter is broken down into the following subsections. First, we will discuss (i) the state-of-the-art numerics for farm modeling (mostly limited to LES). We will build on (ii) state-of-the art analytical wind farm models such as Frandsen–Calaf theory, (iii) Nishino model, and (iv) LES of infinite wind farms, which serve as building blocks to those models. We further discuss the (v) dynamics of large-scale structures commonly seen in wind farms that play a contributing factor to the power generation. Next, we talk about the analytical framework to model important flow-features of wind farms (vi) wake-deficit region and their superposition effects in farms and (vii) induction region. It must be understood that most of these analytical models are either inspired or built from high-fidelity LES. After that, we talk about models and physics representing the farm layout, e.g., (viii) aerodynamic roughness, (ix) complex terrain. Additionally, we also illustrate potential and existing concepts to reduce farm-level aerodynamic interaction by either modifying (x) the wind turbine design or (xi) wind turbine arrangement in the farm, which is a key to optimizing AEP. Finally, we talk about the state-of-the-art approaches to wind farm control involving (xii) wake steering and the more generic, (xiii) dynamic induction control.

4.2 Length scales in wind farms

The aerodynamics in wind farms (aggregate array of wind turbines) are quite complex since, at the scale of the farms (sprawling several kilometers), the interaction with the atmosphere and the modulation of turbulent coherent structures in ABL become more conspicuous as opposed to the blade-resolved scales (in meters) of individual wind turbine. Figure 4.1 shows various regions of the wind farm influenced by the ABL. Region *I* is the induction region, where the flow suffers significant deceleration owing to the blockage effect of wind farms. Due to the induction-blockage effect, which can influence the flow as large as 2.5–3 D upstream of the first row of turbines [3], these decelerated flow structures also contribute to $\sim 5\%$ loss in wind-plant AEP. Region *II* is the entrance or the development region of the "internal boundary layer" (IBL) of wind farms. The flow in the turbine disk region slows down due to the turbine thrust forces causing increased mean shear in the

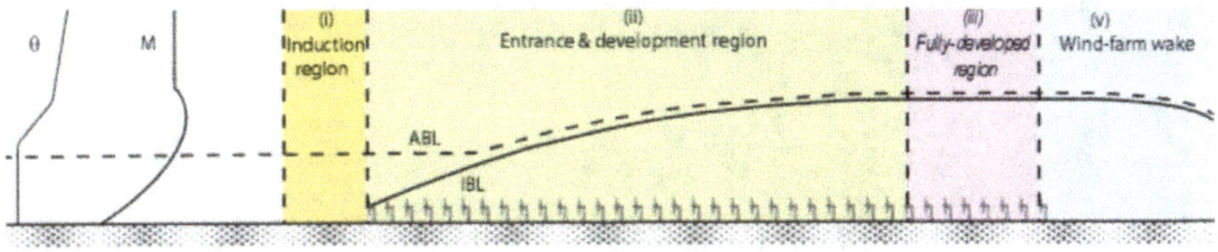

Figure 4.1 *Schematics of wind farm labeling different regions like (i) induction region, (ii) fully developed wind farm region, (iii) wind farm wake region. Figure taken from [2] in accordance with Creative Common Attributions (CC BY) License.*

"wake" region of the turbine, which in turn causes increased turbulence production. These increased gradients promote mixing via vertical entrainment from the flow above, which causes growth of the "IBL." The array of wind farms can also be thought of as added roughness elements to the ground to which the ABL now adapts. The IBL evolves to a fully developed state, which corresponds to a new equilibrium of the wind farm ABL. The new equilibrium is also in part guided by the streamwise distance between the wind turbines, i.e., if the downstream turbine is in the near-wake or far-wake region of the upstream counterpart. Here the turbulent dissipation and the wind turbine power extracted is balanced by the energy coming from the large scale forces. The fully developed state is a new equilibrium that can be described by analytical models (based on energy balance) [4] similar to the equilibrium state of ABL, which corresponds to the balance of turbulence production and dissipation. The details of the analytical model are further described in Sections 4.1 and 4.3. Region *III* is the wind farm wake region, where the streamwise momentum gradually recovers (in the absence of turbine thrust forces). Wind farm wakes, which are essentially an aggregate effect of the superposed individual turbine wake deficits, can be very long and propagate to distances of 20 km or more, depending on the size of the farm layout and atmospheric stability [5]. The dynamics of the wind farm wakes are critical toward optimizing farm to farm interactions when two wind farms are constructed close to each other.

As seen from Figure 4.1, wind farm flows are inherently multiscalar and complex with large scale coherent structures/integral length scales ~0 km to intermediate length scales of several hundred meters ($\sim O(D, z_h)$), where wake dynamics are relevant to Kolmogorov length scales ~0 mm, where turbulent dissipation is important. In addition to the presence of large scale forcing influenced by topography and local weather effects (~0 km), the influence of Coriolis force due to earth rotation and buoyancy effects owing to atmospheric stability adds more complexity to the flow system. These make numerical modeling of wind farm flows an extremely challenging task. Subsequently, the state-of-the-art numerical methodologies rely on high-fidelity LES modeling to capture or resolve the physics of length scales ~0 (10 m) while modeling turbulence phenomenon of smaller scales with supposedly universal behavior and minimally influenced by the large-scale forcing. RANS modeling is popular to model steady-state mean momentum equations in wind farms. However,

(a) (b)

Figure 4.2 (a) Three-dimensional spatiotemporal wake evolution in a typical wind farm in LES simulations. https://stevensrjam.github.io/Website/. (b) Wakes in HornsRev wind farm, North Sea in Denmark. https://www.rechargenews.com/wind/denmarks-horns-rev-1-wind-farm-rendered-inoperative-by-fault/1-1-859985.

using RANS to capture rotor interaction in large wind farms is mostly limited to the spatially varying mean wind and turbulence intensity (TI) in the farms influenced by the farm layout and topography (e.g., hills and terrains) [6, 7]. Also, RANS is considered a tool of lower fidelity than LES. Hence, it is outside the scope of the current chapter. Figures 4.2a and b illustrate the multiscalar nature of large wind farms and wake dynamics as depicted in high-fidelity LES modeling as well as a real snapshot of a production scale of off-shore wind farm. These figures provide a high-level comparison of LES simulations in wind farms and realistic flow depiction. They also serve as a stepping stone toward motivating the readers to discuss the high-fidelity LES simulation modeling framework for wind farms.

4.3 State-of-the-art numerics for farm modeling

4.3.1 Large eddy simulation

The LES involves solving spatially filtered three-dimensional Navier–Stokes (NS) equation (partial differential equation or PDE). For most of the simulations involving ABL and wind farms, the spatial filtering is imposed by the coarse-grid sizes, which acts as a cut-off filter to the turbulent eddies smaller than the grid size.

The spatially filtered NS equations (momentum) can be written as

$$\underbrace{\frac{\partial \widetilde{\mathbf{u}}}{\partial t}}_{\text{I}} + \underbrace{\widetilde{\mathbf{u}}\nabla\widetilde{\mathbf{u}}}_{\text{II}} = \underbrace{-\frac{1}{\rho}\nabla\widetilde{p^*}}_{\text{III}} + \underbrace{\widetilde{\mathbf{F}}_{turb}}_{\text{IV}} + \underbrace{\mathbf{u}\times\mathbf{\Omega}}_{\text{V}} + \underbrace{\frac{\widetilde{\theta}-\widetilde{\theta_0}}{\widetilde{\theta_0}}g}_{\text{VI}} + \underbrace{\nu\nabla^2\widetilde{\mathbf{u}}}_{\text{VII}} - \underbrace{\nabla\cdot\boldsymbol{\tau}}_{\text{VIII}}. \tag{4.1}$$

Similarly, the enthalpy equation can be written as

$$\underbrace{\frac{\partial\widetilde{\theta}}{\partial t}}_{\mathrm{IA}} + \underbrace{\widetilde{\mathbf{u}}\nabla\widetilde{\theta}}_{\mathrm{IIA}} = \underbrace{k\nabla^2\widetilde{\theta}}_{\mathrm{IIIA}} - \underbrace{\nabla\cdot\boldsymbol{q}}_{\mathbf{IVA}} \tag{4.2}$$

Note the use of potential temperature θ in the enthalpy equation. Potential temperature θ is defined as the temperature of a parcel of fluid at temperature T and hydrostatic pressure p if it is adiabatically brought to a reference pressure of p_0. For air with ideal gas approximation, we obtain $\theta = T(p/p_0)^{\gamma-1/\gamma}$, $\gamma = 1.41$. The concept of potential temperature applies to any stratified fluid and is common in atmospheric sciences. It is used because changes in pressure can result in warmer fluid residing under colder fluid—examples being dropping air temperature with altitude. When the potential temperature is used instead, these apparently unstable conditions vanish as a parcel of fluid is invariant along its isolines, and changes in potential temperature can only occur due to the presence of any heat source or any thermodynamic work done.

We use (x, y, z) Cartesian coordinate system, where x corresponds to a streamwise direction, y to a spanwise direction, and z to a vertical (wall-normal) direction. The term **I** is the temporal derivative of the velocity vector contributing to the acceleration, **II** is the nonlinear advection term, and **III** represents the pressure gradient forces (deviation from the hydrostatic term as well as the trace of the subgrid-stress tensor (pertinent to LES)). **IV** represents the thrust, lift, and drag forces due to turbine rotations. Term **V** corresponds to the Coriolis acceleration forces due to the earth's rotation, whereas term **VI** represents the buoyancy term due to density/potential temperature gradient effects, and term **VII** involving the Laplacian reflects the diffusion effect due to molecular viscosity. The final term **VIII** arises due to the numerical effect of spatial filtering in LES (grids). This term is also known as the gradient of the subgrid stresses (SGSs) and arises due to the noncommutativity of filtering and differentiation in the nonlinear advection term in the NS equation. Similarly, the terms IA and IIA in (4.2) represent the temporal variation and the advection of the potential temperature, respectively. Term **IIIA** represents the diffusive effects of potential temperature due to the conduction, and **IVA** represents the subgrid effects of potential temperature.

In the atmospheric turbulence community, the SGS tensor in (4.1), $\boldsymbol{\tau} = \widetilde{\mathbf{u}\mathbf{u}^T} - \widetilde{\mathbf{u}}\widetilde{\mathbf{u}}^T$, is typically modeled using an eddy viscosity closure [8].

$$\boldsymbol{\tau}^{SGS} - \frac{1}{3}\boldsymbol{\tau}^{SGS}\delta_{kk} = -2\nu_T\nabla^s\widetilde{\mathbf{u}} \tag{4.3}$$

Here, τ^{SGS} is the modeled SGS tensor. The eddy viscosity $\nu_T = (l_f)^2\nabla^s|\widetilde{\mathbf{u}}|$ is based on a mixing length hypothesis [9] , δ_{kk} is the Kronecker delta tensor and the filter length scale $l_f = C_s\Delta$, where $\Delta = (\Delta_x\Delta_y\Delta_z)^{1/3}$ is a typical measure of grid size conventionally used by the community. Also, $\nabla^s\widetilde{\mathbf{u}} = 1/2(\nabla\widetilde{\mathbf{u}} + \nabla\widetilde{\mathbf{u}}^T)$ and $|\nabla^s\widetilde{\mathbf{u}}| = (2\nabla^s\widetilde{\mathbf{u}} : \nabla^s\widetilde{\mathbf{u}})^{1/2}$. In the Smagorinsky model, the eddy viscosity closure is done by modeling the constant C_s either dynamically (at each simulation time-step) or predefining C_s statically at each grid location. The Smagorinsky model was originally derived (for isotropic turbulence) assuming that the modeled rate of transfer of energy from large to small scales $-(\boldsymbol{\tau}^{SGS} : \nabla^s \widetilde{\mathbf{u}})$ is equal to the overall rate of the dissipation ϵ. The Smagorinsky model assumes an between kinetic energy flux across scale and the large scales of turbulence. Similarly for potential temperature, $q = \widetilde{\mathbf{u}\theta} - \widetilde{\mathbf{u}}\widetilde{\theta}$ can analogously be modeled using an eddy viscosity closure.

The static or algebraically damped Smagorinsky model is usually criticized for being over-dissipative in the near-wall region. The Lily–Germano model, which has been used as a dynamic model assuming the scale invariance of C_s at different grid-scales Δ, and$\Delta' = \alpha\Delta$, i.e., $C_s(\Delta) = C_s(\alpha\Delta)$ also breaks down in the near-wall region for being under-dissipative. Porté-Agel *et al.* [8] devised a scale-dependent dynamic model that used a power-law profile of C_s at different scales Δ instead of invariance. This has been primarily the state-of-the-art model for simulating ABL flows in the LES framework for the last decade. To address the Smagorinsky model's limitations, i.e., the existence of equilibrium between kinetic energy flux across small and large scales of turbulence, the atmospheric science community has also computed eddy viscosity closures by a separate transport equation of subgrid scale kinetic energy ($k = \text{trace}(\boldsymbol{\tau}^{SGS})$), but such models are not extensively used (until recently [10]) by the wind farm community because it is computationally much more expensive than the scale-dependent dynamic Smagorinsky model. Moreover, if the subgrid scales are properly designed, even the wall-damped Samgorinsky model is capable of providing reasonable results. For example, Figure 4.3 shows the velocity, turbulent stresses, and the integral length scales of neutral ABL computed in the authors' previous work [15] using algebraic wall-damped Smagorinsky model in a high order exponentially accurate spectral element code Nek5000. The results compare reasonably well with previous LES as well as wind tunnel/field experimental results. Note that the current chapter is limited to describing LES results in the neutral ABL, since the literature on wind farms is mostly diverse and well established for neutral stratification of the atmosphere. Even though there are a lot of emerging studies focusing on the atmospheric stability of wind farms, the authors have restricted the discussion of the stability/thermal stratification effect of wind farms as a small discussion in this section and consider it beyond the scope of the current book. The major effect of atmospheric stability is related to the wake-recovery being an inverse function of stability. Consequently, the convectively unstable boundary layers (during daytime) would report fast wake recovery and (owing to higher turbulent intensity) more power production than stably stratified boundary layers. Another aspect to consider is the incoming wind profile at night time, in particular low-level-jets, which has a high-velocity region near the ground. If the nose of the jet is around the wind turbine disk region, it can impact the wake recovery and hence the power production in wind turbines. For more details refer to [16]. Another important phenomenon with the atmospheric stability involves the formation of atmospheric gravity (AG) waves arising from the blockage effect of large wind farms [17]. The AG waves reorganize the pressure distribution in the farms and are also responsible for significant induction losses (flow deceleration) upstream of the wind farm.

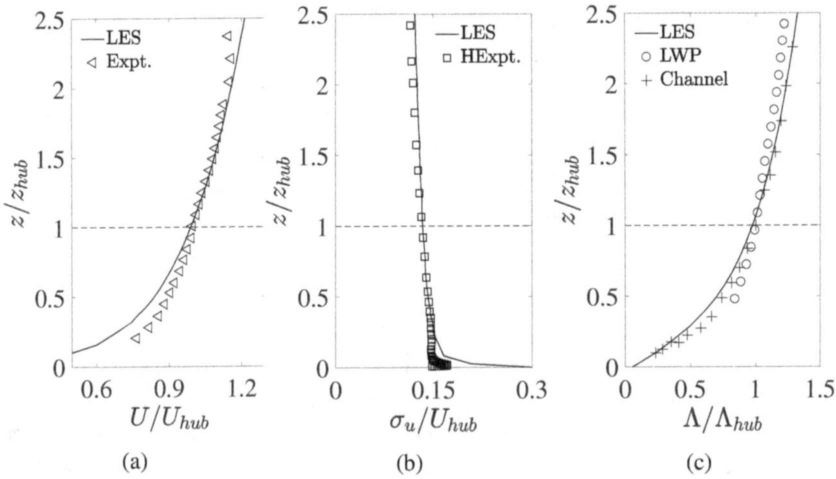

Figure 4.3 *Flow characteristics and validation with experiments of a simulated neutral ABL profile ($Re_\tau \sim 5 \times 10^8$) using wall-modeled LES. (a) Normalized streamwise mean velocity, U/U_{hub}, (b) TI, σ_u/U_{hub}, and (c) normalized integral length scale $\Lambda^u/\Lambda^u_{hub}$. The horizontal lines represent turbine hub-height location. Expt-Hayat* et al. *[11], ($Re_\tau \sim 9.3 \times 10^3$); HExpt, wind tunnel experiment from Hutchins* et al. *[12] ($Re_\tau \sim 1.9 \times 10^4$), channel-channel flow DNS by Sillerro* et al. *[13] ($Re_\tau \sim 2 \times 10^3$), and LWP—Leipzig wind profile, Lettau [14].*

4.3.1.1 Near-wall model

At the bottom surface, boundary conditions involve wall stress vector $\boldsymbol{\tau}_s$ [9] being incorporated through the horizontal velocity vector data $\widehat{\tilde{\boldsymbol{u}}}_h$ at the first grid point using the standard Monin–Obukhov similarity law [18] along with no-penetration conditions of large eddies, $\tilde{w} = 0$,

$$\frac{1}{\rho}\boldsymbol{\tau}_s = -\kappa^2 \frac{\widehat{\tilde{\boldsymbol{u}}}_{h,\frac{\Delta z}{2}}(x,y,t)\left|\widehat{\tilde{\boldsymbol{u}}}_{h,\frac{\Delta z}{2}}\right|(x,y,t)}{\left[\log\left(\frac{z}{z_0}\right)\Big|_{\frac{\Delta z}{2}} - \Psi_M\right]^2}, \quad \Delta z/z_0 \gg 1 \tag{4.4}$$

where $\widehat{\tilde{\boldsymbol{u}}}_{h,\frac{\Delta z}{2}} = \widehat{\tilde{u}}_{\frac{\Delta z}{2}}\vec{e}_x + \widehat{\tilde{v}}_{\frac{\Delta z}{2}}\vec{e}_y$ ($\vec{e}_x$ and $\vec{e}_y$ are unit vectors in the x and y direction), $|\widehat{\tilde{\boldsymbol{u}}}_{h,\frac{\Delta z}{2}}| = \sqrt{\widehat{\tilde{u}}^2_{\frac{\Delta z}{2}} + \widehat{\tilde{v}}^2_{\frac{\Delta z}{2}}}$ and Δz_0 is the aerodynamic roughness length. Here, tilde (~) represents the LES filtering from coarse grids and the "hat" ( ), an explicit filter (may be a low-pass filtering or simple spatial averaging). Ψ_M represents the stability parameter. The variables, $\widehat{\tilde{u}}_h, \widehat{\tilde{u}}, \widehat{\tilde{v}}$ are defined at first grid point from the wall

$z = \Delta z/2$ (for mixed pseudo-spectral staggered finite-difference finite-volume codes, Δz = height of cell, [8]) or half of the distance from first grid point for collocated codes ([15, 19]). For neutrally stratified ABL, $\Psi_M = 0$, such that the log-trend of the mean velocity profile can be obtained. For convectively unstable or stably stratified flows, $\Psi_M \neq 0$, manifesting deviation from the log-law trends. Note, for non-neutral conditions, Ψ is a strong function of $\vec{e}_x$ and $\vec{e}_y$, where L is the Obukhov length [18].

The near-wall boundary conditions are representative of the mean velocity profile, $\overline{U}$ for various atmospheric stability conditions.

$$\overline{U}(z) = \frac{u_\tau}{\kappa} \log_e\left(\frac{z}{z_0}\right) + \frac{u_\tau}{\kappa}\psi_M(\frac{z}{L}) \tag{4.5}$$

Here, u_τ, κ are the friction velocity scale and Von Karman constant, respectively, and ψ_M is the integral to the stability parameter, Ψ_M. The Obukhov length L is a length parameter for the surface layer in the boundary layer, which characterizes the relative contributions to turbulent kinetic energy from buoyant production and shear production. When $L < 0$, the surface layer is statically unstable, and when $L > 0$, the surface layer is statically stable. The absolute magnitude of L indicates the deviation from statically neutral state, with smaller $|L|$ values corresponding to larger deviations from neutral conditions.

For $L \to \infty$, the log trends of the mean velocity profile for neutral boundary layer can be achieved, $\overline{U}(z) = \frac{u_\tau}{\kappa} \log_e\left(\frac{z}{z_0}\right)$.

Additionally, as far as the enthalpy/potential temperature boundary condition is concerned, a heat flux boundary condition is set up at the "bottom wall," with $q_{flux} = 0$ for neutral conditions, $q_{flux} < 0$ (cooling) for stably stratified flows, and $q_{flux} > 0$ (surface heating) for convectively unstable ABL flows.

The "hat" represents an additional explicit filtering carried out in addition to the LES filtering imposed by the grid. The $\widehat{\boldsymbol{u}}_{h,\frac{\Delta z}{2}}$ is calculated at the cell-center in finite volume or staggered-mixed pseudo-spectral methods or as an interpolation between "wall" and Δz (distance between the bottom boundary and the first grid point from the boundary in the vertical direction).

In this chapter, we have presented some of the authors' validation results of the neutral ABL simulations in Figure 4.3 that serve as a representative benchmark simulations for ABL by the wind farm community. The results illustrate that our high-fidelity spectral element simulations (exponentially accurate), in terms of the mean streamwise velocity, streamwise variance of turbulent fluctuations, and integral length scales compare very well with wind tunnel laboratory scale, numerical simulations as well as field experiments.

4.3.2 Wind turbine model—actuator line and actuator disk

The current state-of-the-art framework for wind turbine models are (i) actuator line model (ALM) and (ii) actuator disk model (ADM). The ALM models the blades of the wind turbines as actuator lines with the rotating effects. The momentum deficit in the flow is modeled through the lift and drag forces in the actuator lines (airfoil models) as well as the thrust forces in the turbines. The ADM models the wind turbines

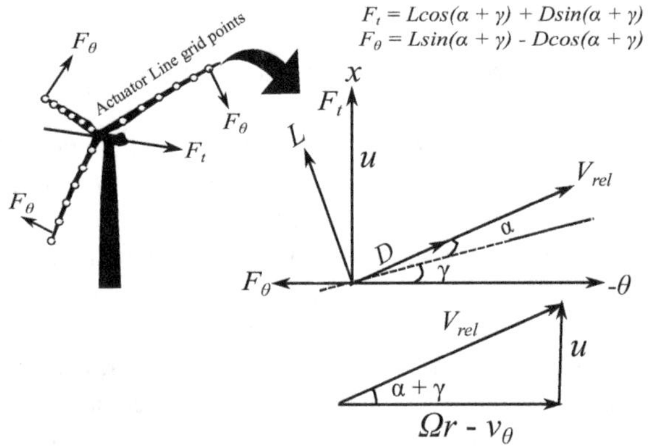

Figure 4.4   Schematic of ALM for wind turbines

as disks without the rotating effects and the momentum deficit in the flow is modeled through the turbine thrust forces.

In an ALM ([20]), the blades with the airfoil cross sections are divided into elements and the local lift (L) and drag (D) forces experienced by each element are calculated as $(L, D) = \frac{1}{2}C_{(l,d)}(\alpha)\,\rho\, V_{rel}^2\, c\, w_d$, where c and w_d are the chord length and the blade width, respectively. The local angle of attack, α, is computed from the relative velocity, V_{rel}, streamwise velocity, u ($\tilde{u}$ in LES), and the pitch angle, γ (see Figure 4.4). The local aerodynamic force is calculated as $\vec{f} = L\vec{e_L} + D\vec{e_D}$ (here $\vec{e_L}$ and $\vec{e_D}$ are the unit vectors in the direction of the local lift and drag, respectively). The total reaction force summed from all the N blade elements experienced by the fluid is distributed smoothly onto several mesh points and is given as the actuator line force in the NS (4.1):

$$\widetilde{\mathbf{F}}_{AL}(x, y, z, t) = -\sum_{i=1}^{N}\vec{f}(x_i, y_i, z_i, t)\,\eta_\epsilon(|\vec{r} - \vec{r}_i|), \tag{4.6}$$

$\vec{r} = (x, y, z)$and (x_i, y_i, z_i) are the coordinates of the centers of the actuator blade elements at time t, and $\eta_\epsilon(d)$ is a Gaussian function in the form of $\eta_\epsilon(d) = 1/\epsilon^3\pi^{3/2}\exp\left[-\left(d/\epsilon\right)^2\right]$. We set $\epsilon = 2\,w_d$ as per previous literature [21] and [20] for optimum results. Recent research has also improved the ALM to an "actuator surface model" (ASM) to incorporate the variation of lift and drag forces over the chord length and blade width by incorporating V_{rel} as a function of c, w_d. However, wind farm research incorporating ASM is limited and hence outside the scope of the current book chapter.

The actuator disk can be modeled in multiple ways, e.g., fixing the rotation effect (ADM-R) of actuator lines and distributing the AL forces over the disk region, as is done by many high-performance code bases, e.g., LESGO [22] and Nalu-Wind [10]. However, in this chapter we will define actuator disk method in the most primitive form (without modeling the rotating effects, ADM-NR) as it was introduced in the LES wind farm/wind turbine community [4]. The ADM forces essentially model the thrust effects and can be written as

$$\vec{f}(x_i, y_i, z_i, t) = -\frac{1}{2}\rho C'_T \langle \widetilde{\mathbf{u}} \rangle^2_{disk} \frac{\pi}{4} D^2 \quad (4.7)$$

where $\langle \widetilde{\mathbf{u}} \rangle_{disk}$ is the rotor-averaged velocity at the wind turbine locations, and $C'_T = C_T/(1-a)^2$, with C_T being the thrust coefficient and a, the axial induction factor. With the thrust force, Note, ρ *and* D are the air density and diameter of the turbine, respectively. The effect of wake rotation can be introduced by noting the wind turbine rpm at the particular wind speed and incorporating a torque at the actuator disk from the wind turbine power. The primitive form of ADM method is popularly used in RANS calculations [7] in wind farms, particularly because of its cheap computational expense. Note that the mathematical formulation of ADM-R [23] is computationally cheaper, yet very similar to the ALM model, in terms of computing both lift and drag forces and then distributing them over the actuator disk based on the local blade and flow characteristics. The major computational advantage compared to ALM comes from the fact that the actuator node locations need not be updated at every time step and interpolate velocity and compute forces on those updated locations. This has been the state-of-the-art ADM models for high-fidelity open-source wind turbine simulator Nalu-Wind [10].

In general, it can be said that the AL model is more advanced than the actuator-disk model [4, 24–26] commonly used in numerical computations of wind farms, in its capability to capture the tip vortices being shed in the near wake [19, 20].

4.4 Analytical models

4.4.1 *Frandsen–Calaf theory*

Analytical wind farm models for asymptotic large wind farms can be derived by considering that an organized layout of wind turbines can augment roughness to the ABL with baseline aerodynamic surface roughness. In this theory, two equilibrium layers are postulated. The top layer is situated above the wind turbine canopy and parametrized by friction velocity $u_{*,hi}$ and equivalent roughness height $z_{0,hi}$. The top equilibrium layer is directly impacted by wind farm layout, farm density, turbine diameter and hub-height, and so on. For infinite wind farms driven by a pressure gradient, the equilibrium law occurs between the pressure force and the friction force at the wall.

$$\frac{1}{\rho}\frac{dp_\infty}{dx} = \frac{u^2_{*,hi}}{H} \quad (4.8)$$

Thus for the top-layer, $u^2_{*,hi} = \sqrt{H\rho^{-1}dp_\infty/dx}$. Note, $\frac{dp_\infty}{dx}$ is the free-stream pressure gradient for the farm layout, ρ is the air density, and H is the height of the ABL. For, the bottom equilibrium layer, the friction velocity, $u_{*,lo} = \sqrt{\tau_w/\rho}$, where τ_w is the bottom wall-shear stress. The roughness height $z_{0,lo}$ is the aerodynamic roughness induced by the vegetation canopies. For very high Reynolds number flows as in ABL, equilibrium layer suggests that $u_{*,hi} \approx -(\langle \overline{u'w'} \rangle)^{1/2}$ for $H \gg z \gtrapprox z_h$, and $u_{*,lo} \approx -(\langle \overline{u'w'} \rangle)^{1/2}$ for $z \lessapprox z_h$, z_h is the hub height of the turbine. This model gives rise to two separate stress layers and hence two logarithmic trends above and below the wind turbine canopy as follows.

$$\langle \overline{u} \rangle_{lo}(z) = \frac{u_*, lo}{\kappa} * \log_e\left(\frac{z}{z_0, lo}\right), \; z_0, lo < z << z_h \tag{4.9}$$

$$\langle \overline{u} \rangle_{hi}(z) = \frac{u_*, hi}{\kappa} * \log_e\left(\frac{z}{z_0, hi}\right), \; z >> z_h \tag{4.10}$$

Note, are the mean-streamwise velocity above and below the wind turbine canopy. The Reynolds averaged streamwise u (time-averaged, mean) momentum equation for fully developed and statistically stationary asymptotic wind turbine array boundary layer can be written as

$$\overline{u}\frac{\partial \overline{u}}{\partial x} + \overline{v}\frac{\partial \overline{u}}{\partial y} + \overline{w}\frac{\partial \overline{u}}{\partial z} = -\frac{1}{\rho}\frac{\partial p}{\partial x} - \frac{\partial}{\partial x}\left(\overline{u'u'}\right) - \frac{\partial}{\partial y}\left(\overline{u'v'}\right) - \frac{\partial}{\partial z}\left(\overline{u'w'}\right) + \overline{f}_x \tag{4.11}$$

Note, in this formulation for ABL flows for high Reynolds number, the viscous stresses have been neglected compared to the turbulent stresses, since the mixing imposed by the turbulent stresses are a few orders of magnitude higher than their viscous counterpart. $\overline{u}\frac{\partial \overline{u}}{\partial x} + \overline{v}\frac{\partial \overline{u}}{\partial y} + \overline{w}\frac{\partial \overline{u}}{\partial z}$ are the x, y, z convective transport of momentum, $\frac{1}{\rho}\frac{\partial p}{\partial x}$ is the streamwise pressure gradient. The fluctuating terms, e.g., streamwise velocity, $u' = u - \overline{u}$ denotes deviation from the temporal mean, $\overline{u'v'}$, $\overline{u'w'}$ denotes the so-called Reynolds stresses, and $-\frac{\partial}{\partial x}(\overline{u'u'}) - \frac{\partial}{\partial y}(\overline{u'v'}) - \frac{\partial}{\partial z}(\overline{u'w'})$ is the term contributing to Reynolds stress transport or turbulent mixing (analogous to viscous mixing but much faster and spatially bigger eddies are involved). The term $\overline{f}_x$ denotes the forcing due to the aggregated streamwise effect of turbine. In its simplest form, $\overline{f}_x$, denotes the thrust effects on the wind turbine disk region which gives rise to the down-stream "wake" regions owing to the velocity-deficit.

The mean velocity field is a complex three-dimensional field, including wakes, shear in boundary layer, etc. A simpler understanding of the model invokes horizontal averaging ($\langle\rangle$ denotes x, y averaging) of the Reynolds averaged mean velocity equation.

Under horizontal statistical homogeneity, $\langle \overline{v} \rangle$, $\langle \overline{w} \rangle = 0$ as well as the gradients, $\partial/\partial x \langle \overline{u} \rangle = 0$.

$$-\frac{1}{\rho}\frac{dp_\infty}{dx} - \frac{\partial}{\partial z}\left(\langle \overline{u''w''} \rangle\right) + \langle \overline{f}_x \rangle = 0 \tag{4.12}$$

where $-\langle \frac{\partial}{\partial z}\left(\overline{u'w'}\right) \rangle = \frac{\partial}{\partial z}\left(\langle \overline{u''w''} \rangle\right)$. The fluctuations, $u'' = u - \langle \overline{u} \rangle$ and $w'' = w - \langle \overline{w} \rangle$.

With further simplification, the temporal/Reynolds and horizontal averaged streamwise momentum equation can be written as

$$-\frac{1}{\rho}\frac{dp_\infty}{dx} - \frac{\partial}{\partial z}\left(\langle \overline{u'w'} \rangle + \langle \overline{\overline{u}''\overline{w}''} \rangle\right) + \langle \overline{f}_x \rangle = 0 \tag{4.13}$$

Note, apart from the conventional time-horizontally averaged Reynolds stress $\langle \overline{u'w'} \rangle$ due to the correlation of streamwise and vertical velocity fluctuations, there is an additional “dispersive stress” (time-horizontally averaged), due to the horizontal heterogeneity of the three-dimensional flow owing to the presence of turbines. In the absence of wind turbines, i.e., just periodic ABL flows, $\overline{u} = \langle \overline{u} \rangle$, $\overline{w} = \langle \overline{w} \rangle$. Hence, the dispersive stresses $\langle \overline{u}''\overline{w}'' \rangle = 0$ for the ABL flows.

With the thrust force, $\langle \overline{f_x} \rangle = \frac{1}{2}\rho C_T U_\infty^2 \frac{\pi}{4} D^2$

$$u_{*,hi}^2 = u_{*,lo}^2 + \frac{1}{2} C_T \left(\frac{\pi}{4} D^2\right) \left[\frac{u_{*,hi}}{\kappa} \log_e \left(\frac{z_h}{z_{0,hi}}\right)\right] \left(\frac{1}{s_x s_y D^2}\right) \tag{4.14}$$

This not only manifests the friction velocity, $u_{*,hi} > u_{*,lo}$, but also illustrates that the differences in the friction velocity of the upper portion of the wind farm canopy and its lower portion is a function of the diameter/hub-height of the wind turbine as well as the farm layout, i.e., the inter-turbine distances in the streamwise and lateral direction. Note, s_x *and* s_y are the non-dimensional inter-turbine distances in the streamwise and lateral direction, respectively, and $s_x D \times s_y D = s_x s_y D^2$ is the area “allotted” per turbine in the infinite farm.

$$z_{0,hi} = z_h \left(1 + \frac{D}{2z_h}\right)^{\frac{\nu_w^*}{1+\nu_w^*}} \exp\left(-\left[\frac{c_{ft}}{2\kappa^2} + \left(\log_e \left[\frac{z_h}{z_{0,lo}}\left(1 - \frac{D}{2z_h}\right)^{\frac{\nu_w^*}{1+\nu_w^*}}\right]\right)^{-2}\right]^{-1/2}\right) \tag{4.15}$$

where the “nondimensional” wake eddy viscosity is $\nu_w^* = \frac{\sqrt{1/2 c_{ft}} \langle \overline{u} \rangle D}{\kappa u_{*,hi} z_h}$, and c_{ft} is the friction coefficient for the farm layout.

Taking the dot product of (4.11) with mean velocity, we obtain the following energy balance equation in the mean sense.

$$\mathcal{W}_p + \Phi_p(z_h + D/2) - \Phi_p(z_h - D/2) = \mathcal{D} + \mathcal{W}_T \tag{4.16}$$

$\mathcal{W}_p$ is the work done by the pressure gradient driving force, which is approximately balanced by the mean kinetic energy (MKE) dissipation $\mathcal{D}$. The work done by the turbine thrust forces $\mathcal{W}_T$ is balanced by the MKE flux difference Φ_p between the top tip $z_h + D/2$ and bottom tip $z_h - D/2$. These two equilibriums reflect the double equilibrium log layer found in infinite wind farms.

In the infinite wind farms, the difference in turbulent shear stress (Reynolds + dispersive) flux component of the MKE flux at the top and bottom region of the rotor is responsible for the mean power per unit area generated by the wind turbines in the farm [4, 23, 27]. The mean power per unit area, $p_{mean} = 1/(N_t s_x s_y D^2) \sum_{i=1}^{N_t} \overline{P}_i$, can be calculated by averaging the temporal mean power $\sum \overline{P}_i$ over all N_t turbines in the farm, and diving by the area $s_x D \times s_y D$.

4.4.2 *Large eddy simulation of infinite wind farms*

In this section, we discuss the LES framework for simulating infinite wind farms. For brevity, we only discuss neutral ABL simulations, although the framework has also been tested for different atmospheric stability conditions by the wind farm community. Figure 4.5 shows a typical representative schematic of an infinite wind farm setup for LES simulations. This is typically done by invoking a horizontally periodic

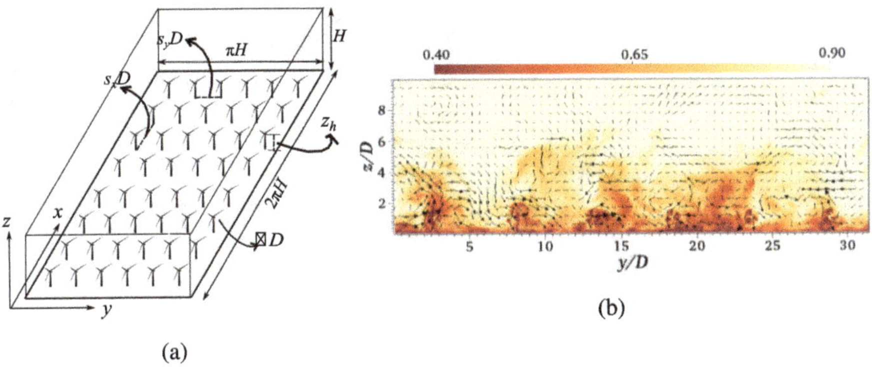

Figure 4.5 *(a) Computational domain $20\pi D \times 10\pi D \times 10D$ for LES showing the 8×6 periodic arrangement of wind turbines for infinite farm setup (hub-height z_h, rotor diameter $D = 0.1H$, H is the ABL thickness). The dashed white arrows indicate the direction of mean wind flow. (b) yz plane at the first row of wind turbines for the 8×6 periodic arrangement. Colors—streamwise velocity, Vectors—in-plane velocity vector (plots constructed from LES simulations data taken from [28]).*

simulation with an organized wind farm layout. Figure 4.5 shows a typical snapshot of LES (at the first row of turbines) conducted by our previous work [28], where multiscalar turbulent features in and around the rotor region are illustrative. Next, we show some of the turbulent statistics and the inference that can be drawn from the theoretical background discussed in Section 4.1.

4.4.2.1 Turbulent statistics

Figure 4.6 shows the mean streamwise velocity profile (validated against previous LES simulations [4]) for a suite of LES cases performed in our previous work in [28]. Cases I, IIa, and IIb have wind turbines placed at hub-height at 10% of boundary layer thickness ($z_h = 0.1\,H$) but with different layouts (8×6, 4×6, and 8×3 array) and hence different streamwise and spanwise distance between the turbines. Case III has the same layout/turbine arrangement as Case I but at a hub-height of 30% of the boundary layer thickness ($z_h = 0.33\,H$). Table 4.1 provides the details regarding the layout of the different asymptotic wind farms discussed in this section. The table is adapted from the authors' previous work [28].

The streamwise velocities manifest a double log trend as predicted by the theory in Section 4.1. The log-trends in Cases IIa, IIb exactly fall on the top of each other indicating that while the bottom log-layer is the manifestation of the ABL aerodynamic roughness, the top log-layer is guided by the roughness of the wind turbine canopy, $z_{0,hi}$, a strong function of the normalized wind turbine distance products $s_x s_y$, which are the same for Cases IIa, IIb. Additionally, the log-layers in Cases IIa, IIb are both shifted in the upward direction, indicating that the canopy roughness in

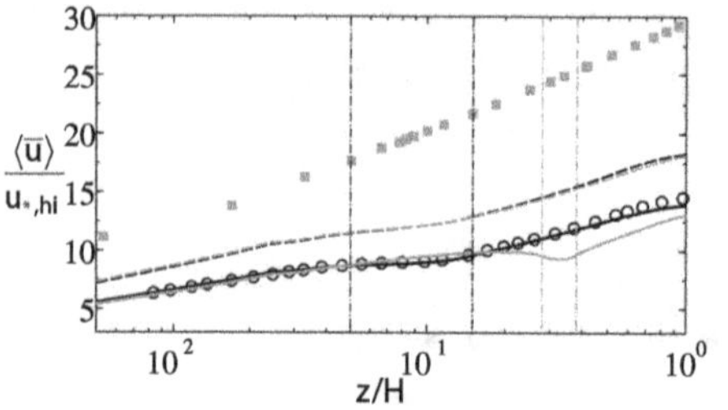

(a)

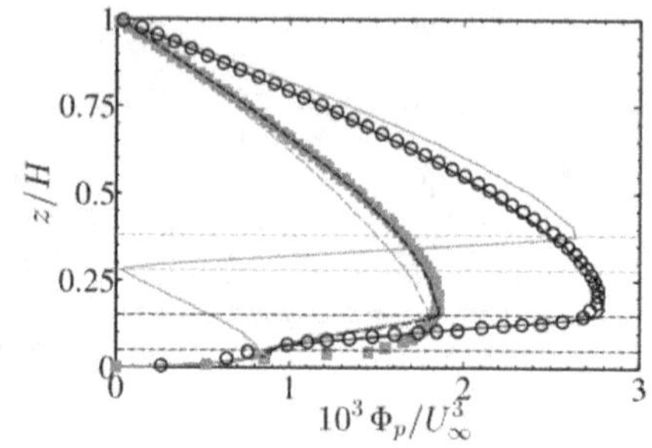

(b)

Figure 4.6 *(a) Mean streamwise velocity $\langle \overline{u} \rangle$ normalized by $u_{*,hi}$ vs z/H. (b) MKE flux Φ_p vs z/H. Solid black—Case I, dashed black—Case IIa, dashed gray—Case IIb, solid gray—Case III. Solid gray □ in (a)—neutral ABL; open ○ in (a), (b)—Calaf* et al. *[4] (same wind farm setup parameters as in Case I). Black, gray dashed dotted vertical (a) and horizontal (b) lines: rotor swept area of Cases I–III.*

Case I is higher than that in IIa, IIb. The bottom log-layer for Case III is similar to Case I, which interestingly indicates that the bottom roughness friction velocity, u_{*lo} (rather the ratio $u_{*,hi}/u_{*,lo}$) is also affected by the layout and hub-height of wind turbines and is a strong function of $(s_x s_y, z_h)$. The difference in the top log-layer between Cases I and III is a manifestation of the location of wake mixing layer in the farm and the differences in $z_{0,hi}$. Similar trends are observable in Figure 4.6, which illustrates the MKE flux Φ_p, a strong function of $(s_x s_y, z_h)$. Thus we see fairly similar trends in $\Phi_p(z)$, and in particular $\Phi_p(z_h + D/2) - \Phi_p(z_h - D/2)$ for Cases IIa, IIb having the same $s_x s_y$. Figures 4.7a and b illustrate the variation of integral length scales and streamwise rms fluctuations for the different asymptotic wind farm layouts I, IIa, IIb, and III. Interestingly, we observe that the major discrepancies in the

Table 4.1 *LES cases for the wind farm simulations. Domain size is $2\pi H \times \pi H \times H$, and $D = 0.1H$, fixed for all cases. N_{turb} is the number of turbines in streamwise-spanwise arrangement. S_x and S_y are the normalized (by* D*) inter-turbine streamwise and spanwise distances between turbines.* z_h and $z_{0,Lett}$ *are the hub-height and estimate of the augmented roughness height (Lettau roughness) of the turbines/wind farms.*

Case	N_{turb}	s_x	s_y	$S_x \times S_y$	S_x/S_y	$Z_{0,Lett}/Z_0$	z_h
I	8 × 6	7.85	5.23	41.05	1.5	9.54	0.1 H
IIa	4 × 6	15.7	5.23	82.11	3.0	4.77	0.1 H
IIb	8 × 3	7.85	10.47	82.11	0.75	4.77	0.1 H
III	8 × 6	7.85	5.23	41.05	1.5	31.5	0.33 H

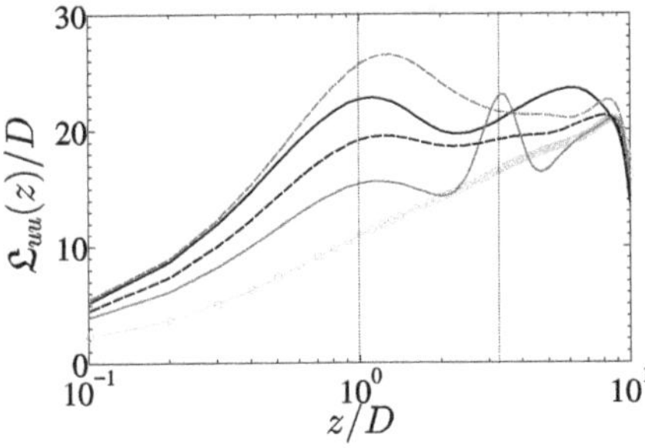

(a)

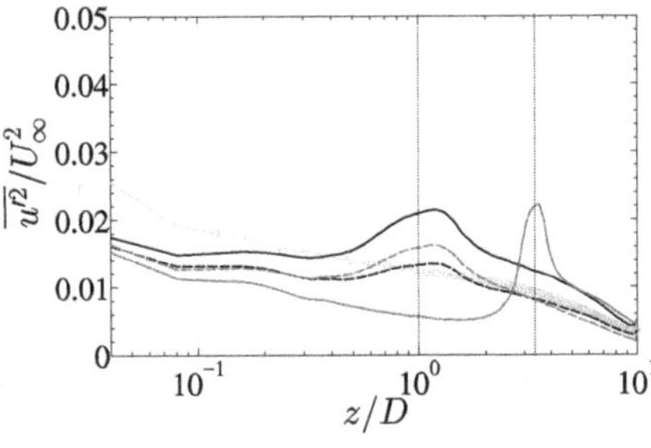

(b)

Figure 4.7 *Vertical variation of spanwise, time averaged, (a) integral length scales $\mathfrak{L}_{uu}$ based on streamwise fluctuations, (b) normalized mean-squared streamwise fluctuations $\overline{u'^2}/U_\infty^2$ (averaged over xy planes), with and without the presence of wind turbines. Solid black—Case I; dashed black—Case IIa; dashed gray—Case IIb; solid gray—Case III; light gray—neutral ABL, without wind turbines. Dotted lines—hub-heights, $z_h = 0.1H$ (Cases I, IIa, IIb), $z_h = 0.33H$ (Case III)*

integral scales and the rms fluctuations at different farm layouts occur at the inner layer in the region around and below the turbine disks while they mostly collapse to a single curve in the far outer layer manifesting minimal influence of farm layout in that region. The mean velocity and the rms fluctuations manifest a strong function of $(s_x s_y, z_h)$ while the integral length scales depend on $(s_x s_y, s_x/s_y, z_h)$. From these results, it is imperative to understand that the turbulent statistics and hence mean power per unit area of the farm depend on $S_x S_y$, Z_h, i.e., the farm density and hub-height of the turbines. Contrary to that, the integral length scales or the turbulent flow organization not only depends on the farm density $s_x s_y$ but also on the farm layout shape s_x/s_y. Since the authors' previous work has illustrated the contribution of organized coherent motions in the wind farm power production [28], the information about the layout shape can be crucial toward harvesting power from these organized coherent motions by placing smaller wind turbines under the shade of already constructed wind farm.

4.4.2.2 Spectra

This section extends our understanding of asymptotic wind farms in the spectral (wave number/wavelength) domain by studying the premultiplied energy spectra in Figure 4.8 excerpted from our previous work in asymptotic wind farms. In the near-wall regime, robust scaling laws of the *attached eddies* [29, 30] can be observed in wind farms. For example, Case I in Figure 4.8a shows the existence of k_x^{-1} scaling (λ_x^0 scaling in premultiplied spectra) in the u energy spectra, which are manifestations of the overlap region of the scales of attached eddies with the integral length scales [12, 15, 31].

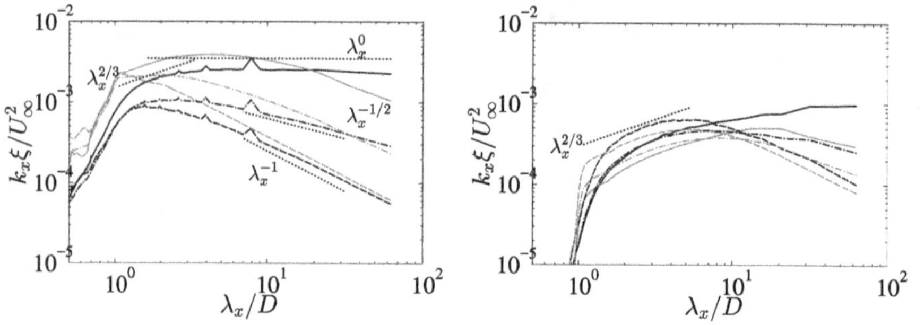

(a) (b)

Figure 4.8 *Premultiplied 1D spectra $k_x\xi$ vs normalized streamwise wavelengths λ_x/D for (a) $z = 0.25D$, (b) $z = 8.75D$. Black lines, case periodic wind farms with 8×6 periodic arrangement; gray lines, neutral ABL. k_xE_{uu} (–), k_xE_{ww} (––), $k_x\phi_{uw}$ (–.–). All spectra normalized by U_∞^2. Gray patch corresponds to the resolved part of the spectra directly influenced by SGS viscosity.*

Neutral ABL simulations indicate similar scaling laws for domains with and without wind turbines. However, the attached-eddy scaling laws extend to larger length scales ($\lambda_x > 10D$) within the wind turbine domain. This observation indicates that the three-dimensional turbulent motions also increase in size near the wall, possibly due to the vertical entrainment of the *MKE flux* in wind farms, which requires *downdrafts* or vertical motions from the high-speed regions above the turbine rotor toward the wall [32, 33]. Note that the higher locations in the boundary layer, e.g., at $z = D$ and $z = 8.75D$, do not show this feature, indicating that the expansion of the wave number region demonstrating k_x^{-1} law is mainly caused by the downdraft effects of the MKE flux. The near-wall structures become more organized, which can be potentially harvested by smaller utility-scale turbines under the shade of the large farm (see the authors' previous work [34] for more details). This idea has been exploited in vertically staggered/multiscale wind farms where the smaller turbines, if properly placed, were shown to be benefited from their larger counterparts.

4.4.3 Two-scale momentum theory

While rigorous theoretical framework based on aerodynamics (actuator disk theory) for the upper limit on the efficiency exists for a single wind turbine, known as Betz limit, developing a similar framework at the farm level is extremely complicated and requires not only an understanding of aerodynamics at the turbine level but also the interaction of wind turbines with the ABL. The present method [35] has been used to develop a theoretical framework to predict the practical upper limit of the efficiency of very large wind farms. Subsequently, we will refer to it as a Nishino model.

$$\langle \tau_w \rangle S + T = \langle \tau_{w0} \rangle S = \text{const.} \tag{4.17}$$

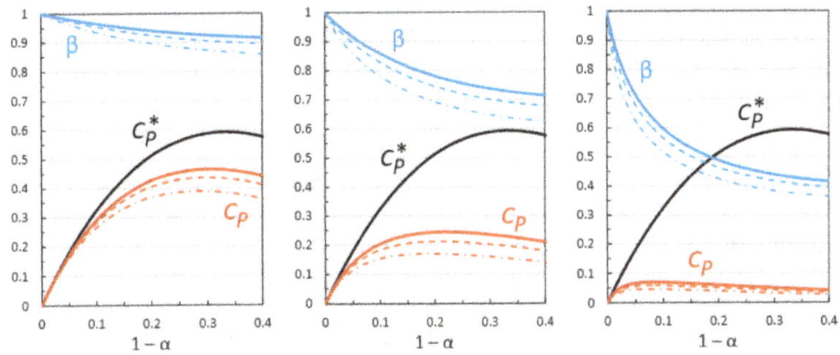

Figure 4.9 Example solutions of the two-scale momentum theory for ideal, very large wind farms, (a) $\lambda/C_{f0} = 2$, (b) $\lambda/C_{f0} = 1.5$, and (c) $\lambda/C_{f0} = 0.2$ (solid lines for $\gamma = 1$; dashed lines for $\gamma = 5$; dashed dotted lines for $\gamma = 1$). $\beta = U_F/U_{F0}$ and $C_P = \beta^3 C_P^*$, where C_P, C_P^* are the power coefficients based on the velocities U_F, U_{F0}, respectively. Figure taken from [35] in accordance with the Creative Commons Attribution (CC BY) License.

where $\langle \tau_w \rangle$ is the wall-shear stress in the ground in wind farms, T is the thrust generated by the wind farms, and $\langle \tau_{w0} \rangle$ is the natural wall-shear stress before the construction of wind farm (ABL). S is the area "allotted" to each turbine (for n turbines in the wind farm, the total wind farm layout area is nS).

U_{F0} and $U_{T0}$are the undisturbed velocity at the farm layer (before farm reconstruction) at a predefined height and averaged over the rotor area "A," respectively. U_F and U_T bear the similar definitions as with U_{F0} and U_{T0} but with the presence of wind turbines in ABL (post-farm construction).

From the actuator disk theory, the aerodynamic thrust by the turbines can be written as

$$T = \frac{1}{2}\rho U_F^2 A \cdot 4\alpha(1-\alpha) \tag{4.18}$$

where $a = (U_F - U_T)/U_F$ is the local axial induction factor and $\alpha = 1 - a$.

$$1 - \frac{\langle \tau_w \rangle}{\langle \tau_{w0} \rangle} = \frac{\lambda}{C_{f0}} \beta^2 \cdot 4\alpha(1-\alpha) \tag{4.19}$$

With $\alpha = U_T/U_F$, area ratio $\lambda = A/S$, and natural friction coefficient based on the undisturbed farm layer wind speed U_{F0}, $C_{f0} = \tau_{w0}/1/2\rho U_{F0}^2$. Additionally, the ratio of the wall-shear stresses of pre- and post-construction wind farms have been defined as $\frac{\langle \tau_w \rangle}{\langle \tau_{w0} \rangle} = \beta^\gamma$, $\beta = U_F/U_{F0}$, $\gamma < 2$.

Figure 4.9 illustrates how the Nishino model works for large wind farms for parameters, $\lambda/C_{f0} = 0.2, 1$, and 5. When $\lambda/C_{f0} = 0.2$ (small), relatively small number of wind farms are installed in the site, and the post-construction farm wind speed U_F is almost equal to the pre-construction wind speed U_{F0} and $\beta \approx 1$. This is reflected in the closeness of C_P^* (local power coefficient) and farm power coefficient

$C_P = \beta^3 C_P^*$. As more wind turbines in the land are installed, λ/C_{f0} increases (the wind farm density increases) and $\beta < 1$ starts dropping further at higher axial induction/rotor loading $(1 - \alpha)$. This is also reflected in $C_P \ll C_P^*$. It must be noted, however, that the normalized power $\eta = \lambda C_p/C_{f0}$ increases with λ/C_{f0}. This is inline with the Frandsen–Calaf theory, which indicates that the MKE flux difference $\Phi_p(z_h + D/2) - \Phi_p(z_h - D/2)$ across the rotor increases with lowering $s_x s_y$ or the wind farm density [28].

4.5 Interaction of large scale turbulence in wind farms

Wind farms in the ABL pose a complex dynamical system with turbulent phenomenology occurring at multiple length scales, mainly due to the interaction of ABL turbulent eddies [36] and the turbulence generated by the wind turbine wakes [4, 37, 38]. Recent works by [28] have shown the importance of coherent turbulent structures serving as mediators of large scale and very large scale motions in the ABL that not only carry around 30–40% of the turbulent kinetic energy but also contribute significantly (around ~ 50%) to the power generated by wind farms. Subsequently, many recent fundamental studies of wind farms have devoted their efforts to understanding the dynamics of these large scale structures to control them for optimizing power harvest in farms.

Modal decomposition methods such as proper orthogonal decomposition (POD) have revealed the presence of counter-rotating energetic roller mode structures presence in ABLs [26, 39]. The presence of the wind turbines/wind farms modulates these large scale structures and leaves it footprint at length scales of the streamwise distances between subsequent wind turbines in farms. The counter-rotating roll-cell structures in the rotor plane are illustrated in Figure 4.10.

The projection coefficients corresponding to the lower order modes (0–3) show low-frequency fluctuations compared to the projection coefficients corresponding to modes four or more, indicating that larger scales of motion are supposedly comprised of eddies of larger time scales. Additionally, the power spectral density of

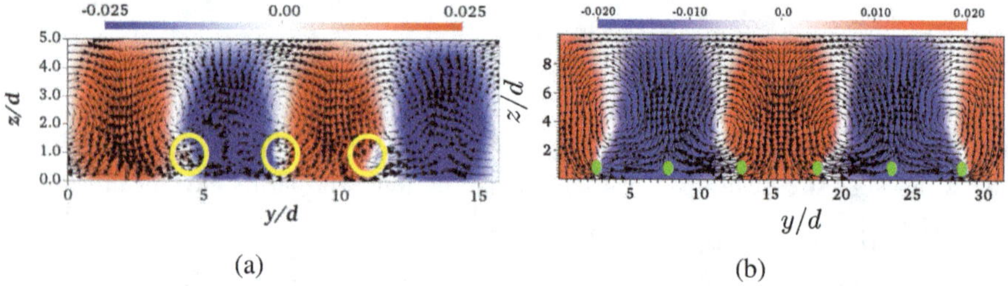

Figure 4.10 *First POD mode containing counter-rotating roll-cells of a (a) 3 × 3 finite-sized wind turbine array (b) 8 × 6 periodic wind farm. Color contours are* u *velocity mode and vectors in-plane* v, w *velocity mode.*

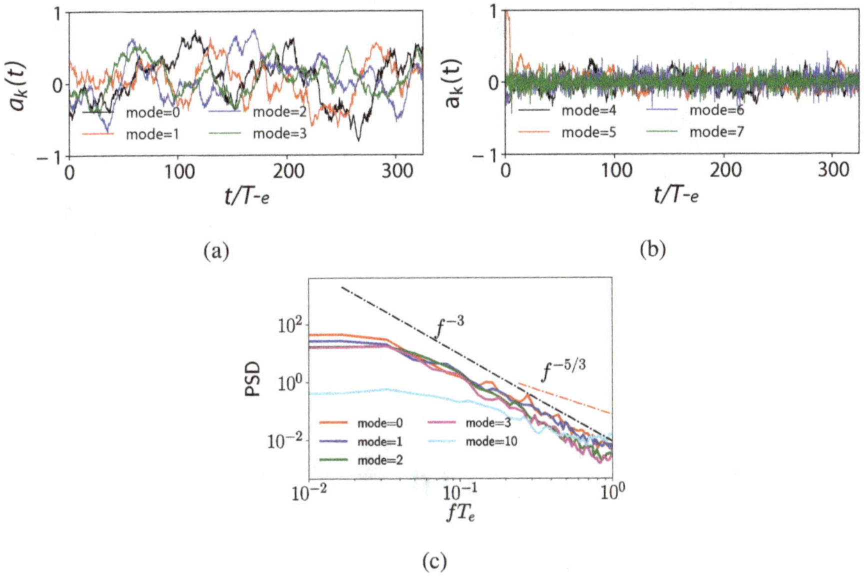

Figure 4.11 *(a), (b) Projection coefficient of three-dimensional POD modes, $a_k(t)$ vs time normalized by flow-through time, $T_e = 15\pi D/U_\infty$. (c) Power spectral density of the projection coefficients vs normalized frequency, fT_e.*

the projection coefficients (in frequency space) of the large scale modes ($m < 5$) manifested an f^{-3} scaling law (f is the frequency), which was attributed to the phenomenon of "merging of eddies" by previous literature [40, 41] (Figure 4.11c). Note, the $f^{-5/3}$ law for modes $m = 9$ (and higher) manifesting energy cascade was also observed. Consequently, it can be argued that the low time scale dynamics of the structures comprising the lower three-dimensional POD modes are predominantly governed by the merging of eddies. More details related to the large-scale dynamics and eigenfrequency of the POD mode shapes can be found in [39].

In the subsequent sections, we will discuss the analytical framework to model various flow features (wakes, induction) in wind farms responsible for power production losses.

4.6 Wake-deficit models

4.6.1 Wake superposition

Wakes generated from the wind turbines account for about 10–20 percent loss of the total power generated by the wind farms [42, 43]. Consequently, the interpretation and modeling of wake deficits are critical toward mitigating farm power losses.

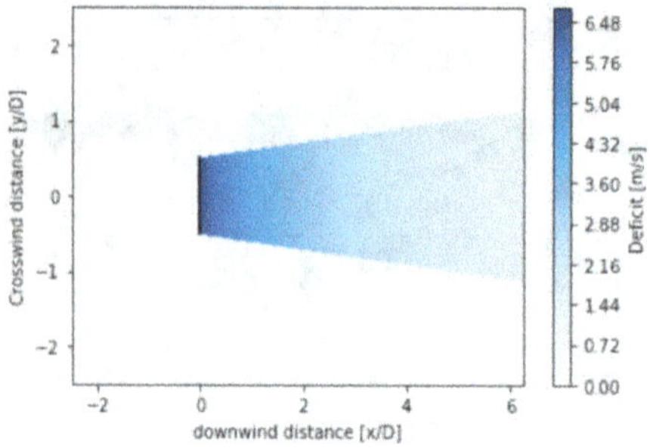

(a)

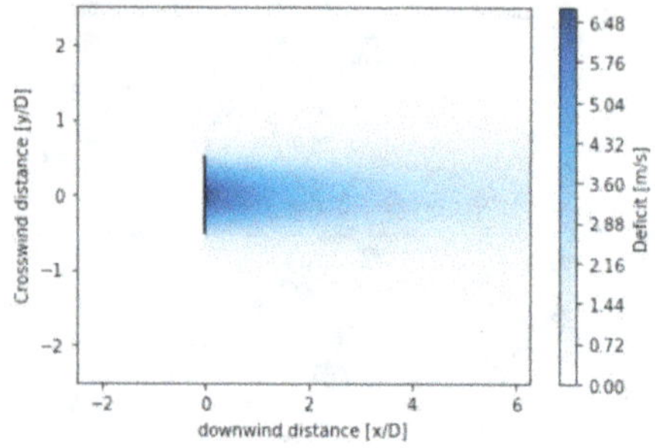

(b)

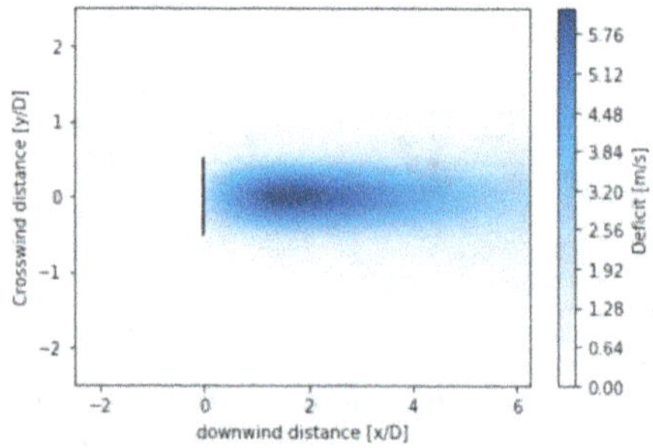

(c)

Figure 4.12 *(a) Top-hat Jensen model, (b) Bastankhah Gaussian profile model, and (c) Zhong model [46] (accounting for the momentum conservation in wakes). Calculations done in open-source AEP calculator PyWake [45].*

Analytical models serve as a powerful reduced-order tool for measuring wind turbine wakes. Analytical models are primarily derived from the first principles using momentum balance but are also inspired by high-fidelity LES results. A comprehensive summary of all the wake-deficit models starting from the top-hat Jensen model or Park model [44] to a more recent Gaussian profile-based Bastankhah model can be found in [1]. The readers are also encouraged to refer to the open-source python-based AEP calculator PyWake tool [45] to learn more about the different wake and wake-superposition models used by the wind energy community. A wind turbine's wake predicted by each of these models is illustrated in Figure 4.12.

The equation of the thrust forces T obtained from the momentum balance is

$$T = \frac{1}{2}\rho C_T U_\infty^2 \frac{\pi D^2}{4} = \rho \int u(U_\infty - u)dA \tag{4.20}$$

where C_T is the thrust coefficient, U_∞ is the free-stream velocity, and D turbine rotor diameter.

This gives rise to the top-hat Jensen model

$$\frac{\Delta u}{U_\infty} = \frac{1-\sqrt{1-C_T}}{(1+2k_f x/D)^2} \tag{4.21}$$

where $k_f = 0.5/\log(z_h/z_0)$ is the expansion rate of the turbine wakes. Z_h and Z_0 are the wind turbine hub-height and the aerodynamic roughness, respectively. A more realistic wake-deficit model is the Gaussian wake model with the assumption that different expansion rates in lateral y- and vertical z-direction.

$$\frac{\Delta u}{U_\infty} = \left(1 - \sqrt{1 - \frac{C_T}{8(\sigma_y \sigma_z/D^2)}}\right) \exp\left(-0.5\left[\left(\frac{y}{\sigma_y}\right)^2 + \left(\frac{z-z_h}{\sigma_z}\right)^2\right]\right) \tag{4.22}$$

where the terms in the equation are modeled as $\sigma_y/D = k_y x/D + \epsilon$, $\sigma_z = k_z x/D + \epsilon$, $\epsilon = 0.2\sqrt{\beta}$, and $\beta = \frac{1+\sqrt{1-C_T}}{2\sqrt{1-C_T}}$.

While the above discussion encompasses the state-of-the-art models for the wake evolution of the wind turbines based on empirical relations as well as the aerodynamics of the wind turbines using actuator disk theory and momentum conservation, the analytical modeling of the wake effects in the farm-scale relies on a robust theory of the superposition of the wakes. This section discusses the recent methods used for the wake superposition of multiple turbines. The wake deficit can be mathematically formulated as

$$\Delta u(x,y,z) = U_\infty - f(\Delta u_i(x,y,z)) \tag{4.23}$$

where Δu is the farm-level wake-deficit, U_∞ is the free-stream wind speed, and $f(\Delta u_i)$ is a function (superposition) of the wake-deficits of the individual turbines, Δu_i. The superposition of the different wakes in the reduced-order wind farm models (algebraic) can be broadly classified based on (i) method of superposition and (ii) definition of the velocity deficit at the ith turbine location. A recent paper [47] discusses these two methods.

The method of superposition popularly used is the linear superposition of the deficits $f(\Delta u(x,y,z)) = \sum_{i=1}^{N} \Delta u_i(x,y,z)$ or the linear superposition of the energy deficits (quadratic superposition) $f(\Delta u(x,y,z)) = \sqrt{\sum_{i=1}^{N} [\Delta u_i(x,y,z)]^2}$, with the velocity-deficit definition defined as $\Delta u_i = U_\infty - u_i(x,y,z)$. Additionally, the velocity deficit could also be defined as $\Delta u_i = u_{in,i} u_i(x,y,z)$, where $u_{in,i}$ is the incoming flow upstream of the ith turbine.

Figure 4.13 taken from [47] illustrates the wind farm power for dominant wind directions computed from the high-fidelity LES models as well as the analytical wake-deficit and wake-superposition model. Similarly, Figure 4.14 shows the power variation in the different rows of Horns Rev wind farm for LES and the analytical models. The results clearly illustrate that if properly designed, the analytical wake-deficit models can be competent to their high-fidelity LES counterparts.

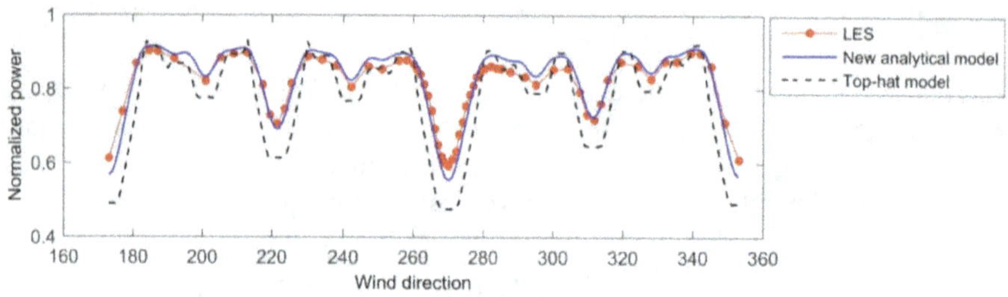

Figure 4.13 *Distribution of the normalized Horns Rev wind farm power output obtained with the new analytical model and LES for different wind directions. Figure taken from [47] in accordance with the Creative Commons Attribution (CC BY 4) License.*

4.6.2 Curled wake model

The curled wake model was recently studied as an analytical model inspired by LES simulations to account for kidney-shaped wakes that arise in wind turbines operating in yaw [48, 49]. The curled wake model is a reduced order model and requires solving linearized RANS model equations. The streamwise mean velocity U comes from the base flow, and the base cross-wind components V and W in the disk rotor region are defined using a superposition of N Lamb-Oseen vortices of circulation strength $\Gamma_i \ \ \forall i = 1, \ldots N$ being shed by the rotor plane. Γ_i is a function of the rotor loading/thrust coefficient C_T and yaw angle θ_{yaw}. The curled wake model plays a critical role

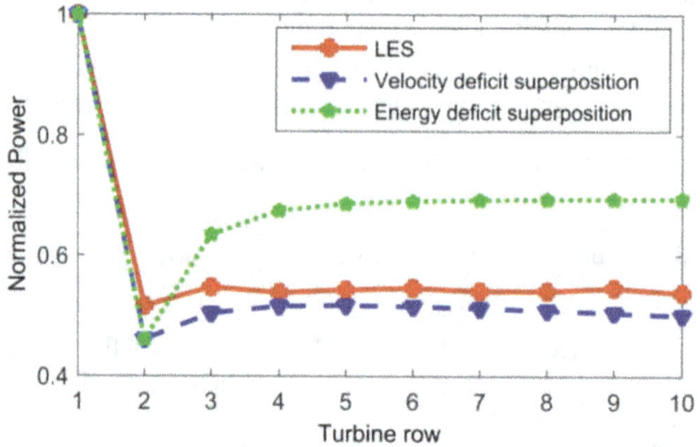

Figure 4.14 *Comparison of the wind farm power output for wind direction of 270° obtained using LES as well as the analytical model with both energy and velocity deficit superposition. Figure taken from [47] in accordance with the Creative Commons Attribution (CC BY 4) License.*

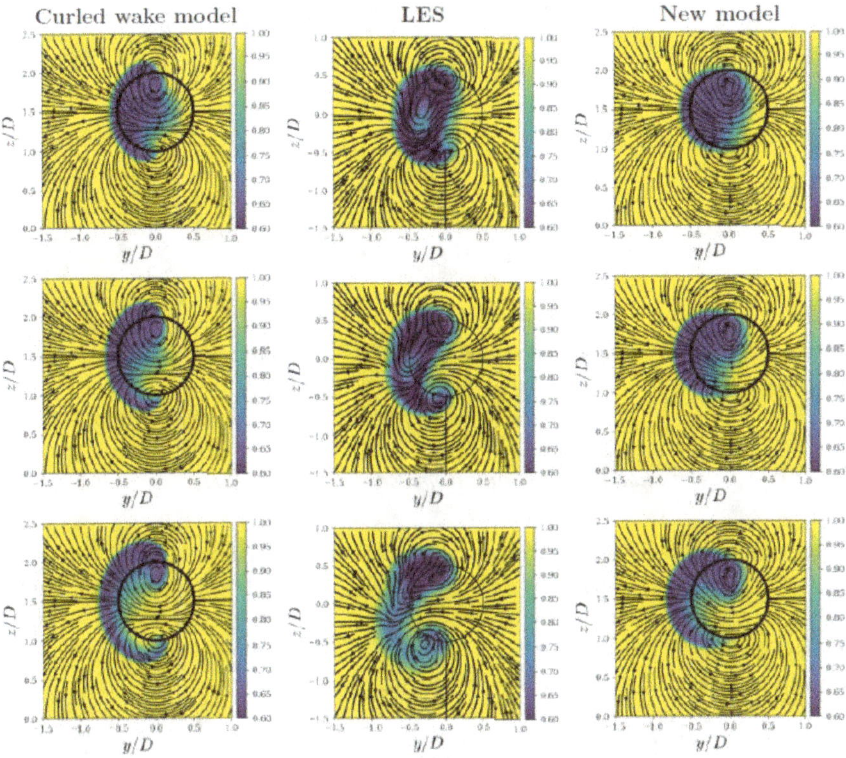

Figure 4.15 *Comparison of streamwise velocity contours between LESs of a wind turbine in* 30° *yaw and engineering models. The streamlines show the in-plane velocities. From top to bottom: downstream locations of* 3*D,* 5*D, and* 8*D. From left to right: the curled wake model [49], LES, and the recent vortex cylinder model [48]. Figure taken from [48] in accordance with the Creative Commons Attribution (CC BY 4) License.*

in reduced order/analytical wind farm models, where the dominant wind direction can change significantly, resulting in wind turbine yaw.

Figure 4.15 compares the behavior of the model against LES simulations and illustrates a lot of promise in the model, especially in the near-wake region and also identifies the potential room for improvement.

4.7 Induction

While the wake-deficit and wake-superposition models have received much attention in the wind energy community, the turbine upstream effects due to blockage are still not properly understood and is an area of active research. The presence of

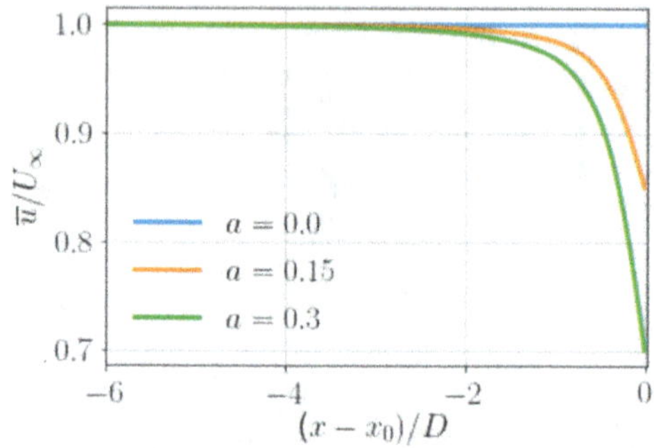

Figure 4.16 *Medici induction model (2011), for different levels of axial induction* a = 0, *(no wind turbine) 0.15, and 0.3. Note for* $a < 0.5$, $a = \frac{1}{2}(1 - \sqrt{1 - C_T})$ *is directly related to rotor loading.* x_0 *is the turbine location. The current figure is recreated by the author using the analytical function from Medici model (2011).*

a wind turbine affects the airflow both upwind and downwind of the turbine. The upwind region affected by the turbine is called the induction region. Prior studies [50, 51] have shown that the main impact of the turbine on that region is a reduction in wind speed, which can be estimated acceptably with the following simple relationship based on the vortex sheet theory [51].

$$\frac{\overline{u}}{U_\infty} = 1 - a\left(1 + \frac{2x}{D}\left(1 + 1 + \left(\frac{2x}{D}\right)^2\right)^{-0.5}\right) \tag{4.24}$$

$\overline{u}$is the streamwise velocity component along the rotor axis (the overbar denotes time averaging), x is the streamwise position (being zero at the turbine and negative upwind), U_∞ is the streamwise velocity component far upwind, D is the rotor diameter, and a is the rotor induction factor. Figure 4.16 manifest the plot of the upstream variation of the wind speed as per (4.24).

The induction/blockage effects have become particularly interesting to the industry since it causes speed reduction not only for upstream of the wind turbine, but for wind farms with rotors closely spaced to each other, e.g., 2.5D apart [52]; the induction can also cause local acceleration (speed-up) around the sides of the wake deficit and serves as a potentially promising way of optimizing power harvest in large wind farms [53].

Figure 4.17a illustrates the effect of superposition of induction and wake regions in wind farms. One of the methods, as discussed in Figure 4.16, is to linearly superpose the induction fields in wind farms (simply adding them), as they have possibly weaker effects than the superposition of wake deficits. The blockage model is developed by iteratively solving for the wakes (upstream to downstream turbines) and induction field for the entire farm and then merging them as $U = U_{base} + u_{wake} + u_{ind}$, where U_{base} is the background base flow. Figure 4.17b illustrates the comparison of the wake and induction region of the vortex cylinder model [53]

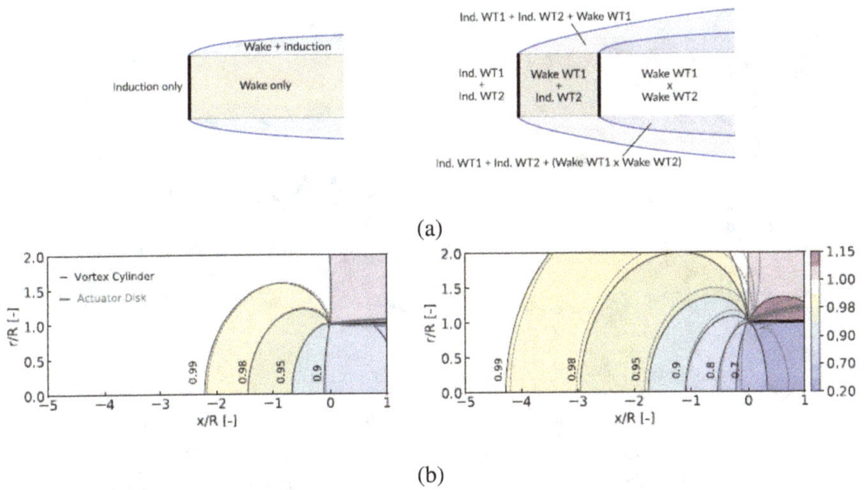

Figure 4.17 *(a) Schematic of the induction region in wind farm, (+): linear superposition and (x): merging. (b) Axial velocity contours normalized by the free-stream velocity (U_x/U_∞), as obtained by the vortex cylinder model (solid) and actuator disk simulations (dashed) for C_T = 0.4 (left) and C_T = 0.95 (right). Figure taken from [53] in accordance with the Creative Commons Attribution (CC BY 4) License.*

compared to actuator disk simulations at different disk-loading conditions (C_T). The contours match reasonably well, illustrating promise that the computationally cheaper engineering model can detect the farm flow at comparable level of fidelity as actuator disk methods.

In the following sections, we highlight several aspects of farm conditions (aerodynamic roughness, terrain effects) on the aerodynamics of wind turbine flows.

4.7.1 Aerodynamic roughness in wind farms

Several studies conducted in the past reveal that the wake recovery rate in wind turbine/wind farms is much faster with higher surface roughness and hence contributes to higher power generation with higher aerodynamic roughness. This, in part, suggests a greater plant capacity factor in onshore wind farms compared to their offshore counterparts (which has lower aerodynamic roughness). A larger aerodynamic roughness causes a more significant TI, which leads to the higher wake recovery rate and hence more efficient power generation by wind farms. Figure 4.18 shows LES simulations of the mean streamwise velocity and TI upstream and downstream of a wind turbine for different aerodynamic roughness. The computational results indeed illustrate that the wake recovery rate is a strong function of the aerodynamic roughness. The incoming turbulent intensity also increases due to higher aerodynamic roughness.

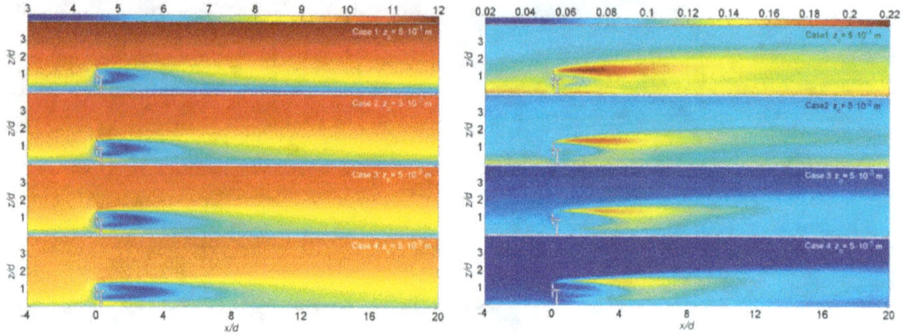

(a) (b)

Figure 4.18 LES study showing contours of the (a) mean streamwise velocity, u *(m/s) and (b) streamwise TI $I_u = \sigma_u/u_{hub}$ in the middle vertical plane perpendicular to the turbines installed over flat surfaces with different roughness lengths. Figure taken from [54] in accordance with the Creative Commons Attribution (CC BY) License.*

The presence of wind turbines/wind farms augments the turbulent intensity in the ABL. Mathematically speaking, the added turbulent intensity can be defined as

$$I_+ = \sqrt{I_{wake}^2 - I_0^2} \tag{4.25}$$

where I_0 is the incoming turbulent intensity in the wind farm whereas I_{wake} is the turbulent intensity in the wake region of the wind turbines. Several empirical/semiempirical relations exist for I_+ as power law coefficients of thrust coefficients, C_T, incoming turbulent intensity to wind turbine, I_0, and the distance from the wind turbine normalized by near-wake length, x/x_n.

$$I_+ = AC_T^{\beta} I_0^{\alpha} (x/x_n)^{-k} \tag{4.26}$$

From the empirical relationship, it is apparent that the higher incoming TI I_0 contributes directly (power-law relation) to the augmented intensity and in general to the augmented turbulent kinematic shear stress, which is primarily responsible for the vertical entrainment of energy from the high-speed region above the turbines thus energizing the turbulent wakes and speeding its wake recovery. Figure 4.18 shows that the turbulent intensity reaches its peak at around 3–4D, after which the intensity starts decaying. Further, Figure 4.19 illustrates the evolution of integral length scales (metric of turbulent coherent structures) in utility-scale wind farms in a wind tunnel facility for low and high aerodynamic roughness [55]. It was illustrated that while both roughness layouts show similar growth trends in the integral length scales, the higher aerodynamic roughness or high incoming turbulent intensity layout manifests better organization of the turbulent structures, which also contributes to its faster wake recovery.

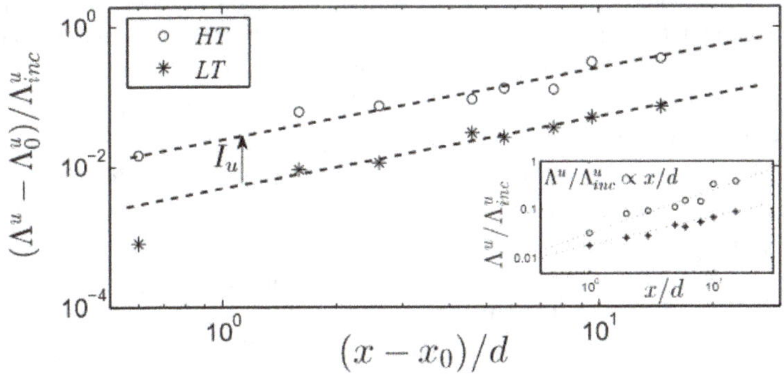

Figure 4.19 *Log–log illustration of the integral length scales along the rotor axis for high free-stream turbulence (HT) and low free-stream turbulence (LT). Here,* $x_0/D \sim 1/8$ *denotes the location where the integral length scales* Λ_u *for both cases are equal. Figure taken from [55] in accordance with the Creative Commons Attribution (CC BY) License.*

4.8 Wind farms in complex terrain

Modeling wind farms in complex terrain serves as a crucial step toward understanding farm-level interaction with the atmosphere in many US sites for onshore wind-farm applications. However, modeling remains a great challenge, especially from the perspective of modeling the "non-flat" ground topography (Figure 4.20). While many RANS approaches [56] exist in simulating farms in complex terrain, recent LES approaches have fundamentally tackled the problem of immersed boundary methodology [57] as well as help develop analytical model of wind farms over hilly topography [58].

One of the key aspects of modeling challenges in wind farms in complex terrain is a lack of a universal superposition method of individual turbine wakes in the farm. This

Figure 4.20 *Wind farms in a complex terrain*

is primarily attributed to wake recovery, which is strongly influenced by (a) nonzero pressure gradient, (b) variable elevation of wake-center trajectory from the ground, and (c) flow separation in hilly topography. For example, recent LES studies by [58] reveals that the favorable pressure gradient on the windward side of the hill accelerates the turbine wake recovery, and the adverse pressure gradient on the leeward side decelerates it. The wake trajectory for a hill of the same height as the hub-height of the turbine is found to closely follow the hill profile on the windward side, but it maintains an almost constant elevation (a horizontal line) downstream of the hilltop. Similar analysis, illustrated in Figure 4.21 was carried out using LES simulations in [57]. The figures clearly indicate the modulation of wind turbine wakes by the favorable pressure gradient on the windward side of the two-dimensional hill. The study further reveals that steep hills have a significant impact on the power production of turbines, and turbines that are significantly taller than the hill benefit from the speed-up of the flow over the hill. For shorter turbines, the power production strongly depends on the turbine location with respect to the hill, but it saturates to a maximum on the hilltop, whereas locations just downstream of the hill are negatively impacted by the hill topography.

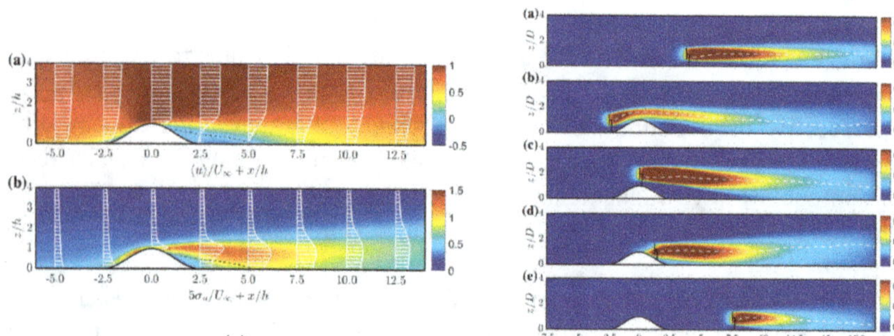

(a) (b)

Figure 4.21 *(a) Time and spanwise-averaged streamwise velocity component and its variance for the ABL flow over a two-dimensional steep hill. The white lines indicate the profile at various downstream locations and the black dashed line indicates the edge of the recirculation zone. (b) Normalized velocity deficit u_{def}/u_d in the mid-plane of the turbine for, (i) the flat terrain case, and (ii)–(v) for flow over a steep two-dimensional hill with the turbine located at various locations at windward and leeward side of the two-dimensional hill. The white dashed lines denote the location of the wake-center trajectory. Figure taken from [57] in accordance with the Creative Commons Attribution (CC BY) License.*

4.9 Reducing farm-level aerodynamic interaction

The aerodynamic interaction at the farm level can be mainly achieved in two ways: (i) modification of the farm's layout and land and (ii) modification of the wind turbines. The basic idea in all these methods is to accelerate the wake recovery of the upstream turbines before they hit the downstream rotors to minimize wake impingement. Some of the state-of-the-art methods are discussed here.

4.9.1 Modification of wind turbines

Some of the existing concepts of the modification of the wind turbines include using (i) two-bladed and three-bladed turbines in the farm and (ii) winglets for truncated-blade wind turbines. The methodology using two-bladed and three-bladed wind turbines [11] is potentially promising since two-bladed turbines have quicker wake recovery compared to three-bladed turbines, and thus can produce potentially more power compared to a farm with all three-bladed turbines. The truncated-blade rotor [59] is again a relatively new concept, which has the potential to improve the farm power harvest. The truncated rotors produced jet-like flow in the wake, inducing large-scale instability and distinct small-scale vortical motions. The mean power reduced significantly with increased blade length ratio L/R. L is the length of the blade that produced nonzero lift and R is the radius of the rotor. Typically winglets are added at the tip of the blade to reduce negative end-effects. However, standard turbines ($L/R = 1$) (no winglets) placed downwind of the truncated counterparts ($L/R < 1$) produced more power than those in the wake of other standard units.

4.9.2 Modification of farm layout: multiscale wind farms

One of the relatively new explored ways of reducing turbine wake interactions is using vertical staggering of wind farms. While horizontal staggering of wind farms have been popularly used by the wind farm community for optimizing power harvest in wind farms [60], the horizontal staggering is effective in sites where there is a dominant wind direction. If the wind direction varies significantly during the year, the layout that is horizontally staggered in one direction might be perfectly aligned for other wind directions. In this regard, the vertical staggering of wind farms offsets the hub-height of the wind turbines allowing no or only partial wake mixing of turbines in the streamwise direction. Xie *et al.* [61] used a vertically staggered wind farm layout, with clusters of small vertical axis turbines (VAWTS) placed around a large HAWT. The authors showed that the wake recovery of the HAWT has significantly accelerated owing to the presence of VAWTS, which has led to more than 10% power gain in the HAWT. In wind tunnel experiments, Chamorro *et al.* [62] used wind turbines of similar capacity, with slightly different hub-heights in a vertically staggered wind farm layout. Vested *et al.* [63] documented experiments with a model of a vertically staggered wind farm featuring HAWTs of two different hub-heights but identical rotor sizes, and reported a 25% increase in the power of shorter wind turbines operating behind the taller ones, as compared to the base configuration

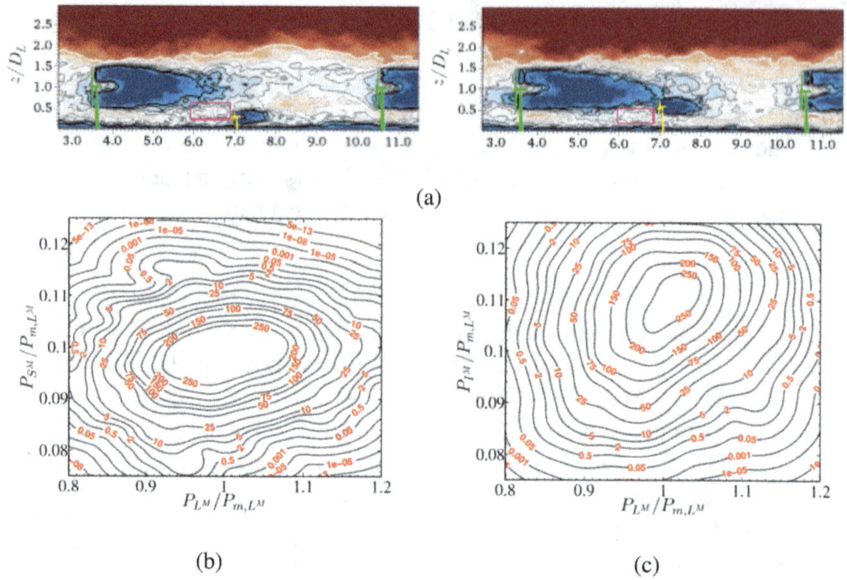

Figure 4.22 *LES study showing contours of the (a) mean streamwise velocity* u *(m/s) of two different configurations of vertically staggered wind farm and (b) joint probability distribution of power of large turbines and the small turbines in the shade of large ones. Figure taken from [65] in accordance with the Creative Commons Attribution (CC BY) License.*

with short wind turbines only. The potential of vertical staggering has further been illustrated in the recent works of Vasel-Be-Hagh and Archer [64].

The authors of the chapter have also demonstrated the potential of such layouts in increasing the overall power yield in these staggered wind farms. Our recent work has extended this concept to a multiscale wind farm, where wind turbines of completely different capacities (MW vs kW) have been used. The small KW-scale wind farms are placed under the shading of the large MW-scale turbines. The small-scale turbines extract power from the organized, coherent motions (modulation of ABL structures by the presence of MW scale HAWTs) below the rotor tip of large turbines. The work demonstrated that if the small turbines are placed at some specified heights where no wake overlap with the large turbines occur, significant power benefits of the small turbines ($\sim$ 5%) can occur along with a drastic reduction in the power variability ($\sim$ 30%).

Figure 4.22a shows the snapshots of the streamwise vertical plane of the multiscale wind farms. The left plot shows that the small wind turbines do not live in the wake shadow of the massive wind farms (favorable layout), whereas in the right plot, there is considerable mixing of the wakes between the large and small wind turbines (unfavorable layout). Figures 4.22b (favorable layout), 4.22c (unfavorable

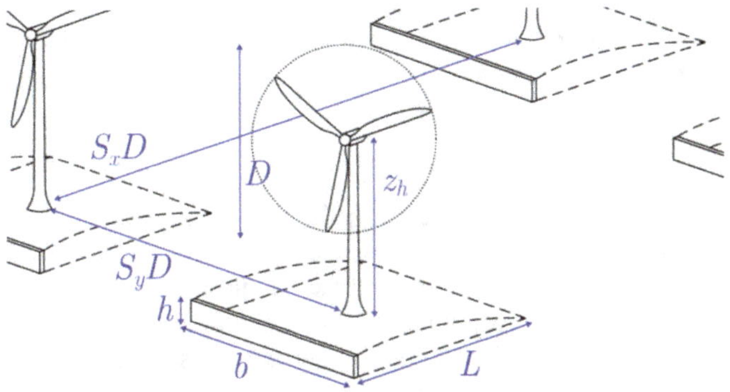

Figure 4.23 Conceptual schematic of the layout of an infinite wind farm with windbreaks. Dashed lines show the region with reversed flow/ circulation region. Figure taken from Tobin and Chamorro [66] in accordance with the CC BY License.

layout) further illustrate the joint probability distribution of the power generated by the massive and small wind turbines. For the unfavorable layout, large variation in the small turbine power corresponding to the large turbine power is illustrated. For favorable layout, the small turbine power remains relatively flat even with large wind turbine power variation, demonstrating that the small turbines are harvesting organized turbulent structures.

4.9.3 Modification of farm land: windbreak models

Tobin *et al.* [66] recently demonstrated the usefulness of windbreaks (typically, rows of trees/canopies or fence structures) in diverting low-level flow to the level of a wind turbine rotor, where a significant increase in power production was shown to agree well with linear theory. The researchers have further demonstrated that the power production gain with windbreaks is dependent on the porosity of the brakes (blockage effect) and is more promising with small wind farms. A typical schematic of windbreaks is illustrated in Figure 4.23.

In the final section, we focus our attention on discussing wind farm control methods.

4.10 Wind farm control

Wind farm control is an active area of research with the sole motivation of optimizing the power harvest/AEP or minimizing the LCOE of the wind farms. Some of the state-of-the-art farm control techniques are discussed in this section.

4.10.1 Wake steering

The wake steering control scheme [67] is solely motivated at reducing the farm power loss due to upstream wake impingement. Mathematically, the wake steering can be formulated

$$\max_{\gamma} \sum_{i=1}^{N_t} P_i, \quad \text{subject to } \gamma \in [\gamma_{min}, \gamma_{max}] \tag{4.27}$$

where P_i is the power of ith turbine and γ is the yaw angle. Despite not being a convex problem, it can be optimized using a genetic algorithm. The wake steering method was tested in an array of six utility-scale turbines, where it increased the power production for wind speeds near the site annual average between 7 and 13 percent and decreased variability by up to 72 percent, for selected wind directions at night. These improvements can contribute to the increasing ability of wind farms to provide reliable, low-cost, and efficient base energy load.

4.10.2 Dynamic induction control

The dynamic induction control is more generic approach of wind farm control [68] showing potential power gains of up to 20 percent. This is more generic than the yaw control directed at steering wakes from impinging downstream turbines, and it involves influencing the induced wake velocity deficit or the axial induction factor.

The optimization is performed using a receding-horizon control framework, as shown in the schematic Figure 4.24. In this framework, wind farm controls are optimized for a finite time horizon T, involving a sequential set of LES and adjoint LES simulations. The adjoint PDE is used to compute the sensitivities of cost functional. Upon convergence of the optimization, optimized control signals are applied in a flow advancement simulation for a time T_A, after which a new optimization window is initiated. Mathematically, the dynamic induction control within each optimization window can be formulated as

$$\min_{\phi,q} = -\sum_{i=1}^{N_t} P_i, \text{ subject to} \tag{4.28}$$

$$\tau \frac{\hat{C}'_{T,i}}{dt} = C'_{T,i} - \hat{C}'_{T,i} \quad i = 1 \ldots N_t \text{ in } (0,\ T], \tag{4.29}$$

$$0 \leq C'_{T,i} \leq C'_{T,\max} \quad i = 1 \ldots N_t \text{ in } (0, T] \tag{4.30}$$

Additionally, the NS equation is also a constraint in the minimization problem. The control variables $\phi = [C'_{T,1}, C'_{T,2}, C'_{T,3} \ldots C'_{T,N_t}]$ are the time-dependent thrust coefficients set points for every turbine and the state variables are the filtered velocity, pressure in NS equation for LES solver as well as the time-filtered thrust coefficients (time constant τ) $[\widetilde{u}_i, \widetilde{p}, \widehat{C}'_{T,1}, \widehat{C}'_{T,2}, \widehat{C}'_{T,3} \ldots \widehat{C}'_{T,N_t}]$.

4.11 Conclusion

In the current chapter, we summarize the recent state-of-the-art advances in the aerodynamics of wind farms deployed in ABLs. This chapter particularly discusses the progress in high-fidelity numerical framework, in particular LESs and the wind

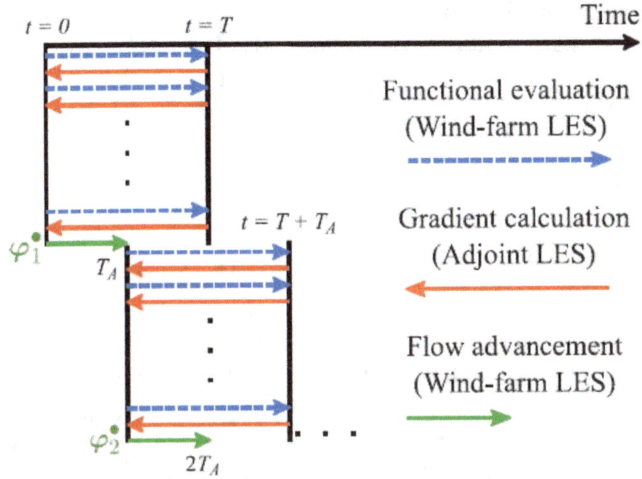

Figure 4.24 *Receding horizon framework subdividing time into discrete flow advancement windows of length* T_A *with prediction horizon T. Each arrow represents a forward or adjoint LES. Every window consists of an optimization stage (blue and red lines) followed by a flow advancement stage with optimal controls (green lines). Figure taken from [68] in accordance with the Creative Commons Attribution (CC BY) License.*

turbine models for computing large wind farm layouts. We have reviewed the LES framework for asymptotically large wind farms, the analytical models relying on momentum/energy equilibrium framework to describe such infinite layout including Frandsen–Calaf theory and the two-scale momentum theory (Nishino model). We have also discussed the analytical models used in the farm layout to describe induction/blockage effect, wake deficits in wind turbines, and the superposition of wakes in the farm. This chapter also discusses about the effect of direction variability, complex terrains in wind farm layout as well as minimizing interaction of wind turbine wakes in farms. From the recent advances in the literature, it is apparent that well established theories of wind farms exist in the asymptotic limit (infinite wind farms) for different atmospheric stability conditions. These theories have been well validated by high fidelity, computationally expensive LES models. From the perspective of very large wind farms, these theories indicate that the addition of wind farms can be treated as an additional roughness increment in ABL. Consequently, the development of a wind farm boundary layer from an undisturbed ABL shows similarities with the flow structure after a surface roughness transition, where the sudden change of surface drag causes the growth of an IBL. These theories certainly have limitations with respect to the environmental conditions, e.g., atmospheric stability and farm layout. For more complex situations, i.e., flow over complex terrains or wind farms of finite scale, the theory of wind farms is still developing, and more studies are required to create a framework of physics-based theories for such complex flow interactions. Finally, some aspects of wind farm control from the numerical perspective have also been discussed in this chapter. These farm control strategies play a crucial role in optimizing power harvest and turbine thrust loads.

References

[1] Porté-Agel F., Bastankhah M., Shamsoddin S. 'Wind-turbine and wind-farm flows: a review'. *Boundary-Layer Meteorology*. Sep 2019. Available from https://doi.org/10.1007/s10546-019-00473-0.

[2] Porté-Agel F., Bastankhah M., Shamsoddin S. 'Wind-turbine and wind-farm flows: A review'. *Boundary-Layer Meteorology*. 2020;**174**(1):1–59.

[3] Bleeg J., Purcell M., Ruisi R., Traiger E. 'Wind farm blockage and the consequences of neglecting its impact on energy production'. *Energies*. 2018;**11**(6):1609.

[4] Calaf M., Meneveau C., Meyers J. 'Large eddy simulation study of fully developed wind-turbine array boundary layers'. *Physics of Fluids*. 2010;**22**(1):015110.

[5] Lundquist J.K., Duvivier K.K., Kaffine D., Tomaszewki J.M. 'Costs and consequences of wind turbine wake effects arising from uncoordinated wind energy development'. *Nature Energy*. 2018;**1**:2058–7546.

[6] Avila M., Gargallo-Peiró A., Folch A., Faulch A. 'A CFD framework for offshore and onshore wind farm simulation'. *Journal of Physics: Conference Series*. 2017;**854**:012002.

[7] van der Laan M.P., Hansen K.S., Sørensen N.N., Réthoré P.-E. 'Predicting wind farm wake interaction with RANS: An investigation of the Coriolis force'. *Journal of Physics: Conference Series*. Jun 2015;**625**:012026

[8] Porté-Agel F., Meneveau C., Parlange M.B. 'A scale-dependent dynamic model for large-eddy simulation: application to a neutral atmospheric boundary layer'. *Journal of Fluid Mechanics*. 2000;**415**:261–84.

[9] Mason P.J. 'Large-eddy simulation of the convective atmospheric boundary layer'. *Journal of the Atmospheric Sciences*. 1989;**46**(11):1492–516.

[10] Shreyas Ananthan and coworkers. *Nalu-Wind Theory Manual [online]*. 2020. Available from https://nalu-wind.readthedocs.io/en/latest/source/theory/.

[11] Hayat I., Chatterjee T., Liu H., Peet Y.T., Chamorro L.P. 'Exploring wind farms with alternating two- and three-bladed wind turbines'. *Renewable Energy*. 2019;**138**(9):764–74

[12] Hutchins N., Nickels T.B., Marusic I., Chong M.S. 'Hot-wire spatial resolution issues in wall-bounded turbulence'. *Journal of Fluid Mechanics*. 2009;**635**:103–36.

[13] Sillero J.A., Jiménez J., Moser R.D. 'Two-point statistics for turbulent boundary layers and channels at Reynolds numbers up to $\delta^+ \approx 2000$'. *Physics of Fluids*. 2014;**26**(10):105109.

[14] Lettau H. 'Note on aerodynamic roughness-parameter estimation on the basis of roughness-element description'. *Journal of Applied Meteorology.* 1969;**8**(5):828–32.
[15] Chatterjee T., Peet Y.T. 'Effect of artificial length scales in large eddy simulation of a neutral atmospheric boundary layer flow: a simple solution to log-layer mismatch'. *Physics of Fluids*. 2017;**29**(7):075105.
[16] Gadde S.N., Stevens R.J.A.M. 'Interaction between low-level jets and wind farms in a stable atmospheric boundary layer'. *Physical Review Fluids.* 2021;**6**(1).
[17] Allaerts D., Meyers J. 'Gravity waves and wind-farm efficiency in neutral and stable conditions'. *Boundary-Layer Meteorology*. 2018;**166**(2):269–99.
[18] Khanna S., Brasseur J.G. 'Analysis of Monin–Obukhov similarity from large-eddy simulation'. *Journal of Fluid Mechanics*. 1997;**345**:251–86.
[19] Chatterjee T., Peet Y. 'Actuator line wind turbine simulations in atmospheric turbulent flows using spectral element method'. *AIAA paper 2015–0727*; Kissimmmee, Florida, 5–9 Jan; 2015.
[20] Troldborg N. Actuator Line Modeling of Wind Turbine Wakes [PhD thesis].Technical University of Denmark; 2008.
[21] Churchfield M.J., Lee S., Moriarty P.J., *et al.* 'A large-eddy simulation of wind-plant aerodynamics'. *AIAA Paper*. 2012;**0537**:2012.
[22] Lesgo: A Parallel Pseudo-Spectral Large-Eddy Simulation Code [online]. 2010. Available from https://lesgo.me.jhu.edu [Accessed 5/18/2021].
[23] Y.-T. W., Porté-Agel F. 'Simulation of turbulent flow inside and above wind farms: model validation and layout effects'. *Boundary Layer Metereol.* 2013;**5**:5340–62.
[24] Tossas Luis Martínez., Leonardi S., Churchfield M., Moriarty P. 'A Comparison of Actuator Disk and actuator Line Wind Turbine Models and Best Practices for Their Use'. *AIAA Journal*. 2012;**16**.
[25] Meyers J., Meneveau C. 'Flow visualization using momentum and energy transport tubes and applications to turbulent flow in wind farms'. *Journal of Fluid Mechanics*. 2013;**715**:335–58.
[26] VerHulst C., Meneveau C. 'Large eddy simulation study of the kinetic energy entrainment by energetic turbulent flow structures in large wind farms'. *Physics of Fluids*. 2014;**26**(2):025113.
[27] VerHulst C., Meneveau C. 'Altering kinetic energy entrainment in large eddy simulations of large wind farms using unconventional wind turbine actuator forcing'. *Energies*. 2015;**8**(1):370–86.
[28] Chatterjee T., Peet Y.T. 'Contribution of large scale coherence to wind turbine power: a large eddy simulation study in periodic wind farms'. *Physical Review Fluids*. 2018;**3**:034601.
[29] Perry A.E., Henbest S., Chong M.S. 'A theoretical and experimental study of wall turbulence'. *Journal of Fluid Mechanics*. 1986;**165**(-1):163–99.
[30] Townsend A.A. 'Equilibrium layers and wall turbulence'. *Journal of Fluid Mechanics*. 1961;**11**(1):97–120.

[31] Nickels T.B., Marusic I., Hafez S., Chong M.S. 'Evidence of the law in a high-Reynolds-number turbulent boundary layer'. *Physical Review Letters*. 2005;**95**:074501.

[32] Chatterjee T., Peet Y. Spectra and large eddy structures in the double log-layer in a high Re wind turbine array boundary layer. ASME Turbo Expo GT2016-56359; 2016.

[33] Peet Y.T., Chatterjee T. 'The contribution of large scale structures in the power generation of finite scale wind farms using large eddy simulation'. *Proceedings of the 10th International Symposium on Turbulence and Shear Flow Phenomena (TSFP10)*; Chicago, IL; Jul 2017.

[34] Chatterjee T., Peet Y. 'Exploring the benefits of vertically staggered wind farms: understanding the power generation mechanisms of turbines operating at different scales'. *Wind Energy*. 2019;**22**(2):283–301.

[35] Nishino T. 'Two-scale momentum theory for very large wind farms'. *Journal of Physics: Conference Series*. 2016;**753**:032054.

[36] Fang J., Porté-Agel F. 'Large-eddy simulation of very-large-scale motions in the neutrally stratified atmospheric boundary layer'. *Boundary-Layer Meteorology*. 2015;**155**(3):397–416.

[37] Chamorro L.P., Lee S.-J., Olsen D., *et al.* 'Turbulence effects on a full-scale 2.5 MW horizontal-axis wind turbine under neutrally stratified conditions'. *Wind Energy*. 2015;**18**(2):339–49.

[38] Önder A., Meyers J. 'On the interaction of very-large-scale motions in a neutral atmospheric boundary layer with a row of wind turbines'. *Journal of Fluid Mechanics*. 2018;**841**:1040–72.

[39] Chatterjee T., Peet Y.T. 'Dynamics of large scale turbulence in finite-sized wind farm canopy using proper orthogonal decomposition and a novel fourier-pod framework'. *Energies*. 2020;**13**(7):1660.

[40] Kraichnan R.H. 'Inertial ranges in two-dimensional turbulence'. *Physics of Fluids*. 1967;**10**(7):1417–23.

[41] Tung K.K., Orlando W.W. 'The and the energy spectrum of atmospheric turbulence: Quasigeostrophic two-level model simulation'. *Journal of Atmospheric Sciences*. 2003;**60**:824–35.

[42] Barthelmie R.J., Frandsen S.T., Nielsen M.N., Pryor S.C., Rethore P.-E., Jørgensen H.E. 'Modelling and measurements of power losses and turbulence intensity in wind turbine wakes at Middelgrunden offshore wind farm'. *Wind Energy*. 2007;**10**(6):517–28.

[43] Barthelmie R.J., Pryor S.C., Frandsen S.T., *et al.* 'Quantifying the impact of wind turbine wakes on power output at offshore wind farms'. *Journal of Atmospheric and Oceanic Technology*. 2010;**27**(8):1302–17.

[44] Barthelmie R.J., Jensen L.E. 'Evaluation of power losses due to wind turbine wakes at the nysted offshore wind farm'. *Wind Energy*. 2010;**13**:573–86.

[45] Rinker J. Pywake. 2010. Available from https://topfarm.pages.windenergy.dtu.dk/PyWake/ [Accessed 18 May 2021].

[46] Zong H., Porté-Agel F. 'A momentum-conserving wake superposition method for wind farm power prediction'. *Journal of Fluid Mechanics*. 2020;**889**:A8.

[47] Niayifar A., Porté-Agel F. 'Analytical modeling of wind farms: a new approach for power prediction'. *Energies*. 2016;**9**:741.

[48] Martínez-Tossas L.A., Branlard E. 'The curled wake model: equivalence of shed vorticity models'. *Journal of Physics: Conference Series*. 2020;**1452**:012069–138.

[49] Martínez-Tossas L.A., Annoni J., Fleming P.A., Churchfield M.J. 'The aerodynamics of the curled wake: a simplified model in view of flow control'. *Wind Energy Science*. 2019;**4**(1):127–38.

[50] Simley E., Angelou N., Mikkelsen T., Sjöholm M., Mann J., Pao L.Y. 'Characterization of wind velocities in the upstream induction zone of a wind turbine using scanning continuous-wave lidars'. *Journal of Renewable and Sustainable Energy*. 2016;**8**(1):013301–697.

[51] Medici D., Ivanell S., Dahlberg J.-Å., Alfredsson P.H. 'The upstream flow of a wind turbine: blockage effect'. *Wind Energy*. 2011;**14**(5):691–7.

[52] Nishino T., Draper S. 'Local blockage effect for wind turbines'. *Journal of Physics: Conference Series*. 2015;**625**:012010.

[53] Branlard E., Alexander R. 'Meyer Forsting. assessing the blockage effect of wind turbines and wind farms using an analytical vortex model'. *Wind Energy [online]*. 2020. Available from https://onlinelibrary.wiley.com/doi/abs/10.1002/we.2546.

[54] Wu Y.-T., Porté-Agel F. 'Atmospheric turbulence effects on wind-turbine wakes: an LES study'. *Energies*. 2012;**5**(12):5340–62.

[55] Jin Y., Liu H., Aggarwal R., Singh A., Chamorro L. 'Effects of freestream turbulence in a model wind turbine wake'. *Energies*. 2016;**9**(10):830.

[56] Schmidt J., Stoevesandt B. 'Modelling complex terrain effects for wind farm layout optimization'. *[online] Journal of Physics: Conference Series*. 2014;**524**:012136.

[57] Liu L., Stevens R.J.A.M. 'Effects of two-dimensional steep hills on the performance of wind turbines and wind farms'. *Boundary-Layer Meteorology*. 2020;**176**(2):251–69.

[58] Shamsoddin S., Porté-Agel F. 'Wind turbine wakes over hills'. *Journal of Fluid Mechanics*. 2018;**855**:671–702.

[59] Cheng S., Jin Y., Chamorro L.P. 'Wind turbines with truncated blades may be a possibility for dense wind farms'. *Journal of Renewable and Sustainable Energy*. 2020;**13**:1810.

[60] Stevens R.J.A.M., Gayme D.F., Meneveau C. 'Effects of turbine spacing on the power output of extended wind-farms'. *Wind Energy*. 2016;**19**(2):359–70.

[61] Xie S., Archer C. 'Self-similarity and turbulence characteristics of wind turbine wakes via large-eddy simulation'. *Wind Energy*. 2015;**18**(10):1815–38.

[62] Chamorro L.P., Tobin N., Arndt R.E.A., Sotiropoulos F. 'Variable-sized wind turbines are a possibility for wind farm optimization'. *Wind Energy*. 2014;**17**(10):1483–94.

[63] Vested M.H., Hamilton N., Sørensen J.N., Cal R.B. 'Wake interaction and power production of variable height model wind farms'. *Journal of Physics: Conference Series*. 2014;**524**:012169.

[64] Vasel-Be-Hagh A., Archer C.L. 'Wind farm hub height optimization'. *Applied Energy*. 2017;**195**(5):905–21.
[65] Chatterjee T. *Understanding the Interaction of Wind Farms and Turbulent Atmospheric Boundary Layer in a Large Eddy Simulation Framework: From Periodic to LIDAR Based Data Driven Cases [online]*. 2018. PhD thesis. Available from http://login.ezproxy1.lib.asu.edu/login?url=https://search-proquest-com.ezproxy1.lib.asu.edu/docview/2041933239?accountid=4485.
[66] Tobin N., Chamorro L. 'Windbreak effects within infinite wind farms'. *Energies*. 2017;**10**(8):1140.
[67] Howland M.F., Lele S.K., Dabiri J.O. 'Wind farm power optimization through wake steering'. *Proceedings of the National Academy of Sciences*. 2019;**116**(29):14495–500.
[68] Munters W., Meyers J. 'Towards practical dynamic induction control of wind farms: analysis of optimally controlled wind-farm boundary layers and sinusoidal induction control of first-row turbines'. *Wind Energy Science*. 2018;**3**(1):409–25.

Chapter 5
Wind farm layout optimization

Hua Li[1], Francisco Haces-Fernandez[2], and Ying Chen[3]

5.1 Introduction to wind farm layout optimization

There are many factors affecting the performance of a wind farm, including site location [1] and layout. Wind farm layout optimization has become one of the critical approaches to increase power output and reduce the total cost of a wind farm. Wind farm layout optimization focuses on optimizing the layout of a wind farm using different intelligent algorithms to maximize its power output or minimize its levelized cost of electricity generated. This chapter first reports the changes in wind farm layouts in the last five decades. It then introduces different optimization algorithms and other models and methods used in wind farm layout optimization with case studies.

5.1.1 Wind farm layout change trend

Over the last five decades, wind farm layouts have evolved as the sizes and rated power of wind turbines increase [2]. Additionally, findings of the impacts of wind turbines on other equipment, surrounding communities, and local ecosystems have led to changes in the location, size, and layout of wind farms [3]. The first commercial wind turbines installed in the 1980s were quite small compared to modern equipment; therefore, their aerodynamics impacts on each other and the surrounding environment did not draw much attention. For instance, the first European wind farm started its operation in Kythnos, Greece, in 1983 with five 15 kW turbines [4] with a rotor diameter smaller than 11 m [5]. These old wind farms have the turbines clustered close together because the turbulent wakes generated by their small blades are short and dissipate quickly. Figures 5.1 and 5.2 are developed using geographic information system (GIS) and data from the United States Wind Turbine Database, which were developed by the US Geological Survey, American Wind

[1]Department of Mechanical and Industrial Engineering, Texas A&M University-Kingsville, 700 University Blvd, MSC 191, Kingsville, TX, USA
[2]Department of Management, Marketing and Information Systems, Texas A&M University-Kingsville, Kingsville, TX, USA
[3]School of Management, Harbin Institute of Technology, Harbin, China

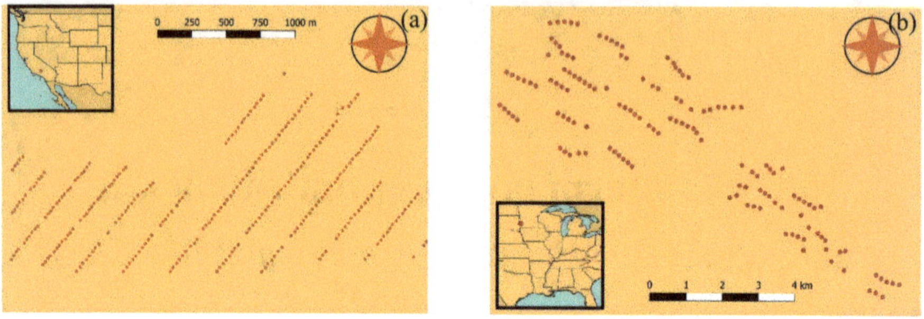

Figure 5.1 *Examples of wind farm layout: (a) Tehachapi Wind Resource Area in Southern California commissioned in 1982, (b) Lake Benton I Wind Farm in Lincoln, Minnesota, commissioned in 1998*

Energy Association, and Lawrence Berkeley National Laboratory [2]. Figure 5.1a shows the layout of Tehachapi Wind, one of the oldest wind farms in the United States. It contains wind turbines with rated power ranging from 40 to 70 kW and rotor diameters of 12–20 m. The separation between wind turbines is about 20–30 m in the crosswind directions and 270–300 m between rows [6]. Figure 5.1b shows the layout of Lake Benton I Wind Farm commissioned in 1998, located in Lincoln, Minnesota. It contains 142 Zond Z-48 wind turbines, which have a rated power of 750 kW and a rotor diameter of 48 m. The separation between wind turbines in crosswind direction increases to 150–180 m and to 1 400–1 800 m between rows [6].

As the size of wind turbines increases, interference caused by the wakes of other turbines, the so-called wake effect, becomes a significant issue in the wind farm operation and its layout design. Orsted, one of the biggest offshore wind operators, recently downgraded the forecasted wind power output for several offshore wind farms because their models underestimated wake interference [7]. The separation between wind turbines continues to increase to reduce wake interference. For instance, Figure 5.2a shows a wind farm with 152 Gamesa G90/2000 wind turbines,

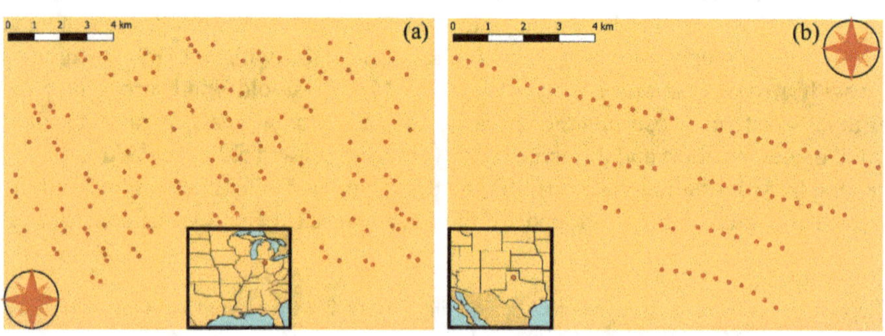

Figure 5.2 *Examples of wind farm layout: (a) Blue Creek wind farm in Van Wert, Paulding, Ohio, commissioned in 2012, (b) FGE Goodnight I in Armstrong County, Texas, commissioned in 2019*

which have a rated power of 2 MW and a rotor diameter of 90 m [2, 8]. In this case, separation in the crosswind direction increases to 250–450 m and to 1 000–1 800 m in the dominant wind direction, which is significantly higher than older wind farms with smaller wind turbines in terms of absolute distance. Addressing the wake effect by further separating the turbines is not feasible since the larger separation would lead to extremely extensive wind farms. Hence, innovative layout designs have been created via wind farm layout optimization to minimize the wake effect. This has become crucial. This is exemplified by FGE Goodnight I, which was commissioned in 2019 with 77 Vestas V136 with a rated power of 3.6 MW and a rotor diameter of 136 m [2, 8]. Its layout is represented in Figure 5.2b, with the separation of the wind turbines reaching 1 600–2 300 m in the dominant wind direction and 340–360 m in crosswind directions.

5.1.2 Wake models and cost models

Wind flowing through a wind turbine causes the wake effect since part of kinetic energy from the wind transfers to the turbine blades, reducing wind speed and generating turbulence after the turbine blades. As wind speed decreases and wind direction changes, it may reduce the power output of the wind turbines in the downwind direction. Different wake models are available for analyzing the wake effect in wind farm layout optimization processes. The most commonly used wake models are Jensen's analytical wake model [9] and its modifications [10]. Various methods have been applied to study wake effects, including mathematical modeling, computational fluid dynamics, remote sensing, and in- situ measurement of turbulence. However, as wind turbines continue to grow and move to diverse geographical locations, a complete understanding of the wake effect will continue to be of interest to both industry and academia.

The costs associated with wind farm development and operation are normally classified in capital expenditures (CapEx) and operational expenditures (OpEx). CapEx, which is generally estimated in dollar per rated power ($/kW), include all the costs incurred before a wind farm initiates its commercial operation, such as the costs of wind turbines (nacelle, rotor, and tower), foundations, electric installations, and all other installation costs [11]. Figure 5.3 indicates the distribution of

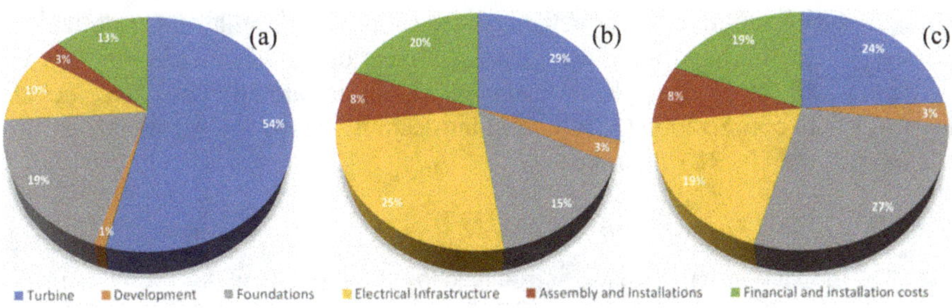

Figure 5.3 CapEx for (a) onshore wind turbine, (b) offshore fixed wind turbine, and (c) offshore floating wind turbine

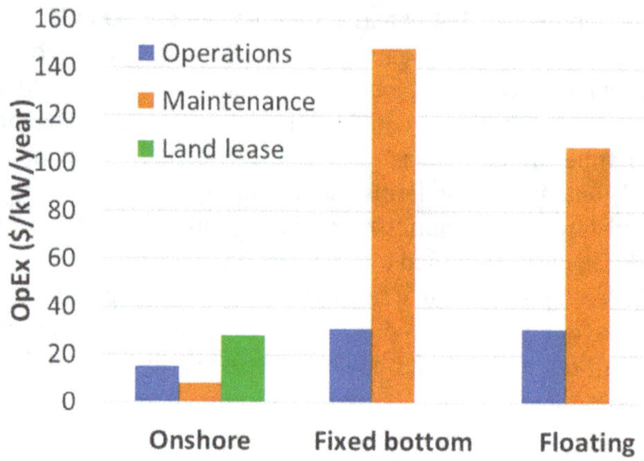

Figure 5.4 OpEx for onshore, offshore fixed bottom, and offshore floating wind turbines

CapEx for onshore wind turbines, offshore fixed bottom wind turbines, and offshore floating wind turbines [12]. The layouts of onshore wind farms are strongly geared toward minimizing wake effects. Minimizing the cost of electrical installation does not have an important weight, considering its low percentage in the CapEx. On the other hand, the cost of subsea cable has a higher weight in determining offshore wind farms' layouts [12].

OpEx are all the cost incurred during commercial wind farm operation. OpEx are estimated in dollars per unit of energy over year or hour ($/kW▪yr or $/MWh) [12]. Figure 5.4 shows that operating a wind turbine in the offshore environment has a much higher cost, with maintenance cost almost 19 times higher than onshore wind farms. The layouts of offshore wind farms are designed to minimize operation and maintenance costs as much as possible, considering the much higher OpEx of these installations.

5.2 Different wind farm layout optimization methods

Different types of optimization methods are used in wind farm layout optimization processes. Heuristic algorithms are the most popular ones. Genetic algorithm (GA), particle swarm optimization (PSO), and other algorithms are discussed here, including ant colony optimization (ACO), teaching-learning-based optimization (TLBO), and random search (RS) algorithm.

5.2.1 Genetic algorithm

GA is a global search heuristic introduced by John Holland in 1960 based on Darwin's natural evolution theory. The GA employs a natural selection process where the fittest individuals are selected for reproduction to produce offsprings of the next generation. Traditionally, a solution of GA is usually represented as a binary

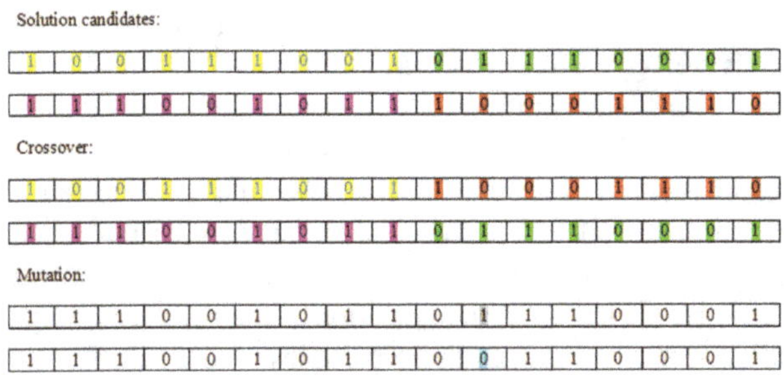

Figure 5.5 An example of crossover and mutation mechanism in GA

string. The initial population of candidate solutions of an optimization problem is improved to better solutions via crossover and mutation, as shown in Figure 5.5.

Wind farm layout optimization processes usually divide a given wind farm into many small sites. Hence, the optimization is to determine whether a small site should host a wind turbine. GA has a natural advantage in addressing such a problem: if a small site requires a wind turbine, "1" can be used for indication; otherwise, "0" will be used. The combination string of "0" and "1" can represent a solution candidate. Mosetti *et al.* [13] are the first to use GA to optimize the layout of a wind farm considering three simplified scenarios. Based on Mosetti *et al.* [13], Grady *et al.* [14] employed more solution candidates and generations in GA for achieving a better layout. After that, Mittal [15] developed the micro-sitting method with GA to find more accurate positions in a wind farm with the cell size of 1 × 1 m, which is much smaller than 200 × 200 m used by Mosetti *et al.* [13] and Grady *et al.* [14]. Acero *et al.* [16] are the first to consider the application of wind turbines with two different hub heights in a one-dimensional wind farm. Based on the idea of Acero *et al.* [16], Chen *et al.* [17–19] further investigated the idea of using wind turbines with different hub heights in wind farm layout optimization. They developed a nested GA to conduct the layout optimization on a two-dimensional wind farm using wind turbines of different hub heights to improve the power output. Chen *et al.* [18] also took advantage of multi-objective GA to optimize the wind farm layout using the real number string to improve the placement accuracy. Some other researchers have also considered the landowners' participation during the implementation of optimization using GA, e.g. [20–22]. Recently, Ju and Liu [23] proposed a self-informed GA with information-guided exploitation to optimize the wind farm layout. Kirchner-Bossi and Porte-Agel [24] utilized GA to optimize a realistic wind farm layout considering a Gaussian wake model. Abdelsalam and El-Shorbagy [25] used a GA-based local search to optimize wind turbines sitting on a wind farm.

5.2.2 Particle swarm optimization method

PSO is a robust stochastic optimization technique, which is based on the movement and intelligence of swarms. PSO applies the concept of social interaction to problem-solving. It was developed in 1995 by Kennedy and Eberhart [26] and Shi and Eberhart [27]. The algorithm uses a number of agents (particles) that constitute a swarm moving around in the search space, looking for the best solution. Each particle in PSO is treated as one wind turbine in a given wind farm. Each particle keeps track of its coordinates in the solution space associated with the best solution (fitness) achieved so far by that particle. This value is usually called personal best. Another best value tracked by the PSO is the best value obtained so far by any particle in the neighborhood of that particle. This value is normally called the global best. Each particle tries to modify its position to get a much better objective result. Particles will remember a better particle position as the personal best position; on the other hand, the swarm consisting of these particles recorded the swarm position as the global best position by recording the particle positions. Then, those steps iterate until the end of the loop. In the end, the swarm captures the best position of each particle as the optimal wind farm layout.

5.2.3 Other algorithms used in wind farm layout optimization

Numerous artificial intelligent algorithms have been used for achieving the optimal wind farm layout design. Besides the GA and PSO, ACO (proposed by Dorigo [28] and extended by Socha and Blum [29]) is another popular heuristic algorithm. ACO consists of several steps: (a) ants first need to explore the area randomly to find food; (b) they leave a chemical pheromone trail on the way when they move the food to the nest; (c) pheromone quantity increases based on the food quantity, and (d) other ants then go to food source following the pheromone trails. Following these four steps, Eroglu and Seckiner [30] are the first to apply ACO to optimize wind farm layout. Particle filtering (PF) introduced by Gordon *et al.* [31] is an approach that aims to achieve reasonable estimates of the state of a stochastic dynamic system based on recursive observations. PF has been widely used in signal processing, computer vision, and control system fields. Eroglu and Seckiner [32] are the first to apply the PF approach to optimize the layout of a wind farm, demonstrating PF can compete with the ACO method.

The TLBO method has a learner phase and a teacher phase: the learner phase is better in exploration than exploitation, whereas the teacher phase is better in exploitation compared to exploration. However, as noted, TLBO is not effective for some multimodal and constrained life problems [33]. Hence, Patel *et al.* [34] proposed an enhanced TLBO to maintain proper trade-off between the exploration and exploitation capability of the algorithm and applied the proposed algorithm to the layout optimization of a wind farm. The results demonstrate that the wind farm's optimum design leads to the economical utilization of wind recourse and the improvement of energy production.

RS algorithm incorporates some kind of randomness or probability in its methodology and does not require the gradient of the problem to be optimized. Such an algorithm moves iteratively to better positions in the searching space, which are sampled from a hypersphere surrounding the current position. In the application of wind farm layout optimization, this algorithm starts from an initial feasible layout and iteratively updates the layout in the feasible solution space conforming to the objective and constraints. Feng and Shen [35] are the first to employ the RS algorithm to optimize the wind farm layout and demonstrate that the proposed RS algorithm performed better than GA in flat terrains. After that, Feng and Shen [36] presented a new random algorithm based on a continuous formulation to optimize the layout of a wind farm on complex terrain. The results show the RS algorithm consistently has a better performance than GA under certain conditions.

Compared to the heuristic algorithms, Park and Law [37] provided a mathematical optimization scheme to efficiently and effectively optimize the locations of many wind turbines, aiming to maximize the wind farm power production. They developed a continuous wake model and formulated the expected wind farm power as a continuous and smooth function with respect to the locations of the wind turbines. They then made use of sequential convex programming for the optimization of the nonlinear mathematical problem. As demonstrated, the optimized wind farm layout within this scheme increases its power efficiency by 7.3 percent.

5.2.4 Wind farm layout optimization with geographic information systems

It is challenging to implement layouts obtained from optimization methods in general scenarios unless real data from the site location is incorporated into the optimization process. For instance, an optimal layout found under uniform surface roughness assumption might be unfeasible in sites with highly variable surface roughness. Integrating layout optimization methods with real site information obtained from GISs will generate optimal solutions for different locations based on their individual characteristics [38]. The initial stage of the GIS-based layout optimization method involves acquiring geospatial data. A wide variety of data are available from the government and private sources, national laboratories, research institutions, and commercial developers. When GIS data from different datasets are used, it is important to have a uniform spatial projection that is adequate for the study location. For instance, for applications requiring the measurement of area, an equal area projection would be suitable, while an azimuthal projection is a good fit for accurate wind directions. It is important to have all data frames in projections with the same measuring units when performing cross-layer analysis. If correct projections are not used, the geospatial analysis will not be possible, or results might not be accurate [39].

GIS allows all the collected data to be integrated into a system that uses location as its unifying criteria. Separate databases such as population density, migratory bird paths, wind speeds, highways, railways, zoning areas, and many others converge on one field: location. It makes it feasible to perform complex analyses, such as how increasing wind turbine development may interfere with migratory bird paths. For each particular location, GIS can perform analysis on average wind speeds, surface roughness, and zoning restrictions, to mention a few, that are very important for understanding and optimizing wind energy harvesting at each location. GIS overlays all the geospatial data into a uniform canvas from which various analyses can be performed [40]. The GIS-based layout optimization model can measure changes in a wide array of factors caused by modifications to the location of each wind turbine. Different surface roughness values, topography, zoning, location of residential areas, wildlife migratory paths, and wind conditions over different layouts will be applied for a particular location [41]. Examples of commercial software that apply GIS specifically for wind farm layout optimization are Wind Atlas Analysis and Application Program developed by the Technical University of Denmark [42] and GE's Digital Wind Farm [43]. The most widely used general application commercial GIS software is ESRI ArcGIS, which can be used as a platform to develop wind farm optimization models.

5.3 Single objective and multiple objectives wind farm layout optimization

Maximizing power output has been traditionally one of the most popular objective functions in wind farm layout optimization, which generally implies a better utilization rate (capacity factor) and potentially increased income from selling electricity to the grid. As the wind energy sector evolves, other factors have drawn more attention and have been considered as the objective function in optimization methods. Cost reduction in both installation and operation has taken a prominent place. However, factors that improve social acceptance of wind turbines, such as noise and visibility reduction, are becoming important elements to improve opportunities for project permit approvals. Aligning wind electricity generation with other factors such as peak consumption or storage technology is becoming more important in wind farm development and operation.

As more and more factors become potential objective functions, it has become essential to evaluate and optimize wind farm layout under two or more objective functions to maximize its potential for permitting approval and successful operation. In locations close to residential areas with a high saturation of wind turbines, two or more objective functions are frequently considered. The first group of objective functions aims to maximize power generation or revenue for the installation whereas the other objective functions seek to minimize social opposition (noise, visibility, and negative impact on other activities or ecosystems), which might lead to permit denials and delays in project development or operation. For remote or arid locations with low risk of social opposition or impact on ecosystems, maximizing power generation while minimizing the distance to transportation infrastructure may be important for the development and construction stages. In addition, minimizing the distance to consumption centers while maximizing power output is also an important consideration for remote locations to reduce transmission losses and new grid developments.

For wind farms with increasingly larger wind turbines, two seemingly opposed objective functions are normally applied, i.e., maximizing the power output while at the same time minimizing the cost of the electric cables used in wind farm. Minimizing this cost will require shorter separation between wind turbines, which in turn increases the negative impact of wake effect on downstream equipment decreasing the total power output. However, it is important to minimize installation and operation costs, especially for large offshore wind farms, where wind turbines with a rotor diameter of more than 200 m are used.

5.4 Case study

A wind farm with 14 wind turbines within a given rectangular area (1 020 × 2 800 m) is chosen as a case study to show the general procedures of wind farm layout optimization. A multi-objective GA is selected as an example to solve this case study. GA has advantages for unrestricted layout approach, as the coordinates of wind turbines can be defined as continuous variables. There are two types of wind turbines available for selection, with each type of turbine has two sizes with lower hub height or higher hub height, as shown in Table 5.1.

In Table 5.1, *S* represents turbine sequence, "type" indicates the actual turbine type, *Rp* represents rated power, *D* represents rotor diameter, *L_h* is the lower hub height of each type turbine, *H_h* is the higher hub height of each type turbine, *In* represents the cut-in the wind speed (m/s), *Out* means the cut-out wind speed (m/s), *Ur* is the rated wind speed (m/s), *A* is the swept area, *C* represents turbine coefficient, *Cp* represents power coefficient, and *a* represents axial induction factor.

The GA toolbox in MATLAB® is used to generate chromosomal strings. Each chromosomal string has 56 bits. The first 14 bits represent the x coordinates of each wind turbine, whereas the second 14 bits represent their y coordinates. The third set of 14 bits indicates the wind turbine types, and the last 14 bits represent the turbine hub height. Two objective functions are chosen, including maximizing the wind farm efficiency and minimizing its cost as defined in (5.1) and (5.2) [17].

$$\text{Farm efficiency} = \frac{\text{power output}}{\text{wind farm installed nameplate capacity}} \tag{5.1}$$

$$\text{Cost} = \text{cost per unit power} \tag{5.2}$$

The parallel multi-objective optimization method is used with equal weight on both objective functions. The initial number of random individuals is 400, and the selection rate is 0.9, with a crossover rate of 1 and a mutation rate of 0.01. The GA runs 1 000 generations to get the optimal results. The fitness convergence curve is shown in Figure 5.6. To choose the optimal layouts from the feasible solution set, the ratio between the values of two objective functions is used as criteria. The higher the ratio means a better layout. The optimal layout is chosen with a farm efficiency of 0.5828 and a cost per unit power of 2.9751. The layout is shown in Figure 5.7.

5.5 Discussion and future trend

Many more factors affect wind farm design, including impact on local communities and the environment. New regulations in Germany restrict the distance between wind farms and residential areas. Concerns about noise, visual impact, and safety have slowed down the approval process of new wind farms, requiring significant modifications. The potential impact on migratory birds and bats has significantly

Table 5.1 Specifications of the GE wind turbines used in the case study

S	Type	*Rp*	*D*	*L_h*	*H_h*	In	Out	*Ur*	*A*	*C*	*Cp*	*a*
0	1.6-82.5	1600	82.5	65	80	3.5	25	11.5	5345.6	0.3212	0.4589	0.1642
1	1.6-100	1600	100	80	96	3.5	25	11	7854	0.2498	0.3569	0.1135

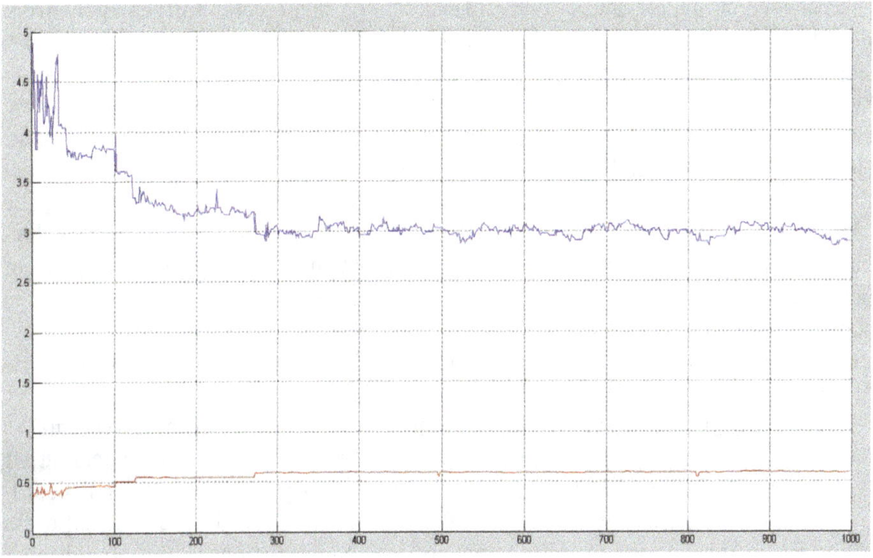

Figure 5.6 Fitness convergence curve: blue for the cost objective function and red for farm efficiency objective function

constrained new wind farm layouts, locations, and equipment [44]. Avoiding congested migratory routes and equipping wind turbines with radar to stop birds and bats are some of the strategies implemented that ultimately significantly impact the wind farm's layout [45].

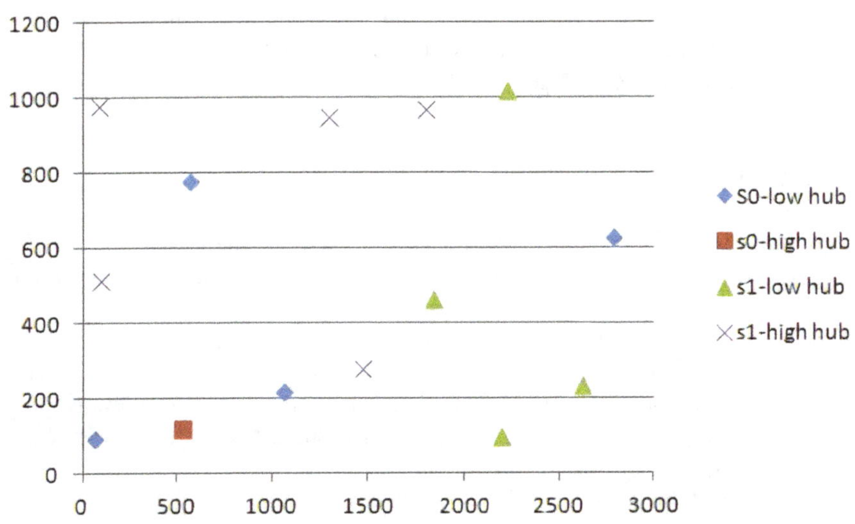

Figure 5.7 Optimal layout selected from feasible solutions

New geographical areas with different topographies and further away from installed infrastructure are also being explored for wind farm development, which significantly impacts wind farm layout design. As wind turbines' sizes increase, it has become more challenging to transport the blades and other wind turbine parts. Having highways or railways close to the proposed site is highly advantageous. However, some potential sites might be remote, limiting the size of the equipment to be installed and influencing the layout of the wind farms [46]. Similarly, the topography of new locations might present additional challenges that affect the layout design. Changing surface roughness in the potential site due to hilly terrain or obstacles creates a high variability on wind turbines' optimal locations. This, in some cases, leads to uneven distribution of wind turbines in the layout, which might increase cost due to more prolonged or more complex service roads or electric transmission lines [47].

Wind turbines have doubled and even tripled the diameter of their rotors in the last two decades. Some offshore wind turbines are reaching rotor diameters larger than 200 m. The separation between wind turbines in layouts has increased accordingly to minimize the impact of the wake effect. However, this has caused areas occupied by wind farms to grow exponentially. Since regions with optimal wind resources are limited, this poses a problem on the saturation of equipment and interference with other economic, transportation, and residential activities. Instead of continually increasing the number and the size of wind turbines in a given wind farm, new optimization methods are being developed to manage and control the operation of these enormous wind turbines in the wind farms under different wind conditions to reduce the wake effect and maximize their power output under any given wind farm layouts. For instance, a new method identified a selected number of wind turbines to be deactivated according to changing wind conditions, improving the wind farm total power output. The method identifies wind turbines that generate greater interference due to the wake effect than the power they can generate. Other equipment stops receiving negative interference by deactivating these turbines and generating additional power output [48].

References

[1] Ramos C.A., Ramirez D., Jin K., Li H. 'Factorial analysis of selected factors affecting the electricity production of onshore wind farms'. *International Journal of Energy for a Clean Environment*. 2011;**12**(1):67–77.

[2] Hoen B.D., Diffendorfer J.E., Rand J.T., Kramer L.A., Garrity C.P., Hunt H.E. *United States Wind Turbine Database [online]*. US Geological Survey, American Wind Energy Association, and Lawrence Berkeley National Laboratory data release: USWTDB V2.3. 2020. Available from https://eerscmap.usgs.gov/uswtdb.

[3] Lundquist J.K., DuVivier K.K., Kaffine D., Tomaszewski J.M. 'Costs and consequences of wind turbine wake effects arising from uncoordinated wind energy development'. *Nature Energy*. 2019;**4**(1):26–34.

[4] Kougias I., Szabó S., Nikitas A., Theodossiou N. 'Sustainable energy modelling of non-interconnected Mediterranean islands'. *Renewable Energy*. 2019;**133**:930–40.

[5] Bauer L., Matysik S. *Wind turbine models database [online]*. 2020. Available from wind-turbine-models.com [Accessed 1 Mar 2020].

[6] Nielsen E.G. Winds of change. 2020. Available from http://www.windsofchange.dk/WOC-usaturb.php [Accessed 1 Mar 2020].

[7] Mathis W., Wienber C. *Offshore wind gets a warning from Its biggest developer [online]*. Bloomberg news. 2019. Available from https://www.bloomberg.com/news/articles/2019-10-29/orsted-sinks-as-company-slashes-outlook-and-warns-of-job-cuts.

[8] Pierrot M. *The wind power database [online]*. 2020. Available from https://www.thewindpower.net/index.php [Accessed 1 Mar 2020].

[9] Jensen N.O. *A Note on Wind Generator Interaction*. Roskilde (Denmark): Risø National Laboratory; 1983.

[10] Archer C.L., Vasel-Be-Hagh A., Yan C., *et al*. 'Review and evaluation of wake loss models for wind energy applications'. *Applied Energy*. 2018;**226**(4):1187–207.

[11] Stehly T.J., Heimiller D., Scott G., Beiter P.C. *2017 Cost of Wind Energy Review (No. DE-AC36-08GO28308)*. CO (United States): National Renewable Energy Lab.(NREL), Golden; 2019.

[12] Stehly T.J., Beiter P.C. *2018 Cost of Wind Energy Review (No. NREL/TP-5000-74598)*. CO (United States): National Renewable Energy Lab. (NREL), Golden; 2020.

[13] Mosetti G., Poloni C., Diviacco B. 'Optimization of wind turbine positioning in large windfarms by means of a genetic algorithm'. *Journal of Wind Engineering and Industrial Aerodynamics*. 1994;**51**(1):105–16.

[14] Grady S.A., Hussaini M.Y., Abdullah M.M. 'Placement of wind turbines using genetic algorithms'. *Renewable Energy*. 2005;**30**(2):259–70.

[15] Mittal A. *Optimization of the layout of large wind farms using genetic algorithm*. MS Thesis. Case Western Reserve University; 2010.

[16] Acero J., Acevedo J., Rendon M., Oleszewski O. 'Linear wind farm layout optimization through computational intelligence'. *Proceedings of 8th Mexican International Conference on Artificial Intelligence*; 2009. pp. 692–703.

[17] Chen Y., Li H., Jin K., Song Q. 'Wind farm layout optimization using genetic algorithm with different hub height wind turbines'. *Energy Conversion and Management*. 2013;**70**(11–12):56–65.

[18] Chen Y., Li H., He B., Wang P., Jin K. 'Multi-objective genetic algorithm based innovative wind farm layout optimization method'. *Energy Conversion and Management*. 2015;**105**(3):1318–27.

[19] Chen Y., Li H., Jin K., Elkassabgi Y. 'Investigating the possibility of using different hub height wind turbines in a wind farm'. *International Journal of Sustainable Energy*. 2017;**36**(2):142–50.

[20] Chen L., McDonald E. 'A new model for wind farm layout optimization with landowner decisions'. *ASME 2011 International Design Engineering*

Technical Conference & Computers and Information in Engineering Conference; Washington DC; 2011.

[21] Chen L., MacDonald E. 'A system-level cost-of-energy wind farm layout optimization with landowner modeling'. *Energy Conversion and Management*. 2014;**77**(5):484–94.

[22] Ju X., Liu Wang F F., Lee W. 'Wind farm layout optimization based on support vector regression guided genetic algorithm with consideration of participation among landowners'. *Energy Conversion and Management*. 2019;**196**:1267–81.

[23] Ju X., Liu F. 'Wind farm layout optimization using self-informed genetic algorithm with information guided exploitation'. *Applied Energy*. 2019;**248**(1):429–45.

[24] Kirchner-Bossi N., Porté-Agel F. 'Realistic wind farm layout optimization through genetic algorithms using a Gaussian wake model'. *Energies*. 2018;**11**(12):3268.

[25] Abdelsalam A.M., El-Shorbagy M.A. 'Optimization of wind turbines siting in a wind farm using genetic algorithm based local search'. *Renewable Energy*. 2018;**123**(1):748–55.

[26] Kennedy J., Eberhart R. 'Particle swarm optimization'. *Proceedings of IEEE International Conference on Neural Networks IV*. 1995:1942–8.

[27] Shi Y., Eberhart R.C. 'A modified particle Swarm optimizer'. *Proceedings of IEEE International Conference on Evolutionary Computation*. 1998:69–73.

[28] Dorigo M. *Optimization, Learning and Natural Algorithms (in Italian). PhD thesis, Dipartimento di Elettronica, Politecnico di Milano, Italy*; 1992.

[29] Socha K., Blum C. 'An ant colony optimization algorithm for continuous optimization: application to feed-forward neural network training'. *Neural Computing and Applications*. 2007;**16**(3):235–47.

[30] Eroğlu Y., Seçkiner S.U. 'Design of wind farm layout using ant colony algorithm'. *Renewable Energy*. 2012;**44**:53–62.

[31] Gordon N.J., Salmond D.J., Smith A.F.M. 'Novel approach to nonlinear/non-Gaussian Bayesian state estimation'. *IEE Proceedings F Radar and Signal Processing*. 1993;**140**(2):107–13.

[32] Eroğlu Y., Seçkiner S.U. 'Wind farm layout optimization using particle filtering approach'. *Renewable Energy*. 2013;**58**(5):95–107.

[33] Patel J., Savsani V., Patel R. 'Maximizing energy output of a wind farm using teachingelearning-based optimization'. *ASME 2015 9th International Conference on Energy Sustainability Collocated with The ASME 2015 Power Conference, The ASME 2015 13th International Conference on Fuel Cell Science, Engineering and Technology, and The ASME 2015 Nuclear Forum*. American Society of Mechanical Engineers. V002T019A004e V002T019A004; 2015.

[34] Patel J., Savsani V., Patel V., Patel R. 'Layout optimization of a wind farm to maximize the power output using enhanced teaching learning based optimization technique'. *Journal of Cleaner Production*. 2017;**158**(11):81–94.

[35] Feng J., Shen W. 'Optimization of wind farm layout: a refinement method by random search'. *Proceedings of 2013 International Conference on Aerodynamics of Offshore Wind Energy Systems and Wakes, 17-19 Jun.* Copenhagen, Denmark; 2013.

[36] Feng J., Shen W.Z. 'Solving the wind farm layout optimization problem using random search algorithm'. *Renewable Energy*. 2015;**78**(**8**):182–92.

[37] Park J., Law K.H. 'Layout optimization for maximizing wind farm power production using sequential convex programming'. *Applied Energy*. 2015;**151**(2):320–34.

[38] Rafiee A., Van der Male P., Dias E., Scholten H. 'Interactive 3D geodesign tool for multidisciplinary wind turbine planning'. *Journal of Environmental Management*. 2018;**205**:107–24.

[39] Kidner D., Sparkes A., Dorey M. 'GIS and wind farm planning'. *Geographical Information and Planning*. Berlin, Heidelberg: Springer; 1999. pp. 203–23.

[40] Abdelaziz A.Y., Mekhamer S.F., Mohamed A.B. 'Geographic information systems (GIS) application in wind farm planning'. *Natural Gas*. 2009;**3**(5):1–4.

[41] Mortensen N.G. *Wind resource assessment using the WAsP software*. DTU Wind Energy. DTU Wind Energy E No. 174; 2018.

[42] Mortensen N.G., Landberg L., Rathmann O., Frank H.P., Troen I., Petersen E.L. 'Wind atlas analysis and application program (WAsP)'. *Wind Energy Department: Scientific and Technical Progress 1999-2000*; 2001.

[43] General Electric. *Digital Wind Asset Performance Management from GE Renewable Energy [online]*. GEA33060 (3/2017). 2017. Available from https://www.ge.com/digital/sites/default/files/download_assets/GE-Digital-Wind-Asset-Performance-Management-Brochure.pdf [Accessed 4 Mar 2020].

[44] Leung D.Y.C., Yang Y. 'Wind energy development and its environmental impact: a review'. *Renewable and Sustainable Energy Reviews*. 2012;**16**(1):1031–9.

[45] Saidur R., Rahim N.A., Islam M.R., Solangi K.H. 'Environmental impact of wind energy'. *Renewable and Sustainable Energy Reviews*. 2011;**15**(5):2423–30.

[46] Serrano-González J., Lacal-Arántegui R. 'Technological evolution of onshore wind turbines—a market-based analysis'. *Wind Energy*. 2016;**19**(12):2171–87.

[47] Liu Y., Chi Y., Wang W., Dai H. 'impacts of large scale wind power integration on power system'. *4th International Conference on Electric Utility Deregulation and Restructuring and Power Technologies (DRPT) (PP. 1301-1305)*. IEEE; 2011.

[48] Haces-Fernandez F., Li H., Ramirez D. 'Improving wind farm power output through deactivating selected wind turbines'. *Energy Conversion and Management*. 2019;**187**(5):407–22.

Part II

Control

Chapter 6

Analyzing data obtained via wind farm supervisory control and data acquisition

Jay Lee[1], Xiaodong Jia[1], Vibhor Pandhare[1], and Marcella Miller[1]

6.1 Introduction

The supervisory control and data acquisition (SCADA) system is a computer system that is designed to monitor and control the wind turbine (WT) and critical subassemblies, including the yaw and pitch, electrical system, rotor and hub, drive train, generator, and braking, to name a few. As shown in Figure 6.1, the SCADA data provides rich operation information from a variety of sensors, which is commonly exploited for performance/condition monitoring, turbine diagnostics and prognostics, intelligent operation, wind farm maintenance planning, and many other applications. The sampling rate of the SCADA system varies from 1/30 Hz to 1/600 Hz based on the hardware capability and the data analysis requirements. The data from the SCADA system record the complete run-to-failure data profiles throughout the turbine uptime, and it holds great value for statistical analysis, data mining, and machine learning.

The condition monitoring system (CMS) plays a pivotal role in the condition-based maintenance (CBM) of the WT drive train, gearbox, rotor, and generator by leveraging the high-frequency waveform measurements from the accelerometers, acoustic emission (AE) sensors, motor current sensors, and so on. In some emerging systems, the temperature of the critical parts of the drive train, oil particulate content, and strain measurements are also included. The sampling rate of vibration data in CMS ranges from 1 kHz to 10 kHz or even higher, which is high enough to capture the detailed vibrational characteristics of high-speed components in the rotating systems. Due to high data velocity, the data acquisition in CMS is triggered by events or routine procedures. Furthermore, various signal processing techniques, such as frequency analysis, envelope analysis, and time-synchronous averaging (TSA), are integrated to process the data and to trigger early warning of electrical and mechanical failures of the system.

[1]University of Cincinnati, Cincinnati, OH, USA

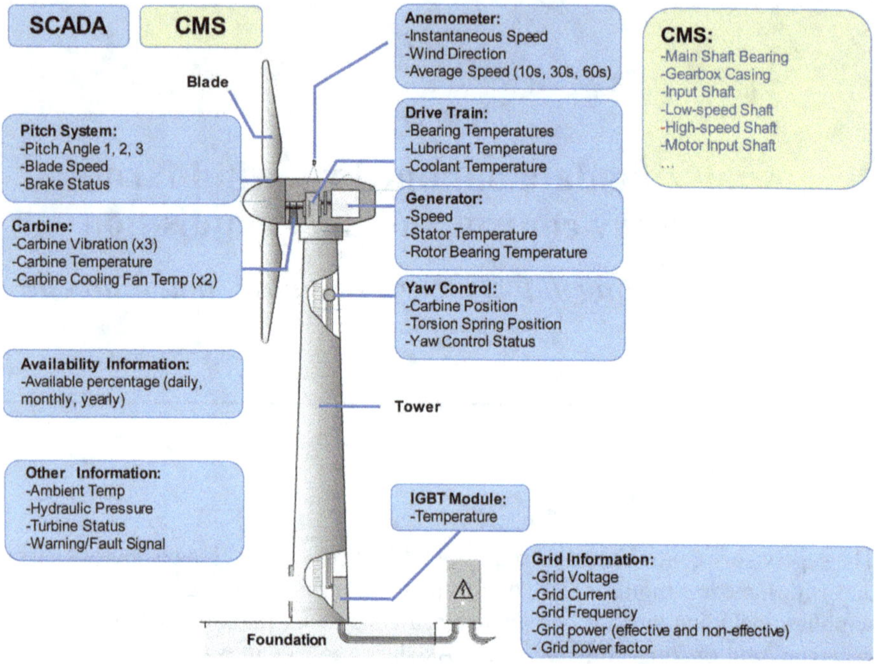

Figure 6.1 *Introduction to the SCADA system and CMS for WTs*

In this chapter, we first discuss the application of SCADA data and CMS data in wind farm operation and maintenance (O&M) by reviewing the commonly used technologies and analytics. Next, the discussion is devoted to a unified framework for building a cyber-physical wind farm, and several case studies are provided to demonstrate the idea. Finally, we highlight the existing challenges and future works.

6.2 Wind farm operation and maintenance with SCADA data and CMS data

6.2.1 Reliability assessment and maintenance strategies

Historical event records in SCADA systems, CMS, and maintenance logs are important data sources to understand the reliability of the WT and critical components. By processing the historical events into annual failure rate and downtime per failure, as shown in Figure 6.2, the reliability analysis will provide critical insights to the turbine design improvement, O&M, and performance assessment. In a recent review study in 2018, E. Artigao *et al.* [1] demonstrate a comprehensive summary and comparison of 13 reliability studies carried out in Europe and the United States. Six of these studies are published in and after 2012. The conclusion indicates the gearbox as a major cause of failure. Other subsystems that merit further attention for condition monitoring include the control systems, blade, generator, and hub. Another

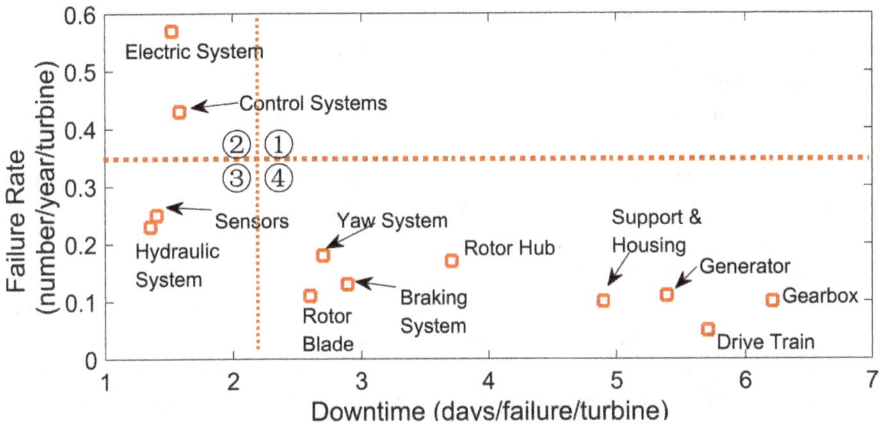

Figure 6.2 *Four quadrant chart for the critical component prioritization. The division of the four quadrants in the figure can be customized based on the cost-benefit analysis.*

more recent review study compares the reliability data from 18 different databases in Europe, Asia, and the United States [2]. This study not only highlights the critical subsystems of WT but also compares the differences between on-shore and off-shore WTs. Furthermore, the uncertainties of failure rate and downtime are explored, and the possible sources of variations are identified.

Understanding the reliability characteristics of WT components and subsystems is critical in designing appropriate maintenance strategies, thus enabling a cost reduction for wind farm O&M. Figure 6.2 shows an example of failure rate and downtime analysis based on the data from the Scientific Measurement and Evaluation Programs in Europe [3]. In Figure 6.2, the dotted lines cut the whole space into four different quadrants. For the components in each quadrant, the recommended maintenance strategies are described as follows:

1. ***Redesign***—components or subsystems in this quadrant have an unacceptable failure rate and downtime, and none of the maintenance strategies is useful to mitigate the operation risk. In this case, the subsystem or component redesign is required to improve reliability and reduce maintenance costs.
2. ***Preventative maintenance***—preventative maintenance is also referred to as time-based or interval-based maintenance in the literature [1]. It consists of regular inspections, commissioning, and replacement of the component or the equipment. The pre-defined inspection interval in preventative maintenance is typically designed based on the reliability characteristic of the component, such as mean time to failure or mean time between failure. Components in this quadrant usually have moderate operation risk with a high failure rate but low downtime.
3. ***Reactive/corrective maintenance***—reactive maintenance is referred to as fail-and-fix, and it happens after the system or component fails. Reactive maintenance

is not favored in practice because it will cause unexpected downtime, and it ignores the opportunities to avoid catastrophic losses at an early stage of the failure. Since the unscheduled maintenance for WTs is very costly, this maintenance strategy is recommended to the components with low-operational risks and failures that are not detrimental to the turbine operation.

4. ***Condition-based maintenance (CBM)***—CBM incorporates real-time condition monitoring and predictive maintenance. Compared with reactive maintenance and preventative maintenance, CBM can effectively reduce the maintenance cost by reducing excessive repair actions and by avoiding unexpected turbine downtime. In CBM, condition monitoring leverages the real-time data from the SCADA system and CMS to evaluate the health status of the equipment and trigger early warnings when certain conditions are met. Based on the real-time data, predictive maintenance modules can anticipate the prospective downtime by forecasting the development of the failure so that the maintenance activities can be planned before the failure happens. Though the CBM is cost-effective, the development of the reliable CBM strategy for the WT and its subassemblies requires tremendous research efforts and a prolonged period of data accumulation. Therefore, it is often applied to the critical components and subsystems which have a low failure rate but a long period of degradation.

6.2.2 Condition monitoring and prognosis with SCADA and CMS data

Driven by the need to reduce the O&M cost of wind farms, SCADA data and CMS data have been widely studied in the modern wind industry to fulfill condition monitoring and fault prognosis. More specifically, the objectives of condition monitoring and fault prognosis include the following:

1. ***Early detection of incipient failures***—early detection is the key enabler of scheduled maintenance. As shown in Figure 6.3, the early detection of system failure is often based on a synthesized health indicator that represents the slow progression of the failure. The health indicator can be simply a discrepancy

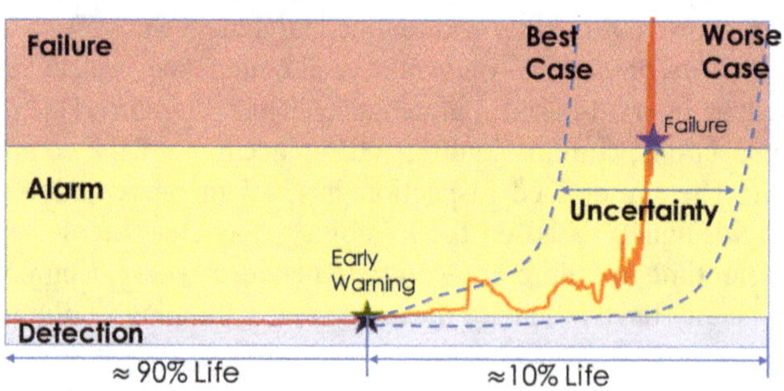

Figure 6.3 Illustration of the objectives of condition monitoring and prognosis

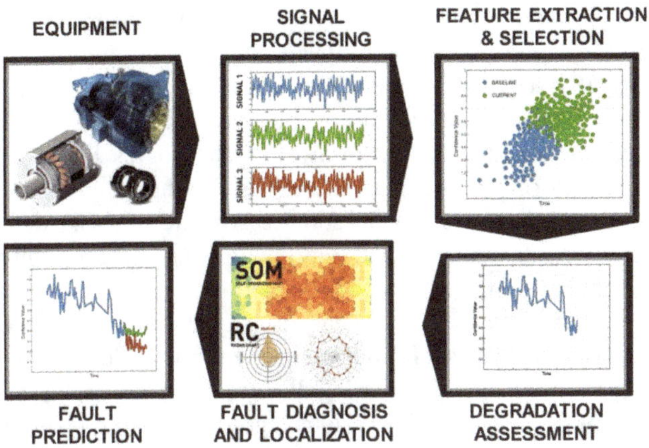

Figure 6.4 Methodology for the condition monitoring and fault prognosis of WT assets

measure that represents the differences between the recent data and the baseline data, or it can have a probabilistic interpretation of quantified operational risk.

2. ***Continuous monitoring of the developing failure***—once an incipient failure is detected, it is important to keep tracking the progression of the failure. By tracking the process, the operator can shut down the system or take timely protective measures before the unacceptable failure threshold is reached. Furthermore, the development of the failure can be predicted based on the degradation trend, which will allow the optimization of the maintenance time.
3. ***Fault diagnosis and localization***—the goal of fault diagnosis is to identify the root cause of the incipient failure, and the purpose of fault localization is to isolate the faulty component and subsystem for repairing actions. The information from fault diagnosis and localization can help understand the failure mechanism and plan necessary resources for maintenance activities.
4. ***Remaining useful life (RUL) prediction***—RUL prediction aims to estimate the remaining time to failure based on the current monitoring data, which is an integral part of fault prognosis [4, 5]. As illustrated in Figure 6.3, the RUL should be predicted once an incipient failure is detected and confirmed. Furthermore, the uncertainly of the predicted RUL should be quantified to allow flexibility in optimizing the maintenance activities.

The methodology for the condition monitoring and fault prognosis based on SCADA data and CMS data is outlined in Figure 6.4. There are six major steps in the pipeline of data analysis:

1. ***Data acquisition***—the data channels that are available in the SCADA system and CMS are detailed in Figure 6.1. Based on the literature, the SCADA is often useful for turbine-level fault detection when the prospective failure could affect the power conversion efficiency of the WT. However, it is

challenging to isolate and diagnose the failure based on the SCADA only due to the limitation of data fidelity. In comparison, the high-velocity data (from the CMS), such as motor current, vibration, and acoustic signal, carries detailed information about the failure and provides reliable evidence for fault localization and diagnosis. Furthermore, technologies like vibration-based condition monitoring and oil data analysis for the rotating machinery are very mature in most CMS [6]. In some cases, the SCADA data, especially the temperature data, are analyzed jointly with CMS data to improve the fault detection accuracy [7].

2. ***Signal processing***—in vibration analysis, signal processing is employed to enhance the fault signatures from measurement noise. For example, the envelope analysis is often used for bearing diagnosis and the TSA is useful to indicate the faulty gears. Similarly, various signal processing techniques are also available for acoustic signals, motor current signals, and so on. A review of signal processing tools is presented in [8]. In the SCADA data analysis, signal processing techniques, such as empirical mode decomposition [9] and wavelet analysis [10], are also applied to the wind speed and other variables to enable further statistical analysis.
3. ***Feature extraction and selection***—the goal of feature extraction is to process the waveform data into structured data tables for further statistical analysis and data modeling. The frequently used features include (1) descriptive statistics based on the time domain waveforms, such as mean, variance, kurtosis, and skewness, (2) signal amplitude and phase information within a pre-selected frequency band, and (3) domain-specific features based on operation experience and physical knowledge, such as turbulent intensity, distribution of wind directions, and so on.
4. ***Degradation (health) assessment***—degradation assessment is also referred to as trend analysis in literature [11]. The goal of degradation or trend analysis is to derive the slow degradation trend, as shown in Figure 6.4. The flowcharts in Figure 6.5 summarize two popular methods for degradation assessment. The residual-based method in Figure 6.5(a) establishes a simulator to estimate the expected system output under normal conditions. By comparing the differences between the expected output and the measured output, a residual-based indicator can be computed to describe the WT degradation. When the system starts to degrade, an increase in the residual is expected. The discrepancy-based method in Figure 6.5(b) estimates the difficult-to-measure parameters in the system based on the measurement input and output. The discrepancy between the estimated system parameter and design reference indicates the severity of the degradation. In general, the baseline simulator and the system observer can be either physics-based or data-driven, depending on the specific applications. The physics-based models are typically more reliable than data-driven methods due to the complexity of the WT design and the non-stationary operating conditions. However, data-driven models are more cost-effective and easy-to-implement than physics-based models, especially when abundant SCADA data and CMS data are available.

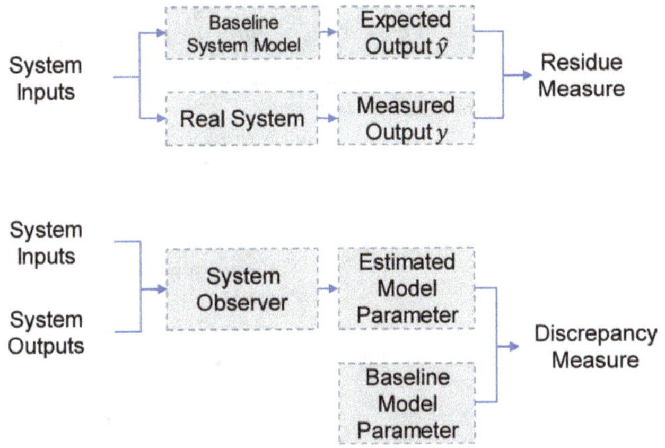

Figure 6.5 (a) Residual-based method and (b) discrepancy-based method for degradation assessment

5. ***Fault diagnosis and localization***—fault diagnosis and localization decisions are often obtained from classifiers and rule-based inference systems in practice. These classifiers and inference systems take the features in step (3) as input and then suggest possible root causes of the failure as output. This information is useful for the maintainers to plan inventories and resources to repair the system.
6. ***Fault prediction (FP)***—FP aims to forecast the future development of the failure and determine the RUL of the system. Typically, probabilistic approaches, such as Kalman filters and random coefficient models, are widely used for RUL prediction since the prediction uncertainties can be quantified in these methods. Neural networks, especially deep neural networks (DNN), recently reported excellent prediction accuracies, although the model explainability is still a concern, and the prediction uncertainty cannot be quantified in these methods.

Among the six processing steps in the methodology in Figure 6.4, the signal processing and feature extraction steps are critical to the success of the models. The development of effective features for subsequent modeling requires the convergence of domain knowledge, operation experience, and statistical insights. Also, the quality of the extracted features must be strictly validated and verified. The rest of the analysis procedures, including feature selection, degradation assessment, fault diagnosis and localization, and FP, can be automated by utilizing the off-the-shelf algorithms for machine learning and data mining. According to [12–14], the learning problems that can be addressed by machine learning are summarized as follows:

- ***Classification***, to obtain a learning function that separates pre-defined data classes;
- ***Clustering***, to obtain a descriptive function that groups data into several clusters;
- ***Regression***, to obtain a learning function that relates the inputs to the outputs;
- ***Detection***, to discover the significant changes in the data set;

Table 6.1 Learning problems in condition monitoring and fault prognosis

Machine learning problems	Condition monitoring tasks	Example algorithms
Prediction	FP	Auto-Regressive Moving Average (ARMA) Stochastic processes Proportional Hazard Model DNN
Detection	(Early) fault detection	K-Nearest Neighbor Principal Component Analysis (PCA)-T^2/SPE (squared prediction error)
Inference	(Early) fault detection (Early) fault diagnosis	Fuzzy Inference Systems (FIS) Bayesian Belief Networks Expert Systems
Classification	(Early) fault diagnosis	Support Vector Machine (SVM) Artificial Neural Networks (ANN) Decision Tree (DT) Gaussian Process Classifier
Clustering	(Early) fault diagnosis	Self-Organizing Map (SOM), K-means, Gaussian Mixture Model (GMM)
Regression	Health assessment (HA)	SVM, DT, ANN, Gaussian Process Regression

- ***Dependency inference***, to obtain a model that describes the dependencies between observable values and unobservable states;
- ***Prediction***, to obtain a learning function that predicts future values based on recent observations;
- ***Summarization*** (feature extraction and selection), to find a compact description of data.

Table 6.1 maps the machine learning problems into the research tasks in condition monitoring and fault prognosis. Some example algorithms are also listed. It is important to note that the list in Table 6.1 is not exhaustive. However, these algorithms are well-proven in machine learning studies with successful applications in condition monitoring and fault prognosis. Table 6.2 shows some examples of applying these algorithms for condition monitoring and fault prognosis of WT and subassemblies.

As a summary of the existing studies about the condition monitoring and prognosis of WTs, the SCADA data and CMS data are mainly utilized in the following five different ways:

Table 6.2 Summary of techniques for the condition monitoring and fault prognosis of WT components

Component	Data	Method	Reference
Pitch	SCADA	Adaptive neuro-fuzzy inference system	Chen *et al.* [15, 16]
Pitch	SCADA	Multilayer perceptron	Kusiak and Verma [17]
Gearbox	Oil data	Particle filtering	Zhu *et al.* [18]
Gearbox	AE data	TSA and kurtosis	Li *et al.* [19]
Gearbox	Vibration SCADA Oil data	Multi-sensor fusion	Pan *et al.* [7]
Generator bearing	SCADA	Neural network	Schlechtingen and Santos [20]
Generator brush	SCADA	Multilayer perceptron, SVM, and boosting tree	Verma and Kusiak [21]
Main bearing	SCADA	Multilayer perceptron	Butler *et al.* [22]
Main bearing	Vibration	Kalman filter	Bechhoefer and Schlanbusch [23]
Blade	AE data	Logistic regression	Regan *et al.* [24]
Blade	Vibration	Auto-associative neural network	Dervilis *et al.* [25]
Drivetrain	Vibration	Frequency analysis, TSA, envelope analysis	Siegel *et al.* [26]
Drivetrain	SCADA	SVM	Tang *et al.* [27]

1. ***Normal behavior modeling***—as shown in Figure 6.5, understanding the normal behavior of WTs plays a vital role in detecting anomalies and potential failures at an early stage. A straightforward form of a normal behavior model is using data-driven algorithms, such as Auto-Regressive with eXogenous input [28], ANN [20, 29], and FIS [15, 16], to model the healthy behavior of WTs. This type of method comes with a lower cost and short development cycle. However, the performance of the model varies from case to case since most of the data-driven algorithms are black-box, and it is unknown how to improve the model performance step by step. Another type of behavior model is a physics-based model, which describes the normal behavior of the turbine with a set of physical equations. The physics-based models are reliable and trustworthy. However, these models are only available to critical subsystems, such as pitch mechanisms, yaw systems, and electric systems. Moreover, the development of the physics-based model is expensive and may require sensitive design information.
2. Degradation assessment and prognosis—by analyzing the SCADA parameters that are collected over a long period, many studies have successfully reported a developing trend before the failure event. Feng et al. [30] investigated the trending of gear efficiency decrease and temperature increase based on the SCADA

data and CMS. Their method successfully detected the planetary gear failure in the drivetrain six months in advance. Astolfi et al. [31] studied the temperature against the rated power and detected turbine problems for nine 2-MW on-shore WTs.

Based on the developing trend that is estimated from the SCADA and CMS data, prediction algorithms are then utilized to extrapolate the developing trend and forecast the RUL of the turbine. Commonly used prediction algorithms include ARMA, stochastic process methods, particle filters, DNN, and the proportional hazard model as shown in Table 6.1. Among these methods, the probabilistic algorithms with quantified prediction uncertainties are preferred in practice because it is important to consider the prediction uncertainties when planning the maintenance activities.

3. ***Failure analysis***—the SCADA and CMS data provide rich information to study the physical mechanisms of failure (PoF), which will be useful in designing preventative measures to lower the operation risk. Qiu *et al.* [32] investigated a thermophysics-based method for the fault diagnosis of a WT drivetrain. By comparing the simulation results and SCADA data, their method indicates that the nonlinear correlation between gear oil temperature and wind power provides sound evidence for transmission efficiency degradation. Gray and Watson [33] proposed a novel damage model to estimate the probability of failure for several critical failure modes and components based on the PoF. The method is practiced in a wind farm of 160 1-MW WTs to forecast the gearbox failures.
4. ***Clustering of similar WTs***—clustering techniques are powerful tools to cross-leverage the data from similar WTs in a fleet to improve the performance of prognostics and condition monitoring. Catmull [34] and Kim *et al.* [6] are the early studies that cluster similar WTs by using the SOM. The SOM rearranges neurons with similar input parameters, such as wind speed, applied loads, ambient factors, and others, to form clusters. By clustering the similar WTs and aggregating the data together, the goal is to enhance the model performance and improve FP accuracies. Currently, fleet-level condition monitoring and prognostics are vibrant to explore. However, there is still no clear evidence that the fault detection or prediction accuracy will be significantly improved by clustering the data from similar turbines.
5. ***Smart alarming systems***—the alarming records in the SCADA are useful information resources to study the reliability of the turbine, understand the root cause of the failure, and identify the false alarms. To give a few examples, the study in [35] explored the possibility of using alarming records in SCADA to complement the data scarcity challenge in the reliability assessment of WT. Their conclusion indicates that there are significantly more alarms than actual failures for certain components, and the SCADA can help ameliorate the scarcity of failure data. The study in [36] proposed a novel method to identify the false alarms by performing correlation analysis on all variables in SCADA. They managed to classify the alarming records into true alarm, false alarm, missed alarm, and true no alarm. Chen *et al.* [37] utilized a Bayesian Network to reason the root cause of the failure based on 26 months of historical SCADA data, including both

signals and alarms. The model can infer the probability of the failure event and suggest a clear relationship between cause and effect.

6.2.3 Performance monitoring with SCADA data

The ultimate promise of performance monitoring is to maintain the power conversion efficiency of the WT to a certain level of excellence. In practice, the performance of a WT is directly influenced by the aerodynamic characteristics of turbine blades, condition of the drivetrain system, pitch/yaw control strategy, electric system, and generator efficiency. As the turbine components or subassembly degrade, the overall performance of the turbine decreases, and less power is generated than normal operation. For example, ice and dirt deposits on the blade, drivetrain misalignment, lack of lubrication, or improper pitch/yaw control strategy will all cause turbine performance to drop. Therefore, it is clear that condition monitoring and prognostics of the turbine play a pivotal role in the performance monitoring of the WT.

Beyond the condition monitoring, SCADA data is also frequently exploited for power curve monitoring and production loss quantification to maintain the turbine performance.

1. ***Power curve monitoring***—the power curve is useful to monitor and assess the overall turbine performance. In condition monitoring, power curve degradation is also regarded as a precursor of turbine failure. The power curve is a graphical representation of the functional relationship between turbine power output and wind speed. Typically, the WT power curve is divided into three distinctive regions by three important speed limits, as shown in Figure 6.6. (1) The cut-in wind speed allows the turbine to begin generating electrical power, and the

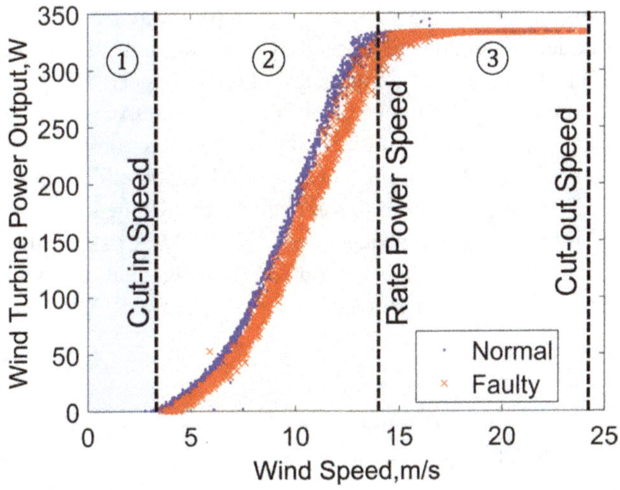

Figure 6.6 Power curve and turbine performance degradation

turbine is not rotating when the wind speed is below the cut-in speed. (2) The WT power output reaches a maximum at the rated power wind speed, and there is no further rise in the power output as wind speed further increases. (3) The rotor braking system will start to engage at the cut-out speed, which prevents the rotor damages caused by overspeeding.

Power curve monitoring focuses on estimating the degradation of the turbine based on the power curve between the cut-in speed and the rated power speed. As shown in Figure 6.6, the performance degradation of the WT will result in an actual power curve below the normal power curve or the design reference. In [38], a PCA-based method is proposed and validated for power curve monitoring. The method is practiced on an on-shore WT, which experienced two major downtime events in two years. The result indicates a noticeable performance degradation before the turbine failure. Similarly, the study in [39] proposed a multi-regime modeling technique based on the GMM to describe the power degradation and obtained good results. Regarding other available methods for power curve monitoring, a comprehensive summary can be found in [40].

2. ***Power loss investigation and quantification***—SCADA data is frequently leveraged to investigate or quantify the power loss induced by blade icing, wake effects, wind uncertainties, system degradation, and many other reasons. Investigation and quantification of power loss will be tremendously helpful in maintaining the turbine efficiency and in providing insights to improve the design and turbine control strategies. Said *et al.* [41] investigated the methods to quantify the power loss due to wake interactions by using the SCADA, meteorological, and wind LiDAR data. Based on the data from an on-shore wind farm, their conclusion indicates that power loss induced by wake interactions is more significant under stable atmosphere conditions than under convective conditions. The power loss for WTs in the second row of the wind farm reaches 40–60% of potential wind power under stable conditions, and this number further increases to 80% for the WTs in the third row. Davis *et al.* [42] proposed a statistical model to estimate the power loss by blade icing based on the 10-minute records of nacelle-measured temperature and wind speed, the power output of the turbine, and quality assurance flags from the SCADA system. There are also several recent studies in [43, 44] focusing on the detection of blade icing based on SCADA data. The study in [45] investigated terrain effects on the wind farm efficiency by exploring the correlation between the Ruggedness IndeX (RIX) of the terrain and the polar efficiency, where the SCADA data is utilized to compute the polar efficiency. Their conclusion indicates that, for one of the wind farms under study, the polar efficiency increases as the RIX increases.

6.2.4 Wind speed/power prediction

Wind patterns are often irregular and intermittent. This poses a significant challenge for wind fields, as converting this sporadic input source into a reliable electric output is a complicated task. In general, the uncertainty of wind as a source leads to higher

Table 6.3 Summary of wind speed and power prediction for the short, medium, and long term [46]

Time horizon	Range (approx.)	Analytics	Applications
Short term	up to 6 h	• Persistence method (naïve predictor) • Statistical approach (time-series models; ARMA, linear predictions, etc.) • Statistical approach (ANN; feed-forward, recurrent, multilayer perceptron) • ANN + fuzzy logic	• Ramp up/ramp down decisions • Economic dispatch planning
Medium term	6 h–1 day	• Numeric Weather Predictors (NWP) + neural network	• Generator decisions • Guidance on bidding strategy in the day-ahead electricity market
Long term	1 day–1 week	• Physical approach (NWP)	• Unit commitment and reserve requirement decisions • Maintenance scheduling

than necessary O&M costs; therefore, wind speed and power prediction can be considered a key input to wind farm dispatching and maintenance planning to adapt to these uncertain conditions and reduce the associated costs. Ultimately, improvements in wind speed and power prediction offer a number of economic and technical benefits; system capacity management and stability, optimal operation conditions for minimal emissions, business operation strategies, and even efficient fulfillment of maintenance requirements are all influenced by the quality of wind-forecasting information that is available.

Two elements are considered in wind prediction algorithms. Wind *speed* predictions are concerned with anticipating the patterns of the upcoming wind source input, including expected direction, frequency, location, and, of course, speed. Wind *power* predictions are concerned with estimating the upcoming expected electric power output of each WT based on the wind speed predictions. Predictions can be made for the short, medium, and long term, each of which can offer unique benefits. Table 6.3 shows an approximate (exact definitions vary) time range for and different analytic methods used to obtain each of these types of forecasts (details regarding the analytic approaches can be found in [46]).

In addition, Table 6.3 includes information on the possible applications best suited to each forecast time horizon. Short-term predictions, for up to 6 h in the future, are ideal for making decisions about the immediate operation of the wind farm. This includes ramping up the load to properly leverage a predicted future upswing in wind speed or ramping down to save on operation costs during a period with little to no wind. Power predictions in the medium range allow wind farms to optimize their bidding strategy for the next day's electricity markets. Long-term predictions can assist wind farm operators in properly allocating reserves to maintain a certain output level during times with a lower input and scheduling maintenance downtime to minimize the costs of opportunity loss. Overall, accurate and reliable wind-forecasting is vital to the successful management of a wind farm.

Wind predictions generate a future estimate based on both past and current conditions. Present conditions can be read from real-time sensors, but information on the past, actual wind patterns that were observed in the wind field, and upon which future predictions and decisions can be made, is found by accessing the SCADA data for each turbine in the field. In addition to providing historical wind information, SCADA data also contains information on historical WT operation patterns and power output trends. Overlaying and comparing the SCADA-based wind data with the SCADA-based turbine operation and output data allows for additional insights to be made into the relationship between wind speed, wind power, and optimal operating conditions, thus strengthening the prediction models and improving the forecast for the realization of further benefits.

6.2.5 Predictive maintenance planning

Wind farms are often erected in remote areas, such as mountainous areas and offshore locations, to obtain a reliable wind source and large open spaces, which are essential for optimal wind energy production. However, one challenge that is caused by the remoteness of the location and the logistical limitations is the high O&M costs. According to the survey in [47, 48], the maintenance costs for wind farms represent as much as 20 percent (onshore) to 35 percent (offshore) of its lifetime cost. Thus, optimal maintenance planning plays a critical role in ensuring reliable operations as well as economic value. This problem is generally referred to as maintenance planning. It is a complex problem to produce an optimized strategy detailing maintenance schedules, technician assignments, and vessel routes. In the literature, many investigations have been performed to obtain solution approaches. Multiple factors, including the operating condition of the WT, availability of maintenance resources, weather conditions for vessel travel, and others, are considered when optimizing the maintenance schedule. Feng [49] proposed an intelligent planning approach considering the significance of production loss before and during maintenance activities with the help of wind speed predictions. Zhang [50] approaches the problem by minimizing the energy loss for downtime. Other approaches involve maximizing the system reliability [51], optimizing the penalty cost for delayed tasks [52, 53], and also minimizing the production loss [54, 55] with a different model formulation from [49].

SCADA systems play one of the most important roles in developing this maintenance strategy by providing data for three critical aspects:

1. ***Reliability assessment and condition monitoring***—the SCADA system enables reporting a fault, once detected and diagnosed, by generating a work order for

maintenance. Once the fault is reported, an estimation of the RUL can be made, which provides a time window recommendation within which maintenance should be performed under ideal conditions. Damage and costs can rise significantly if a fault is not fixed before it turns into a major failure.

2. ***Performance monitoring***—data from SCADA are suitable to monitor the performance of the WT. Deviations in the produced power from ideal produced power under the same conditions can indicate the health of the WT. Such observations can also identify deviations in the operation setting of the WT that may need maintenance actions. Performing such maintenance actions can help sustain the performance of the WT to a certain level of excellence and prevent failures.
3. ***Wind speed/power prediction***—while deciding the schedule to perform a maintenance action, it is beneficial to identify a time window that causes minimum production loss. Such a maintenance strategy is called an opportunistic maintenance strategy. To evaluate this optimal time window, additional factors that contribute to the performance need to be considered. Wind speed is one such critical factor. Long-term wind speed prediction forms an important input in deciding the optimal maintenance timing. Along with wind speed, the predicted wind power can also be incorporated to minimize the power loss caused due to turbine shutdown. Data obtained from SCADA are used to make these predictions.

6.3 A unified framework for cyber-physical wind farms

Developing intelligent wind farm systems offers significant opportunities for improving turbine field performance, decreasing operating and maintenance costs, and increasing overall efficiency; however, implementing such systems highlights some unique challenges beyond what is seen in other industrial applications of performance and health monitoring and maintenance planning. First, the input, wind, is highly variable. Unlike other industrial settings where the machine inputs are known and supplied by a reliable, controllable source, WTs operate in a less orderly environment. Wind speed, direction, energy density, and duration are all dynamic, nonlinear processes. Despite best efforts to predict and mitigate the effects of these fluctuations, they are all still highly variable sources of wear and degradation on the turbine components. In addition, wind farms are typically located in remote, isolated areas by design to not interfere with the day-to-day operations of other infrastructures of civilization. This means that maintenance is more challenging to coordinate, requires more sophisticated tools or methods, and is more costly and time-consuming. Therefore, an innovative and intelligent operating and maintenance planning system is especially valuable in this context.

Overall, an intelligent wind farm architecture must offer insights into a number of key areas that align with the wind farm O&M strategies listed in Section 6.2. The system solution should allow monitoring and management of the turbine and all related components, which includes displaying accurate and easily interpretable details of degradation. This information can be used to inform revision or upgrade decisions for individual components or subsystems. In addition, a health trend analysis and RUL prediction for the entire turbine structure should be readily available. Besides monitoring the status of the WT components, an intelligent approach would also continuously monitor and evaluate the health of the turbine outputs, i.e., the power generation performance of the entire wind farm. To combat the variability of the wind source and to take full advantage of wind availability, any implemented system would require intelligent dispatch capabilities based on wind resource prediction technologies. Finally, all garnered knowledge would need to be applied to optimize the wind farm operation, scheduling, and maintenance practices.

Implementing a cyber-physical system (CPS) allows all collected information, including sensor signals, controller information, maintenance records, and any other digitized data, to be monitored and integrated between the physical world and the cyber computational space. A CPS is an intelligent industrial technology approach that encompasses the entire process of data collection, storage, analysis, prediction, and decision-making. The CPS also accounts for relationships between different processes as well as relevant business considerations. In addition, the CPS offers a closed-loop solution by transforming the collected data into executable knowledge. By leveraging the available SCADA and CMS data, the above-mentioned requisite capabilities for an intelligent wind farm can be built into a CPS using the popular 5C architecture as a functional framework.

Following the 5C framework gives a clear, sequential workflow to construct a CPS. Considerations in the connection, conversion, cyber, cognition, and configuration levels provide step-by-step guidelines for developing and deploying a CPS. The general components of CPS are (1) a physical world that supplies data and receives feedback and (2) a cyberspace that manages data and contains analytic and computational abilities. While these general components incorporate the overarching goals of a CPS, this simple framework alone is vague and does not present enough specific information for implementation purposes. As such, further development of the framework, namely the 5C concept, is required to achieve the desired system that can create value from the incoming data. Detailed explanations of the 5C components in Figure 6.7 are as follows:

1. ***Connection***—the first piece of assembling a CPS is establishing a smart connection to the machines, components, and conditions of interest. Acquiring accurate, reliable, and comprehensive quality data is essential for ensuring the success of the CPS. As previously mentioned, the data can be measured by sensors, obtained through the controller, or accessed through manually kept records. For the specific case of a wind farm, SCADA data, CMS data, and maintenance records are primary sources of information. Sensors must be properly selected and specified for the application. Data of various types are included, so seamless and well-organized data acquisition procedures need to be implemented at this first level to meet the objectives and needs of future information analysis tasks

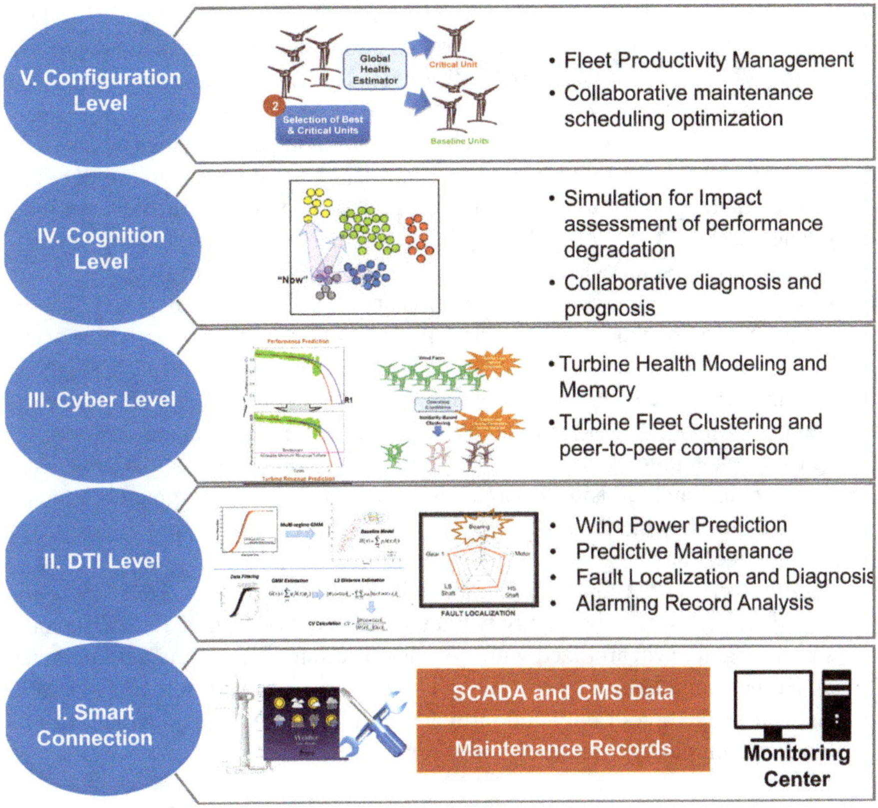

Figure 6.7 A unified framework for cyber-physical wind farms

and activities. This data collection environment serves as the foundation that supports the remaining layers of CPS.

2. ***Conversion***—the second level focuses on converting the data, which has a lower value density, to information, which has a higher value density. This level is sometimes called the data-to-information (DTI) level. Using previously described algorithms and approaches that have been specifically tailored to prognostics and health management applications, this conversion transforms the data into extracted features that can be filtered, classified, and prioritized. Unlike simply collected data, the processed information ensures interpretability and allows for further analysis. Determination of health values, RUL, and other predictive maintenance metrics, wind power prediction, fault localization and diagnosis, and alarming record analysis add an element of self-awareness to the CPS.
3. ***Cyber***—at the cyber level, the focus shifts toward centralizing, sorting, and maintaining the information. The cyberspace modeling and analysis environment synchronizes different information streams, forms associations between related processes, records changes, and performs higher-level analysis and prediction

functions. Going beyond the level of single machine insights obtained in DTI conversion, the cyber interface extracts additional information that can provide an overview of the status of the entire fleet. This also allows self-comparison and peer-to-peer cross-comparison metrics to be generated. A single WT's performance can be rated against the other turbines in the field. Similarly, previous assets (historical information) from turbine health monitoring and memory can be evaluated against present operating conditions to predict future behaviors. At this level, the benefits of CPS truly begin to emerge as the supporting connection and conversion levels are used to enhance fleet-wide analysis.

4. ***Cognition***—pattern recognition and detection of relationships between components, environmental conditions, and maintenance or operation activities occur at the cognition level. Since the system possesses individual machine status information and comparative information, simulations of the impact of continued performance degradation along with a collaborative diagnostic approach can suggest maintenance tasks that are optimally prioritized. At this stage, data visualization and decision optimization tools are vital to completely convey acquired knowledge to the users and to provide users a means of support in making decisions.
5. ***Configuration***—the final configuration layer closes the loop. Here, the optimized results for fleet productivity management and collaborative maintenance scheduling are synchronized with system execution. The turbine field operates under the configuration layer's supervisory control, allowing self-configuration and self-adaptation of the wind farm and enabling a timely deployment of utilized information and acquired knowledge.

The described unified 5C-based framework for a CPS can be successfully deployed in a wind farm application. SCADA and CMS provide information on environmental conditions, input control and system state parameters, work conditions, vibration (and other useful signatures of critical components), power grid dispatching, personnel management, and maintenance. The data streams are comprehensively analyzed and transformed into useful information that feeds the generation of comparison metrics, pattern recognition and relationship establishment protocols, and visualization and decision-making tools. Through these processes, the CPS garners advantages in turbine performance evaluation and performance decline prediction, wind farm operation, and maintenance management. An intelligent wind farm system operating under such a CPS would enjoy the benefits of improved turbine field performance, decreased operating and maintenance costs, and increased overall efficiency.

6.4 Case studies

6.4.1 SCADA-based power curve monitoring

A WT power curve expresses the power output of the turbine as a function of the wind speed. It reflects the power generation capability of the WT under a range of wind speeds. It can be used as an effective indicator for monitoring the health of the WT. It enables characterizing the performance of an individual WT without needing expert knowledge on turbine design, components, or other aspects. All streams of data—especially wind speed and power output—needed to obtain the power curve for a WT and apply the methodology to assess the degradation can be sourced through SCADA.

A typical WT power curve is segmented as described previously (see Figure 6.6) in different zones defined by three key wind speed values: (1) the cut-in wind speed; (2) the rated output wind speed; and (3) the cut-out wind speed. The cut-in wind speed is the minimum wind speed required for the turbine to begin rotating and generating electrical power. After the cut-in wind speed and up to the rated output wind speed, the power curve reveals a quasi-linear relationship between wind speed and output power. The turbine produces the maximum power output at the rated output wind speed. The power output does not increase after this point, even if the wind speed increases. In fact, if the measured wind speed reaches the value of cut-out wind speed, the braking system in the WT assembly is activated to keep the rotor stable, as the rotor may get damaged at higher speeds.

The degradation of an individual WT can be assessed by quantifying the deviation in the power curves between normal conditions and degraded conditions. The power loss due to degradation mainly occurs in the quasi-linear region of the power curve, i.e., between the cut-in wind speed and rated output wind speed. This deviation from the normal operation condition is often presented as a confidence value or health value. The analysis pipeline from obtaining SCADA data to inferring the health value involves two major steps: (1) data preprocessing and (2) performance modeling, as shown in Figure 6.8.

1. ***Data preprocessing***—data preprocessing is a fundamental and essential step in a data-driven method. It is especially important in the current application since the health value is obtained by a direct comparison between normal and degraded turbine conditions without any intelligent model. Thus, the data quality becomes an important factor in deriving the results. Data preprocessing for WT SCADA data, in particular, involves three aspects: (1) control scheme filtering to rule out disturbances of WT operational influence; (2) regime selection

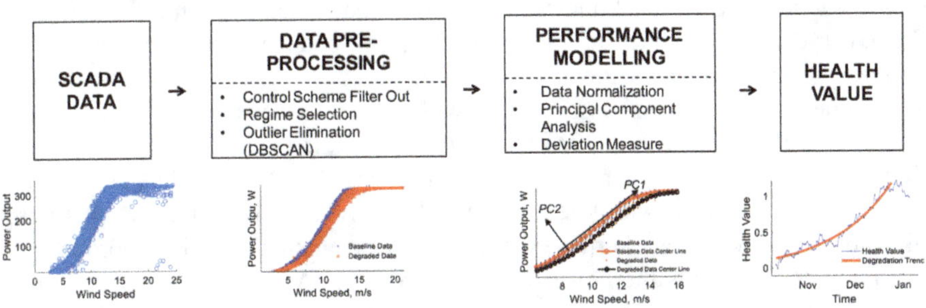

Figure 6.8 Overall approach for monitoring WTs using power curve [38]

for selecting segments that satisfy model assumptions; and (3) outlier elimination for noise reduction.

Loss in power output in a WT can be a result of non-degradation related factors as well, such as certain operational conditions. For example, the blade pitch angle needs to be adjusted to curtail the power output for over-speed protection or other turbine safety considerations. The filtering of samples in the data, which represents the loss in power output due to such factors, is performed in the control scheme filtering rule. As mentioned earlier, since only the quasi-linear region in the power curve carries useful degradation information of WTs, the regime selection process is used to perform the appropriate data segmentation and selection. A GMM approach with three components can be used to separate the standard WT power curve into three different 2-D Gaussian distributions such that one of the distributions covers the quasi-linear region. After regime separation, outlier elimination can be performed to remove noise caused by instrumentation error or natural deviations in the system behavior. The Density-Based Spatial Clustering of Applications with Noise (DBSCAN) algorithm is one widely employed algorithm for outlier removal that can be used here.

2. ***Performance modeling***—the pre-processed power curve data needs to be modeled into a metric reflecting the relative conditions of the WT. The measured deviation in this metric between normal and degraded conditions can then be quantified into a health value that assesses the degradation. Dimension reduction methods can be used to extract a few high-information variables from many low-information variables. These high-information variables can be suited for modeling a system's condition. PCA is one of the most widely used methods for dimension reduction and can fulfill the purpose. PCA is popular for its convenience of implementation and interpretation and its excellent performance when applied appropriately.

For the given application scenario, PCA is adopted as a coordinate transformation technique. The input matrix of PCA is a vertical stack of the baseline power curve data and testing power curve data. The stacked data is then normalized to a mean of zero and unit variance. On applying PCA, the normalized data yields the projection of the power curves in the principal component (PC) space. The first dimension or the first principal component (PC1) points to the direction with maximum variation, and the PC2 direction carries the major degradation information in the power curve. Thus, the variation of data in the secondary PC direction can be used as a metric to quantify the deviation of the test power curve from the baseline. This value of variation can be normalized with respect to the variation in conditions where no degradation is present to obtain a relative health value for better interpretation and visualization of the result. It is worth noting that PCA is a linear approach that works for a system with linear or quasi-linear relationships. For systems with high non-linearity present, kernel-PCA can be an excellent alternative approach.

Figure 6.9 presents the power output from SCADA data of a large-scale onshore WT for a period of almost two years from 2008 to 2010, sampled at 10-minute intervals. It indicates four turbine downtimes during the two years, with one

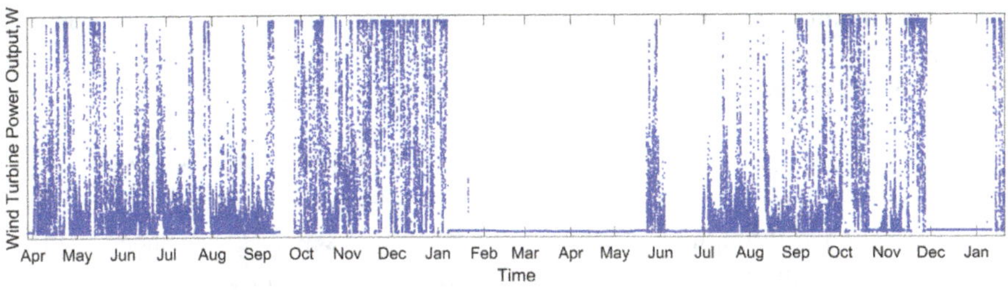

Figure 6.9 Time series data of WT power output in two years [38]

downtime lasting as long as five months. Each of these long periods of downtime implies a major maintenance action. It is expected to observe a significant rise in the health value from the time after maintenance is completed until the next maintenance event. It is important to mention that a week of SCADA data is missing in the first week of September in Figure 6.9. However, according to the maintenance log, no maintenance actions were recorded during that period, and thus, the health value is not expected to rise before this time.

For the given data set, pitch angle and average active power are used to remove the influence of the control scheme on the data while the average wind speed and average active power are inputs into the power curve monitoring algorithm for turbine degradation assessment. Health assessment is performed daily, and one week's data prior to the current date is used as input every time the health assessment algorithm is triggered. Three weeks of data after the first maintenance were chosen as the baseline data based on the assumption that the WT will work under normal conditions after the maintenance action. Using this baseline data, the deviation between the baseline and the following test data is quantified to obtain the health value of the WT using the described algorithm.

The resulting health value of the WT over the entire duration is shown in Figure 6.10 with the convention of a higher value meaning a higher level of degradation. The dotted horizontal line in Figure 6.10 represents the average of the health value under normal conditions. A three-sigma limit is used to set the threshold for degradation detection. The heath values inside the control limits are considered normal, while those outside the control limits indicate the degradation of the WT performance. The health values indicate a significant increase in degradation before the

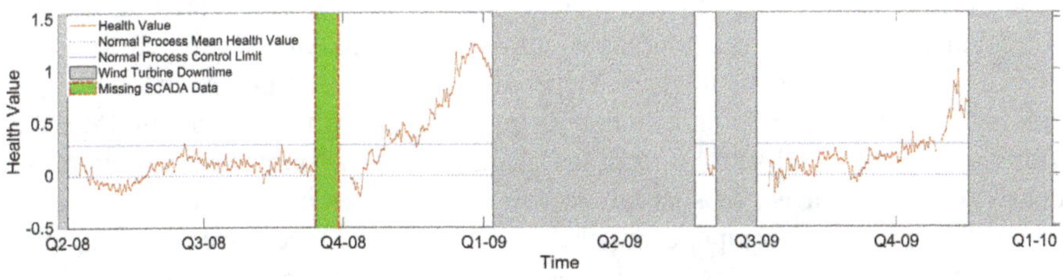

Figure 6.10 Health value of the WT over the period of two years [38]

two major maintenance periods, marked by shaded regions. The results are easy to interpret and include minimal false alarms. The health values also indicate a stable trend of degradation, which makes it suitable for prediction analysis as well.

In comparison with other methods for performance modeling such as GMM with L2 distance, SOMs, and neural networks, the described method has many superior features [37], such as (1) straightforward interpretation of the result; (2) a broad application scope through machine performance curves with quasi-linear shapes, such as a diesel engine or a gas turbine engine; (3) simple implementation without an iterative process or non-convergence issues; and (4) better performance.

6.4.2 Condition monitoring of drivetrain system

The ability to diagnose and locate a faulty component can offer substantial improvements in the reliability and availability of the WT asset. It forms the step after degradation is detected in the health condition of the WT. While low-fidelity data from SCADA can be enough to assess the degradation of the WT, as seen in the previous section, fault localization and diagnosis need high-fidelity specific signals that can depend on components and fault types. CMS is employed to fulfill this requirement and can be triggered for high-velocity data collection once degradation has been detected in the asset.

Considering that it is not feasible to monitor every aspect of a WT, numerous studies have been performed to identify subsystems that are critical to the reliability of the WT and show evidence of high failure rates. WT drivetrains, particularly the gearbox, are one of the most critical subsystems regarding maintenance. Along with the gears, gearbox bearing failures are also a common component failure. Thus, most efforts have been put into improving the capability to identify failures in gears and bearings in the drivetrain system of a WT.

Given the complexity and sophistication of such critical rotating machinery, multiple failure modes can be observed for each component. Identifying different failure modes may require different types of data, and thus different types of sensors and specific signal processing techniques based on the characteristics of the failure mode to be identified. Table 6.4 summarizes the most common failure modes, sensors, and signal processing techniques for critical WT components.

The commonly observed failure mode for gears is scoring and scuffing. This can result from multiple root causes, such as lack of lubrication, misalignment, or trapping of debris at undesired locations, resulting in higher loads and damage at the area affected. Other failure modes include surface fatigue, tooth fracture, fretting corrosion, etc. Vibration is one of the commonly used and effective signals for identifying such failure modes in gears. Depending on the signal processing technique employed, additional rotation data may be required to extract information from the vibration signal, which can be recorded using encoders. Apart from vibration, oil debris and motor current sensors can also be used. For bearings, faults can occur on any of the constituent components—the outer-race, inner-race, cage, or rolling element. Corrosion, overheating, cracks, and assembly breakage are some of the failure modes caused by gradual wear, loose fit, overloading, and lubrication starvation.

Table 6.4 Summary of failure modes, sensors, and signal processing techniques for critical WT components

Component	Failure mode	Sensors	Signal processing analytics
Gear	Scoring and scuffing, surface fatigue, tooth fracture	• Vibration • Encoder • Oil debris • Motor current	• Frequency spectrum • TSA • Phase demodulation • Cepstrum • Wavelet analysis • Residual analysis for auto-regressive filter
Planetary gear			• Instantaneous angular speed estimation • Planet separation method
Bearings	Outer-race, inner-race, cage, rolling elements	• Vibration • Oil debris	• Kurtogram • Envelope analysis • Wavelet analysis
Shaft systems	Imbalance, misalignment	• Vibration	• Frequency spectrum

Vibration is the primary signal employed for bearing fault diagnosis, although oil debris and motor current can also be used. In some situations, shaft systems can also cause failures that can indirectly lead to bearing and gear failures. Primary failure modes in shaft systems are imbalance and misalignment, and the vibration is the most effective signal for identifying this fault category.

As it is evident that vibration is one of the most versatile signals for fault identification in rotating machinery, numerous vibration-based algorithms have been developed over time. In the data analysis pipeline, these algorithms are used to process the vibration signal to extract useful information or features to be applied in a model for fault identification. The signal processing techniques essentially exploit the periodic character of the system to enhance the information in the signal. The geometric parameters of the system, including the number of gear teeth, pitch diameter, and the number of rotating elements in the bearing can be used to extract useful features that can locate the fault and to measure the fault intensity. These features can then act as inputs to classification or anomaly detection models such as SOMs, SVMs, etc., to be able to perform fault detection and diagnosis in practice. Therefore, it is important to note that the performance of the overall model is highly dependent on the quality of the features and thus the signal processing technique. Five widely used signal processing algorithms for gear and bearing faults using vibration signals are summarized in Table 6.4.

1. ***Frequency domain methods***—the fast Fourier transform is employed to transform the time domain signal to the frequency domain. For rotating machinery,

exploration of frequency domain signatures is often a typical first approach prior to applying more advanced processing methods. However, frequency domain signatures by themselves can also serve as effective features in some cases. For meshing gears, signature frequencies are present for gear mesh frequencies and sidebands. These can provide an initial evaluation of the gear wheel's health condition. Although a normal and degraded gear may have a similar gear mesh frequency peak, the magnitude of sidebands can reflect significant differences between health conditions. In addition, the location of the fault across shafts or gears can be indicated by the spacing between the sidebands. For bearings, signature frequencies can be used for fault identification at the inner-race, outer-race, rotating element, and cage. Frequency spectrum analysis is a relatively simple method for extracting gear condition indicators without a tachometer signal. However, as it lacks the capacity to enhance the signal-to-noise ratio with the support of a tachometer, advanced processes are required.

2. ***Cepstrum processing method***—the real cepstrum is calculated as the inverse Fourier transform of the logarithm of the power spectrum. It provides an enhanced method for analyzing a family of harmonics over just the frequency domain representation. For mechanical systems and gears especially, the analysis of sidebands can be more effective and convenient using cepstrum. By comparing the cepstrum for a series of sidebands spaced at a given shaft speed between the baseline and degraded state, the health condition of each gear wheel can be identified.
3. ***Spectral kurtosis filtering***—the vibration signals for damaged gear and bearing components nominally exhibit an impulsive signature. It may be difficult to detect the impulsive signature as other sources of vibration could mask the signature. Optimal resonant band filters, such as spectral kurtosis, Kurtogram, and wavelet filter bank, can help enhance the impulsive fault signature that is contaminated by measurement noise. An advantage of this method is that filtering of signals is possible without any a priori knowledge of the proper frequency band in which to filter; this method is especially useful in selecting an appropriate filter in envelope analysis. The filtered signal can then be used to extract features. However, for the case of detecting faults with non-impulsive consequences such as extended spalls or faults, the method may not be useful.
4. ***Time-synchronous averaging***—TSA allows the extraction of the periodic vibration waveform by averaging multiple waveforms aligned by time buffers. This extracted waveform allows further analysis of the vibration signatures in terms of residuals and improved amplitude modulations. The residual signal for a gear of interest can be extracted by filtering the shaft harmonics and the gear mesh frequency and its harmonics from the periodic vibration waveform. Deviations in this residual signal around a gear mesh frequency could be indicative of a fault in the gear teeth. For detecting defects such as a fatigue crack in a gear wheel, amplitude and phase modulations in the periodic vibration waveforms can be compared with a baseline. A requirement for using TSA is the availability of an accurate tachometer signal, which may not

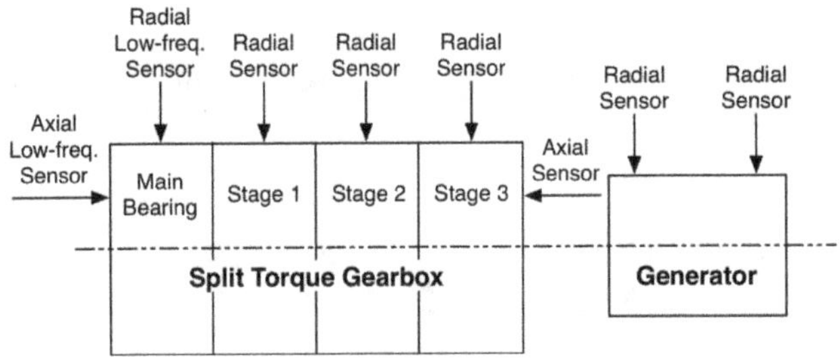

Figure 6.11 Example of a drivetrain system with sensor locations for a WT

always be possible. Additionally, a long acquisition time may be needed for gears with low rotational speeds due to the need for collecting a set of readings for averaging.

5. ***Bearing envelope analysis***—for performing a bearing envelope analysis, primarily a band-pass filter is applied around an exciting natural frequency. After filtering the signal, the envelope signal is extracted using the Hilbert transform. This envelope signal is then further analyzed in the frequency domain. Amplitude modulation at defect signature frequencies is much easier to observe for bearing damages in the envelope spectrum as compared to just the frequency spectrum. However, the selection of the demodulation band is complicated. Also, higher sampling rates may be required depending on the excited system resonance.

Figure 6.11 shows an example of a drive train system for a WT. Different locations have been appropriately identified, and accelerometers are installed to monitor the gearbox. It is critical to identify an optimal location and sensor type to be able to collect useful vibration signals. Once the raw vibration data is acquired, Figure 6.12 describes the steps to perform condition monitoring of the gearbox. TSA can be performed as a signal processing technique to enhance the quality of the signal and extract features. The amplitude of peaks at different critical frequencies can be extracted as the features from the enhanced signal. This set of baseline features can be used to train an SOM to learn the normal condition of the gearbox. As test samples are recorded, the same signal processing technique and feature extraction procedure can be applied to them. Test features thus obtained can be projected on the learned map. The distance between the test features and the best-matching unit of the map can be used to quantify the deviation of the test sample from the baseline sample. This distance is called the Minimum Quantization Error (MQE). A three-sigma limit of the MQE value of the baseline data can be used as a threshold for fault detection. Figure 6.13 shows the MQE values for the gearbox over the period of time when it undergoes failure. It can be observed that the deviation from the baseline could be observed as early as one week before the occurrence of failure.

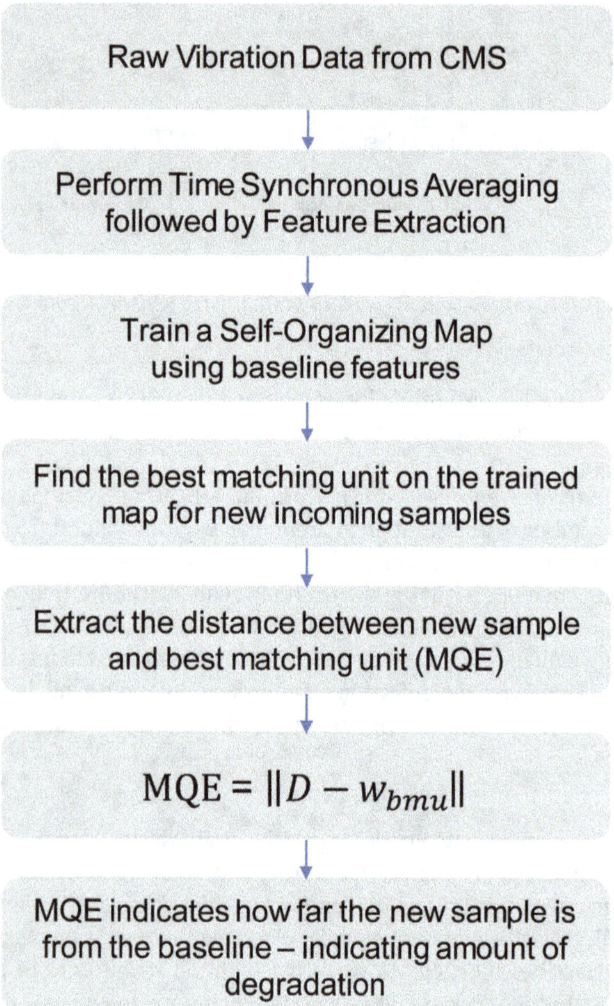

Figure 6.12 Flowchart for condition monitoring of gearbox using self-organizing maps

6.4.3 Intelligent wind field operation and maintenance system

As production demand continuously expands toward the constraints of production capacity, well-informed and executable production plans (i.e., O&M scheduling) that require few, if any, adjustments to be made mid-operation are essential if a wind farm hopes to remain competitive. In addition to the considerable business loss that is caused by a production plan that needs to be adjusted, the human resource requirements and costs, including cooperation between multiple departments, capacity allocation decisions for production execution, and constant communication with workers and outside entities, are also substantial. As the gap between current and optimal

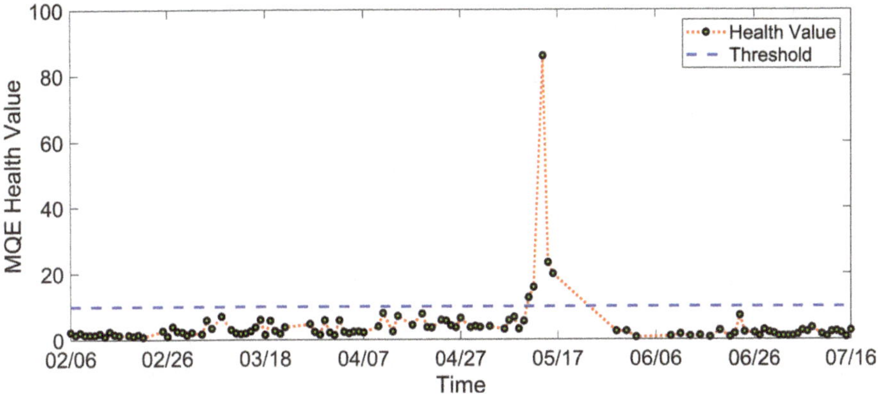

Figure 6.13 Fault detection of gearbox using MQE values

production output closes, the margin of error for production planning decreases, driving the necessity for computational approaches to production planning.

The application of such computational approaches in an intelligent wind field is based on the collection and proper management of a large amount of data. This accumulated data provides insights into the relevant processes, from which a deeper understanding of the wind farm can be obtained. With this enhanced knowledge, decision-making problems can be translated into mathematical models and evaluated using advanced algorithms. Critical objectives and goals can also be quantified and evaluated under this system, allowing the best solutions to be found within the boundaries and rules of each decision. This approach offers a significant improvement over traditional systems, which mainly rely on the knowledge of experienced workers and are therefore singular, stagnant solutions that offer little in the way of flexibility and adaptability.

These traditional systems and standard software-based production planning approaches face two main shortcomings. First, with the race to achieve optimum production, the complexity of decision-making has greatly increased as the issues that must be resolved are continuously evolving. Traditional approaches are already at their limit in processing a single wind field, let alone a multi-field production environment. Second, with the exponential number of variables involved in a WT production setting, the number of possibilities to explore can often reach into the billions. While less complex scenarios can be adjusted and supplemented by expert input and solved via heuristic methods, the current, complex industrial landscape of the wind field means that more advanced algorithms or approaches are required.

To address and avoid these issues, the comprehensive intelligent wind field system incorporates health management, operation, and maintenance tasks that can be separated into two areas of focus, shown in Figure 6.14—(1) predictive analytics that assess turbine performance and (2) dynamic O&M optimization. Through the analysis of real-time data, pattern recognition is implemented to assess the work environment and wind resource cycles, and effective features are extracted to

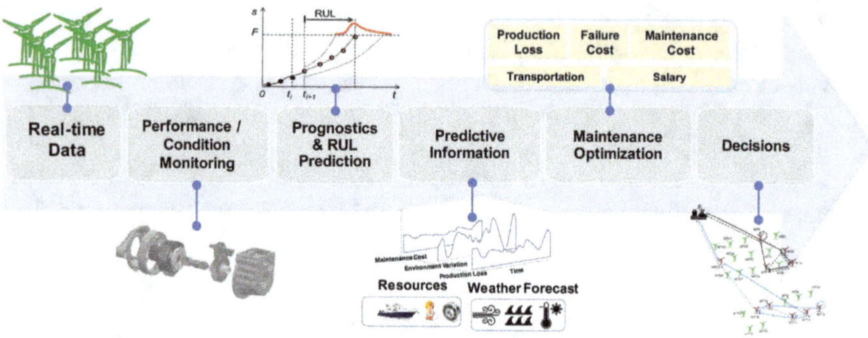

Figure 6.14 Predictive analytics and dynamic operation/maintenance scheduling for intelligent wind farm

establish the health of the turbine and its components. The present state of degradation is also evaluated. After this diagnostic stage, prognostic techniques are used to assess possible failure modes and operation risks, and RUL predictions are made. Careful monitoring of the turbine and balancing of the anticipated conditions allow the maximization of power generation capacity and reduce the system downtime as much as possible.

After the WT data is analyzed and processed, the features and models are uploaded to a cloud service platform. Along with resource information and weather patterns, the trends of the wind field and turbines are tracked in this central location, which helps to unify management and provide further analysis. This cloud platform is used for real-time monitoring, and historical performance information from multiple wind farms can be aggregated to enhance the analysis and prediction capabilities of the system. Comparisons between this library of historical turbine operation data (production loss and maintenance cost trends, environmental variation, etc.) and the uploaded baseline models provide keen, data-driven insight into efficient management of the turbine state. These evaluations can be made across time (self-comparison) or across clusters of like machines (peer-to-peer comparison) using different data resources to produce a complete set of comparative dimensions and insights.

Given these insights, maintenance can be optimized, including minimizing production loss, failure cost, and maintenance cost; increasing safety; and executing the most efficient transportation routes and maintenance schedules [56]. Having a complete portfolio of this information available in an easily digestible format allows the best decision to be made for fleet-wide deployment and beyond. The exhibited modular design of an intelligent WT system and decisions made from these schedules update the data for analysis, making the entire process scalable and customizable for the user while simultaneously allowing intelligent WT upgrades to be completed.

An optimization model for short-term and medium-term O&M planning was developed and specifically tailored to offshore wind farms. This model is based on actual wind farms and design optimization models, and a number of constraints are applied to ensure that the real-world operation of sites and dispatching of maintenance is simulated as accurately as possible. The focus on maintenance tasks for off-shore wind farms accounts for factors such as the ships used and navigation costs incurred, the weather conditions, maintenance personnel requirements, order of completion of the maintenance tasks, and the health status of the turbines. A flexible maintenance fleet was assumed so that, for each necessary maintenance task, there are a number of available maintenance vessels and several maintenance teams. While this approach helps to reduce the costs, it also enlarges the search space for feasible solutions, thereby increasing the complexity of the optimization problem. To solve the optimization problem, a two-layer genetic algorithm was designed. This algorithm is highly scalable, provides stronger computing power than commercial optimization software, and is well-suited to the challenges of maintenance dispatching for off-shore wind farms.

The optimization model is integrated to enable the user interface of the intelligent wind farm O&M system. The interface includes four modules of visual output: optimized strategic route scheduling, cost comparisons for the before and after of optimal strategy implementation, a perspective on maintenance costs, and an analysis of historical source costs. This interface quickly provides users with information to support maintenance scheduling decisions. As an example, 17 maintenance tasks were assigned for an off-shore wind farm. The developed algorithm forecasted a cost for the optimized intelligent scheduling plan that was a 25 percent reduction over the non-optimized scheduling plan. This demonstrates the abilities of the intelligent wind field O&M system to improve the efficiency of operation and reduce the costs of maintenance.

6.5 Summary and future perspectives

As a summary, the data records from the SCADA system contain rich information about turbine O&M, which has been utilized for (1) reliability assessment; (2) condition monitoring and performance monitoring; (3) power loss investigation and quantification for turbine redesign and intelligent control; (4) wind speed/power prediction; and (5) maintenance planning. Among these applications, the reliability assessment is essential for selecting cost-effective maintenance strategies for different turbine parts and subassemblies. Condition monitoring, which is appropriate for the turbine parts with a low failure rate but a long period of degradation, is pivotal to avoid or reduce unscheduled turbine downtime. When developing the CMS for WT, SCADA data are often analyzed in combination with high-resolution data from CMS for (1) normal behavior modeling; (2) degradation (or trending) analysis; (3) failure mechanism analysis; (4) turbine clustering analysis; and (5) alarming code mining. The data from the SCADA system are appropriate for data mining and machine learning, and the data have demonstrated great abilities for anomaly detection and trend analysis at the turbine level. The data from CMS are usually high-frequency waveform data, and proper signal processing techniques are needed to extract meaningful features from the raw data. The CMS data is often used to diagnose turbine failures and study the failure mechanism, which provides supplemental information to the analysis of SCADA data. Beyond the condition monitoring, the SCADA data also provides valuable information for power curve monitoring, power loss investigation and quantification, and wind speed/power prediction. This information will help the operators to develop intelligent power dispatching plans and the designers to improve the turbine design and the pitch/yaw control strategies. Finally, the knowledge learned from SCADA data analysis, including parts reliability, turbine health/performance status, and forecasted wind resources, are aggregated in an optimizer to generate maintenance plans with cost breakdowns for wind farms management and maintenance.

Based on SCADA (and CMS) data analysis, a unified framework of the cyber-physical wind farm is presented to promote the seamless integration of design into turbine intelligent operation, the construction of unified platforms for fleet management and maintenance, and the development of large-scale optimizers for collaborative maintenance planning among multiple wind farms. Three case studies from our previous collaborations with real-world wind farms are utilized to demonstrate the framework and some of the intelligent applications.

Though the analysis has been fruitful, the development of reliable and accurate condition monitoring models for all failure modes of a WT is still challenging. The difficulties that need to be addressed in future works include the reliability and explainability concerns of data-driven models and the cost and deployment issues of physics-based models. Special attention needs to be paid to the non-stationary operation conditions of the WT, the fault isolation for multiple component systems, and collaborative diagnosis and prognosis for turbine fleets. Besides, advanced predictive analytics are critical to overcoming the intermittency of wind resources, and efficient optimizers are also highly desired to promote nation-wide maintenance resource planning and vessel routing.

References

[1] Artigao E., Martín-Martínez S., Honrubia-Escribano A.,*et al.* 'Wind turbine reliability: a comprehensive review towards effective condition monitoring development'. *Applied Energy*. 2018;**228**(2):1569–83.

[2] Dao C., Kazemtabrizi B., Crabtree C. 'Wind turbine reliability data review and impacts on levelised cost of energy'. *Wind Energy*. 2019;**22**(12):1848–71.

[3] Faulstich S., Hahn B., Tavner P.J. 'Wind turbine downtime and its importance for offshore deployment'. *Wind Energy*. 2011;**14**(3):327–37.

[4] Jia X., Cai H., Hsu Y., Li W., Feng J., Lee J. 'A novel similarity-based method for remaining useful life prediction using kernel two sample test'. *Annual Conference of the PHM Society*. 2019;**11**(1):1–9.

[5] Cai H., Jia X., Feng J., Li W., Pahren L., Lee J. 'A similarity based methodology for machine prognostics by using kernel two sample test'. *ISA transactions*. 2020;**103**:112–21.

[6] Kim K., Parthasarathy G., Uluyol O., Foslien W., Sheng S., Fleming P. Use of SCADA Data for Failure Detection in Wind Turbines. *Proceedings of the ASME 2011 5th International Conference on Energy Sustainability. ASME 2011 5th International Conference on Energy Sustainability, Parts A, B, and C*; Washington, DC, USA, ASME, August 7–10, 2011; 2011. pp. 2071–9.

[7] Pan Y., Hong R., Chen J., *et al.* 'Performance degradation assessment of a wind turbine gearbox based on multi-sensor data fusion'. *Mechanism and Machine Theory*. 2019;**137**(1–2):509–26.

[8] Lee J., Wu F., Zhao W., Ghaffari M., Liao L., Siegel D. 'Prognostics and health management design for rotary machinery systems—reviews, methodology and applications'. *Mechanical Systems and Signal Processing*. 2014;**42**(1-2):314–34.

[9] Wang S., Zhang N., Wu L., *et al.* 'Wind speed forecasting based on the hybrid ensemble empirical mode decomposition and GA-BP neural network method'. *Renewable Energy*. 2016;**94**:629–36.

[10] Wang L.-J., Dong L., Liao X.-Z., Gao Y. 'Short-term power prediction of a wind farm based on wavelet analysis'. *Proceedings of the CSEE*. 2009;**29**(28):30–3.

[11] Hameed Z., Hong Y.S., Cho Y.M., *et al.* 'Condition monitoring and fault detection of wind turbines and related algorithms: a review'. *Renewable and Sustainable Energy Reviews*. 2009;**13**(1):1–39.

[12] Kantardzic M. *Data Mining: Concepts, Models, Methods, and Algorithms.*' USA and Canada: John Wiley & Sons; 2011.

[13] Murphy K.P. *Machine Learning: A Probabilistic Perspective*. Cambridge, Massachusetts, London, England: MIT Press; 2012.

[14] Jordan M.I., Mitchell T.M. 'Machine learning: trends, perspectives, and prospects'. *Science*. 2015;**349**(6245):255–60.

[15] Chen B., Matthews P.C., Tavner P.J. 'Wind turbine pitch faults prognosis using a-priori knowledge-based ANFIS'. *Expert Systems with Applications*. 2013;**40**(17):6863–76.

[16] Chen B., Matthews P.C., Tavner P.J. 'Automated on-line fault prognosis for wind turbine pitch systems using supervisory control and data acquisition'. *IET Renewable Power Generation*. 2015;**9**(5):503–13.

[17] Kusiak A., Li W. 'The prediction and diagnosis of wind turbine faults'. *Renewable Energy*. 2011;**36**(1):16–23.

[18] Zhu J., Yoon J.M., He D., *et al.* 'Online particle-contaminated lubrication oil condition monitoring and remaining useful life prediction for wind turbines'. *Wind Energy*. 2015;**18**(6):1131–49.

[19] Li C., Sanchez R.-V., Zurita G., Cerrada M., Cabrera D., Vásquez R.E. 'Gearbox fault diagnosis based on deep random forest fusion of acoustic and vibratory signals'. *Mechanical Systems and Signal Processing*. 2016;**76-77**(1):283–93.

[20] Schlechtingen M., Ferreira Santos I. 'Comparative analysis of neural network and regression based condition monitoring approaches for wind turbine fault detection'. *Mechanical Systems and Signal Processing*. 2011;**25**(5):1849–75.

[21] Verma A., Kusiak A. 'Fault monitoring of wind turbine generator brushes: a data-mining approach'. *Journal of Solar Energy Engineering.* 2012;**134**(2):021001.
[22] Butler S. 'A feasibility study into prognostics for the main bearing of a wind turbine'. *IEEE International Conference on Control Applications. IEEE*; 2012.
[23] Bechhoefer E., Schlanbusch R. 'Generalized prognostics algorithm using Kalman smoother'. *IFAC-PapersOnLine.* 2015;**48**(21):97–104.
[24] Regan T., Beale C., Inalpolat M. 'Wind turbine blade damage detection using supervised machine learning algorithms'. *Journal of vibration and acoustics.* August 2, 2017;**139**(6):061010.
[25] Dervilis N., Choi M., Taylor S.G., *et al.* 'On damage diagnosis for a wind turbine blade using pattern recognition'. *Journal of Sound and Vibration.* 2014;**333**(6):1833–50.
[26] Siegel D., Zhao W., Lapira E., AbuAli M., Lee J. 'A comparative study on vibration-based condition monitoring algorithms for wind turbine drive trains'. *Wind Energy.* 2014;**17**(5):695–714.
[27] Tang B., Song T., Li F., *et al.* 'Fault diagnosis for a wind turbine transmission system based on manifold learning and Shannon wavelet support vector machine'. *Renewable Energy.* 2014;**62**(13):1–9.
[28] Garlick W., Dixon R., Watson S. *A model-based approach to wind turbine condition monitoring using SCADA data.* Loughborough University. Conference contribution; 2009. Available from https://hdl.handle.net/2134/5488.
[29] Zaher A., McArthur S.D.J., Infield D.G., Patel Y. 'Online wind turbine fault detection through automated SCADA data analysis'. *Wind Energy.* 2009;**12**(6):574–93.
[30] Feng Y., Qiu Y., Crabtree C.J., *et al.* 'Monitoring wind turbine gearboxes'. *Wind Energy.* 2013;**16**(5):728–40.
[31] Astolfi D., Castellani F., Terzi L. 'Fault prevention and diagnosis through SCADA temperature data analysis of an onshore wind farm'. *Diagnostyka.* 2014;**15**.
[32] Qiu Y., Feng Y., Sun J., *et al.* 'Applying thermophysics for wind turbine drivetrain fault diagnosis using SCADA data'. *IET Renewable Power Generation.* 2016;**10**(5):661–8.
[33] Gray C.S., Watson S.J. 'Physics of failure approach to wind turbine condition based maintenance'. *Wind Energy.* 2010;**13**(5):395–405.
[34] Catmull S. *Self-Organising Map Based Condition Monitoring of Wind Turbines. EWEA 2011 - 14-17 March 2011.* Brussels, Belgium: European Wind Energy Association; 2015.
[35] Reder M.D., Gonzalez E., Melero J.J. 'Wind turbine failures - tackling current problems in failure data analysis'. *Journal of Physics: Conference Series.* 2016;**753**:072027.
[36] Pliego Marugán A., García Márquez F.P. 'Advanced analytics for detection and diagnosis of false alarms and faults: a real case study'. *Wind Energy.* 2019;**22**(11):1622–35.

[37] Chen B., Tavner P.J., Feng Y., Song W.W., Qiu Y.N. *Bayesian Network for Wind Turbine Fault Diagnosis EWEA 2012*. Copenhagen, Denmark: European Wind Energy Association; 2012.
[38] Jia X., Jin C., Buzza M., *et al.* 'Wind turbine performance degradation assessment based on a novel similarity metric for machine performance curves'. *Renewable Energy*. 2016;**99**:1191–201.
[39] Lapira E., Brisset D., Davari Ardakani H., *et al.* 'Wind turbine performance assessment using multi-regime modeling approach'. *Renewable Energy*. 2012;**45**(3):86–95.
[40] Wang Y., Hu Q., Li L., *et al.* 'Approaches to wind power curve modeling: a review and discussion'. *Renewable and Sustainable Energy Reviews*. 2019;**116**(9):109422.
[41] El-Asha S., Zhan L., Iungo G.V. 'Quantification of power losses due to wind turbine wake interactions through SCADA, meteorological and wind LiDAR data'. *Wind Energy*. 2017;**20**(11):1823–39.
[42] Davis N.N., Pinson P., Hahmann A.N., *et al.* 'Identifying and characterizing the impact of turbine icing on wind farm power generation'. *Wind Energy*. 2016;**19**(8):1503–18.
[43] Liu Y., Cheng H., Kong X., *et al.* 'Intelligent wind turbine blade icing detection using supervisory control and data acquisition data and ensemble deep learning'. *Energy Science & Engineering*. 2019;**7**(6):2633–45.
[44] Pandit R.K., Infield D. 'SCADA-based wind turbine anomaly detection using Gaussian process models for wind turbine condition monitoring purposes'. *IET Renewable Power Generation*. 2018;**12**(11):1249–55.
[45] Castellani F., Astolfi D., Terzi L., Hansen K.S., Rodrigo J.S. 'Analysing wind farm efficiency on complex terrains'. *Journal of Physics: Conference Series*. 2014;**524**:012142.
[46] Soman S.S., Zareipour H., Malik O., Mandal P. 'A review of wind power and wind speed forecasting methods with different time horizons'. *North American Power Symposium*. 2010:1–8.
[47] Sinha Y., Steel J.A. 'A progressive study into offshore wind farm maintenance optimisation using risk based failure analysis'. *Renewable and Sustainable Energy Reviews*. 2015;**42**(204):735–42.
[48] Blanco M.I. 'The economics of wind energy'. *Renewable and Sustainable Energy Reviews*. 2009;**13**(6-7):1372–82.
[49] Feng J. 'An intelligent system for offshore wind farm maintenance scheduling optimization considering turbine production loss'. *Journal of Intelligent & Fuzzy Systems*. 2020:1–13.
[50] Zhang J. 'Optimal preventive maintenance time windows for offshore wind farms subject to wake losses'. *12th AIAA Aviation Technology, Integration, and Operations (ATIO) Conference and 14th AIAA/ISSM*; Indianapolis, IN; 2012.
[51] Perez-Canto S., Rubio-Romero J.C. 'A model for the preventive maintenance scheduling of power plants including wind farms'. *Reliability Engineering & System Safety*. 2013;**119**(4):67–75.

[52] Dai L., Stålhane M., Utne I.B. 'Routing and scheduling of maintenance fleet for offshore wind farms'. *Wind Engineering*. 2015;**39**(1):15–30.
[53] Irawan C.A., Ouelhadj D., Jones D., *et al*. 'Optimisation of maintenance routing and scheduling for offshore wind farms'. *European Journal of Operational Research*. 2017;**256**(1):76–89.
[54] Kovács A., Erdős G., Viharos Z.J., *et al*. 'A system for the detailed scheduling of wind farm maintenance'. *CIRP Annals*. 2011;**60**(1):497–501.
[55] Kovács A., Erdös G., Monostori L., *et al*. 'Scheduling the maintenance of wind farms for minimizing production loss'. *IFAC Proceedings Volumes*. 2011;**44**(1):14802–7.
[56] Lee J. *Industrial AI*. Singapore: Springer Books; 2020.

Chapter 7

Innovative control strategies for wind turbines

Yin Minghui[1], *Chen Zaiyu*[1], *Zhou Lianjun*[1], *Xu Yan*[2], *and Zou Yun*[1]

When operating below its rated wind speed, a variable-speed wind turbine (VSWT) often uses maximum power point tracking (MPPT) to maximize wind energy extraction. However, as wind resources in regions with high wind speeds have been widely exploited, energy extraction from VSWTs is increasingly challenged by unfavorable wind conditions, such as low wind speeds and strong turbulence. Meanwhile, the inertia of wind turbine rotors increases with the scale of wind power generation systems. Consequently, the divergence between turbulent wind speeds, which hinder MPPT, and slow dynamics of the large-scale VSWT rotors becomes prominent, leading to substantial tracking loss [1–3] and even tracking invalidity [4, 5]. The effects of turbulence on wind energy extraction are analyzed in this chapter to address this divergence [6, 7]. Then, a tracking range reduction method is proposed to improve the wind energy extraction efficiency, considering the slow dynamics of wind turbine rotors [8]. This method mainly modifies the reference input for MPPT control. In the wind turbine control and control theory in general, enhancing the control performance of a system depends on improved control strategies. Nevertheless, few studies have focused on modifying the reference for changing the MPPT target and improving the control performance. Based on the concept of tracking range reduction, the effective tracking range (ETR) can be determined and applied to the conventional optimal torque (OT) control for increasing wind energy extraction [9].

7.1 Wind turbine model for MPPT design

Figure 7.1 illustrates a VSWT comprising a mechanical system and an electrical system. The former contains a wind turbine rotor, generator rotor, and gearbox, which is excluded in direct-drive wind turbines. The latter contains generator windings and

[1]School of Automation, Nanjing University of Science and Technology, Nanjing, China
[2]School of Electrical and Information Engineering, Changsha University of Science and Technology, Changsha, China

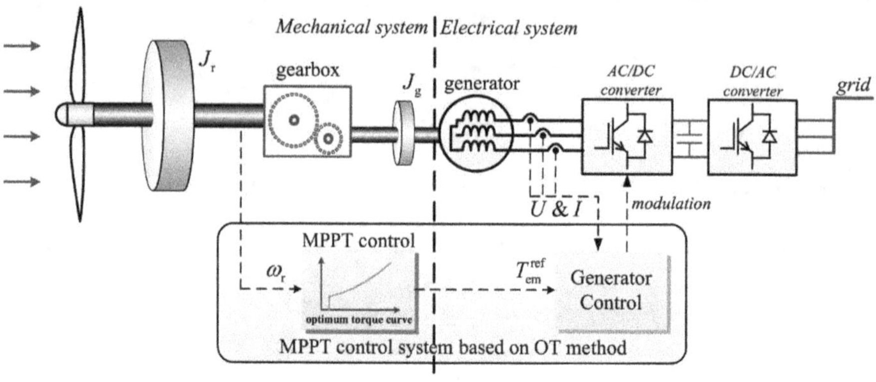

Figure 7.1 Block Diagram of Wind Turbine Model and MPPT System

power converters. Therefore, the mechanical and electromagnetic dynamics should be simultaneously considered when modeling a VSWT.

As the electromagnetic response [10] is generally much faster than the mechanical response [11], the two systems can be decoupled to separately model the mechanical and electromagnetic dynamics. This chapter focuses on the mechanical dynamics of wind turbines. Thus, the generator and converter are not modeled, and the generator is assumed to instantly follow the torque reference requested by MPPT.

7.1.1 Wind turbine rotor aerodynamics

According to Betz's law [12], the aerodynamic power captured by a wind turbine can be expressed as:

$$P_m = 0.5\rho\pi R^2 v^3 C_P(\lambda, \beta) \tag{7.1}$$

where ρ is the air density, R is the radius of the wind turbine rotor, v is the wind speed, and power coefficient C_P is a function of blade pitch angle β and tip speed ratio (TSR) λ, which is given by:

$$\lambda = \omega_r R/v \tag{7.2}$$

where ω_r is the speed of the wind turbine rotor. For any pitch angle β, there exists a unique optimal TSR, denoted by λ_{opt}, which yields maximum power coefficient C_P^{max}.

The aerodynamic torque produced by the wind turbine can be expressed as:

$$T_m = P_m/\omega_r \tag{7.3}$$

In variable-speed MPPT, pitch angle β remains fixed, and thus power coefficient C_P is a function of TSR λ. The C_P-λ curve is determined by the aerodynamic characteristics of the wind turbine. A typical C_P-λ curve is shown in Figure 7.2 [13]. When the TSR reaches optimum λ_{opt}, the maximum power coefficient C_P^{max} is yielded. Usually, the aerodynamic performance of a wind turbine can be represented by the C_P-λ curve.

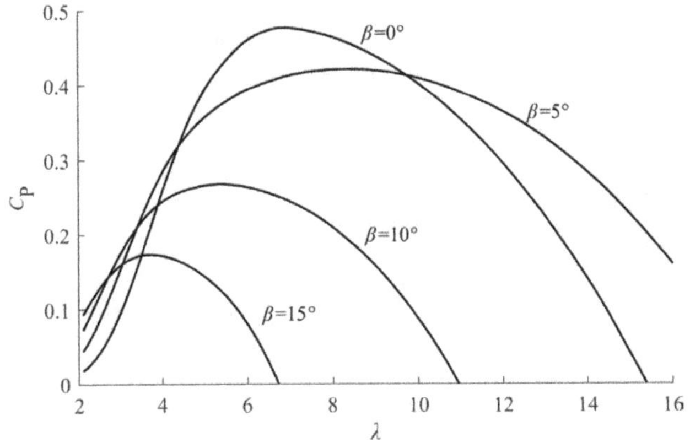

Figure 7.2 Typical C_P-λ Curve

7.1.2 Wind turbine mechanical dynamics

The mechanical system of a wind turbine (Figure 7.1) consists of a wind turbine rotor with large inertia J_r and a generator rotor with small inertia J_g. In the model, the rotors are connected by an ideal gearbox with a gear ratio n_g. The mechanical dynamics of a wind turbine can be described by the following two-mass model [14]:

$$\begin{cases} J_r\dot{\omega}_r = T_m - D_r\omega_r - T_{ls} \\ J_g\dot{\omega}_g = T_{hs} - D_g\omega_g - T_{em} \\ n_g = \omega_g/\omega_{ls} = T_{ls}/T_{hs} \end{cases} \tag{7.4}$$

where ω_g is the generator speed, ω_{ls} is the low-speed shaft speed, T_{ls} and T_{hs} are the low- and high-speed shaft torques, respectively, T_{em} is the electromagnetic torque of the generator, and D_r and D_g are the external damping values of the rotor and generator, respectively.

Assuming a perfectly rigid drivetrain, the mechanical dynamics of a wind turbine can be further represented as a lumped-mass model:

$$J_t\dot{\omega}_r = T_m - D_r\omega_r - T_g \tag{7.5}$$

where $J_t = J_r + n_g^2 J_g$, $D_t = D_r + n_g^2 D_g$, and $T_g = n_g T_{em}$. Additionally, damping can be ignored for simplicity.

7.1.3 MPPT

MPPT aims to maximize wind energy harvesting using wind turbines. As optimal TSR λ_{opt} corresponds to maximum power coefficient C_P^{max}, ensuring that the wind turbine always operates at the optimal TSR under varying wind speed allows maximizing wind energy extraction. The operating point at λ_{opt} and C_P^{max} is called the maximum power point, and maximizing wind energy extraction can be formulated as a problem of rotor speed tracking. Specifically, the rotor speed can be regulated

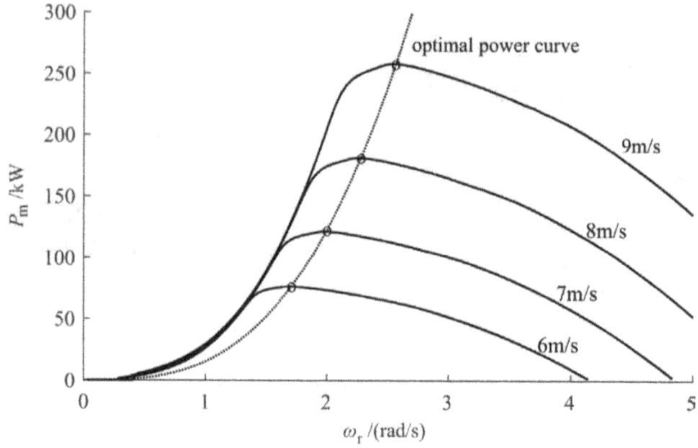

Figure 7.3 Maximum Power Points (circles) at Different Wind Speeds and Optimal Power Curve (Dotted Line)

by torque control to track the optimal speed, which is determined by the instantaneous wind speed [15]:

$$\omega_{opt} = v\lambda_{opt}/R \tag{7.6}$$

According to (7.6), the maximum power point (i.e., optimal rotor speed ω_{opt}) varies according to the wind speed. If all maximum power points are connected, a curve of maximum aerodynamic power versus rotor speed (called optimal power curve for short) can be obtained, as illustrated by the dotted line in Figure 7.3 [13]. Therefore, under wind speed fluctuations, MPPT aims to maintain the wind turbine operating along this curve to maximize the energy extraction efficiency by achieving C_P^{max}.

Several MPPT strategies, including OT control [16–18], TSR control [15, 19, 20], and perturbation and observation control [16, 21] (i.e., hill-climb searching method), have been proposed. Both the OT and TSR control strategies are widely used in megawatt-scale wind turbines with large inertia.

7.1.3.1 OT control

According to (7.1) and (7.6), the optimal power curve of a wind turbine is given by:

$$P_m^{opt}(\omega_r) = K_{opt}\omega_r^3, \quad \omega_r > \omega_r^{bgn} \tag{7.7}$$

where ω_r^{bgn} is the lowest rotor speed for power generation called starting speed and K_{opt} is the gain of the OT curve defined as

$$K_{opt} = \left(0.5\rho\pi R^5 C_P^{max}\right)/\lambda_{opt}^3 \tag{7.8}$$

The OT curve for the model given by (7.4) is derived as [14]

$$T_{em}^{opt}(\omega_r) = \frac{K_{opt}}{n_g}\omega_r^2 - \left(n_g D_g + \frac{D_r}{n_g}\right)\omega_r, \quad \omega_r > \omega_r^{bgn} \tag{7.9}$$

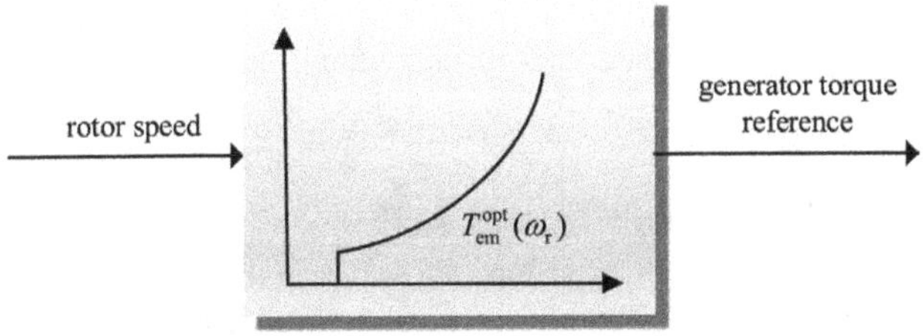

Figure 7.4 Block Diagram of OT Control

To achieve MPPT, OT control adjusts the generator torque along the OT curve according to the measured rotor speed ω_r, as illustrated in Figure 7.4.

7.1.3.2 TSR control

TSR control corresponds to rotor speed feedback control, as shown in Figure 7.5. The optimal rotor speed ω_{opt} determined according to the measured or estimated wind speed is regarded as the reference input [22]. Then, the deviation between the feedback and optimal rotor speed allows adjusting the feedback error to zero by the generator torque. The wind turbine tracks the optimal rotor speed to operate at the maximum power point. Although the principle of TSR control is simple, its dependence on real-time wind speed information hinders its applicability.

7.2 MPPT under turbulence

Turbulence challenges MPPT in large-scale VSWTs, and rapid wind speed fluctuations further hinder MPPT. Moreover, the growing length of wind turbine blades increases the rotor inertia and slows down the dynamic behavior of the wind turbine. Unlike high wind speed with weak turbulence, low wind speed with strong turbulence compromises MPPT in large-scale VSWTs, thus considerably deteriorating the wind energy conversion efficiency. In this section, the tracking loss and tracking

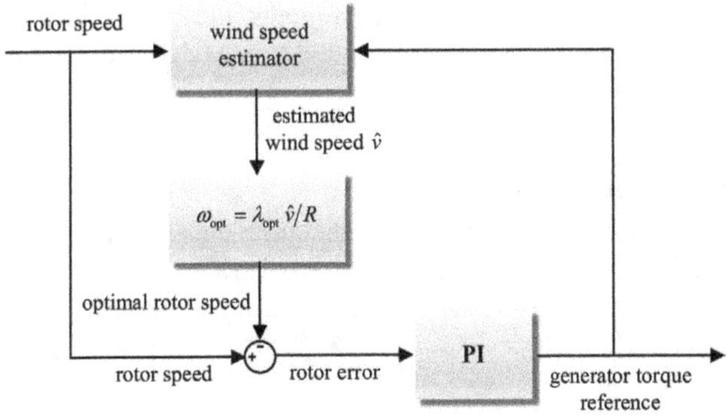

Figure 7.5 Block Diagram of TSR Control

invalidity due to the slow dynamics of large-scale VSWTs under turbulence are considered to provide the foundations for analyzing the influence of turbulence on the MPPT performance.

7.2.1 Turbulence characteristics

Turbulent wind refers to the fluctuation of wind speed in periods usually ranging from several minutes to tens of minutes. It can be modeled by the superposition of the mean wind speed and turbulence [23]:

$$v = \bar{v} + v_t \tag{7.10}$$

where mean wind speed $\bar{v}$ remains constant over the considered period [18, 24]. The turbulence component v_t represents the variation of wind speed around the mean value and can be considered a zero-mean stationary random process [25, 26]. The statistical description of turbulence characteristics mainly includes the mean wind speed, turbulence intensity (TI), integral length scale, power spectral density, and turbulence frequency [18, 25, 27].

The generation of turbulent wind is a complex process. Nevertheless, the time series in minutes of turbulent wind can be established by an autoregressive moving-average model [25, 28] and filtering [29]. Professional simulation software, such as Bladed [30] and TurbSim [31], have been developed to generate the time series of turbulent wind tailored to specific mean wind speed, TI, integral length scale, and power spectral density. These software packages not only ensure that the simulated wind profiles satisfy standard specifications but also simplify the simulation of turbulent wind.

7.2.1.1 Mean wind speed

The mean wind speed indicates the turbulence strength. If the turbulent wind speed over a period T is $v(t)$, mean wind speed $\bar{v}$ can be defined as

$$\bar{v} = \frac{1}{T}\int_0^T v(t)\, dt \tag{7.11}$$

In periods from several minutes to tens of minutes, the mean wind speed variation is small. Thus the mean wind speed is generally considered to be constant within such periods.

7.2.1.2 Turbulence intensity

The TI represents the fluctuation degree of the turbulence component relative to the mean wind speed. It can describe the wind speed distribution around the mean value, as larger turbulence indicates broader wind speed fluctuation. The TI is defined as [32]

$$\mathrm{TI} = \frac{\sigma}{\bar{v}} \tag{7.12}$$

where σ is the standard deviation of the wind speed series over a period:

$$\sigma = \sqrt{\frac{1}{T}\int_0^T [v(t) - \bar{v}]^2 \, dt} \tag{7.13}$$

The TI is related to atmospheric stability, terrain clearance, and terrain roughness [18, 33]. According to standard IEC-61400-1 [32], TI can be divided into three levels: higher (A), medium (B), and lower (C). The TI corresponding to the turbulence level is determined by [32]

$$\text{TI} = \frac{I_{\text{ref}}(0.75\bar{v}+5.6)}{\bar{v}} \tag{7.14}$$

where I_{ref} is 0.16, 0.14, and 0.12 for levels A, B, and C, respectively.

7.2.1.3 Integral length scale

Integral length scale L is a measure of the average size of a turbulent eddy in turbulent wind. It is related to the spatial correlation of the turbulence. Although the movement of the atmosphere is complex in the atmospheric boundary layer, the wind speed shows some regularity in specific time and space up to a certain height above ground. In fact, the wind characteristics at two points within the same eddy are usually similar. The integral length scale of the turbulence is also closely related to the terrain clearance and roughness, being an attenuation function of terrain roughness and increasing with height [33]. The value of the integral length scale generally ranges from 50 to 500 m [24, 34, 35].

7.2.1.4 Power spectral density

The power spectral density of turbulence describes the mean power according to frequency. The change in wind speed can be considered the superposition of various frequency components [24, 33], with low-frequency components being usually larger than the high-frequency ones. The power spectral density is a function of the mean wind speed, turbulence standard deviation, and integral length scale. Two representations are recommendable for the power spectral density, namely, the von Karman spectrum and Kaimal spectrum [36]. The von Karman spectrum at frequency f is expressed as [37]

$$S(f) = \frac{4\sigma^2 L/\bar{v}}{\left(1 + 70.8\left(fL/\bar{v}\right)^2\right)^{5/6}} \tag{7.15}$$

The Kaimal spectrum has a unilateral form defined as [38]

$$S(f) = \frac{4\sigma^2 L/\bar{v}}{\left(1 + 6fL/\bar{v}\right)^{5/3}} \tag{7.16}$$

The von Karman spectrum is theoretically derived using physics and hydrodynamics and has been confirmed on wind tunnel tests. Given its simple analytical expression, it suitably describes turbulence in a wind tunnel, but it may be unrealistic in the atmosphere [39]. The Kaimal spectrum is an empirical description based on full observations of the atmosphere and conforms to atmospheric turbulence in nature [40, 41]. Thus, the Kaimal spectrum is commonly used in wind energy research to describe the power spectral density of turbulence [42].

7.2.1.5 Turbulence frequency

As turbulent wind has various frequency components, the frequency levels among different wind profiles are difficult to distinguish [43]. Nevertheless, wind energy generation is generally concentrated at low frequencies in the wind spectrum. However, the power spectrum represented as a curve is not an adequate quantitative index to describe the frequency characteristics of turbulent wind, unlike single-valued statistical indices such as the mean wind speed, TI, and integral length scale. Thus, a single-valued index to represent the turbulence frequency is necessary to study its effects on the MPPT performance of wind turbines. Equivalent turbulence frequency ω_{eff} has been proposed to describe the frequency characteristics of turbulence [27]:

$$\omega_{\text{eff}} = \frac{1}{\sigma}\left|\frac{\Delta v}{\Delta t}\right|_{\text{rms}} = \frac{1}{\sigma}\sqrt{\frac{1}{N}\sum\left(\left|\frac{\Delta v}{\Delta t}\right|\right)^2} \tag{7.17}$$

where Δv is the deviation of the wind speed during a sampling period, and N is the number of periods Δt within a statistical period.

7.2.2 Turbine tracking loss

With the growing capacity of wind turbines, their rotor inertia increases dramatically. The OT curve given by (7.9) shows that OT control is based on the steady states of wind turbines. Thus, OT control neglects the slow mechanical dynamics of wind turbines and corresponding MPPT dynamics, because only the steady-state operation points are considered according to the wind speed. The inability of VSWTs to accelerate or decelerate instantly in response to wind speed fluctuations reduces wind energy extraction. In addition, the actual power coefficient curve differs from the ideal one, as demonstrated through experiments [1].

As the optimal rotor speed for MPPT depends on the wind speed, fluctuations cause continuous random variations of the tracking reference. Under strongly turbulent wind with high-frequency fluctuations, the highly variable tracking reference can hardly be tracked. Meanwhile, large-scale wind energy exploitation, especially in low-wind areas, has substantially increased the inertia of the wind turbine rotors, consequently undermining dynamic performance. Severe wind-speed fluctuations and slow dynamic response of VSWTs constitute an irreconcilable problem, and wind turbines are subject to MPPT under wind fluctuations most of the time, generally impeding operation at the maximum power point, thereby leading to tracking losses.

The turbine tracking loss of wind energy is caused by the MPPT deviation from the maximum power point. Research on mechanisms has been conducted and improved early in the twenty-first century to mitigate this loss. Tracking loss

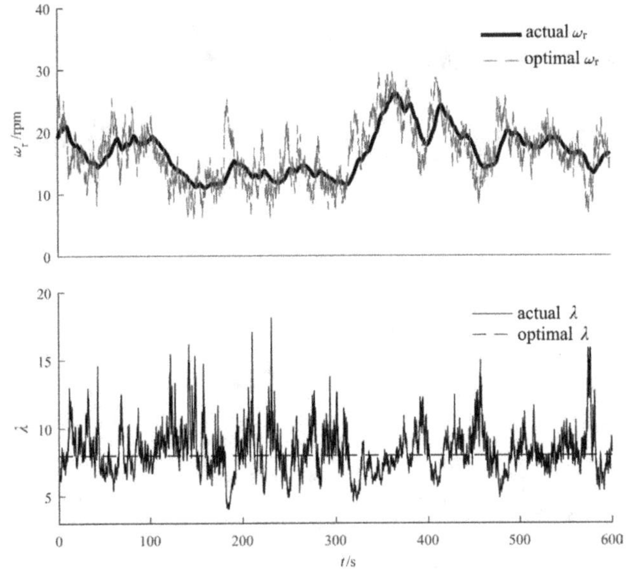

Figure 7.6 Turbine Tracking Loss

originates from the slow dynamics of wind turbines with large inertia, impeding the proper response to wind speed variations.

Figure 7.6 shows MPPT simulation results of wind turbines with large inertia under the turbulent wind. The speed trajectory shows that the rotor speed cannot reach the optimal value most of the time, as indicated by the deviation of the TSR from its optimal value. Therefore, it is difficult for VSWTs to achieve maximum wind energy extraction.

To verify the MPPT performance, wind energy extraction efficiency P_{favg} is defined as a closed-loop performance index of wind turbine operation under turbulent wind conditions [44, 45]. It can be approximated as the ratio of extracted wind power to available power as follows:

$$\begin{cases} P_{favg} = \sum_{i=1}^{n} P_{cap}(i)\,\Delta t / \sum_{i=1}^{n} P_{wy}(i)\,\Delta t \\ P_{cap} = T_g\omega_r + J_t\omega_r\dot{\omega}_r \\ P_{wy} = 0.5\rho\pi R^2 v^3 cos^3\psi \end{cases} \tag{7.18}$$

where n is the number of simulation steps, Δt is the sampling time, P_{cap} is the captured wind power, P_{wy} is the inflow wind power, and ψ is the yaw angle, which can be neglected by assuming its best value of 0°.

7.2.3 Turbine tracking invalidity

Turbine tracking invalidity is a more severe problem of tracking loss, because it renders VSWTs unable to operate at the maximum power point and even lose

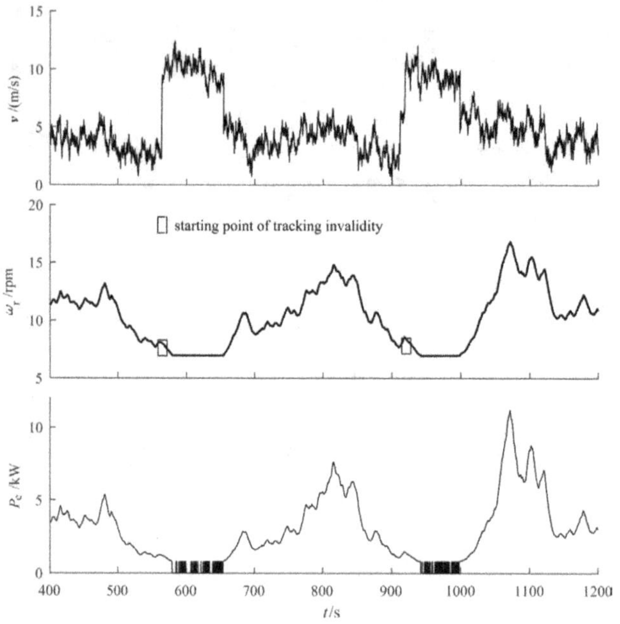

Figure 7.7 Behavior During Turbine Tracking Invalidity [4]

the rotor speed tracking ability in the worst case [2]. An experimental study on generation at low wind speed has shown that rapid wind speed fluctuations may lead to turbine tracking invalidity. During turbine tracking invalidity, the rotor speed cannot track fluctuations that make the wind speed increase rapidly, and the rotor speed continuously falls to its starting value, as shown in Figure 7.7. When the wind speed increases suddenly, the rotor speed cannot be increased by MPPT, greatly reducing the wind energy extraction efficiency.

Figure 7.8 illustrates the occurrence of tracking invalidity [5]. The relation $T_m < T_g$ holds when the VSWT tracks wind gusts that decelerate the rotor. In the figure, the two solid lines represent the aerodynamic torque under wind speeds v_1 and v_2, and the dashed line represents the electromagnetic torque curve. Consider that the wind turbine operates at the optimal rotor speed for v_1 (point A). When the wind speed suddenly rises to v_2, T_g remains at point A because the rotor speed cannot vary abruptly, but T_m changes to point B. Thus, $T_m < T_g$, and the rotor continuously falls to its starting speed instead of accelerating to the optimal rotor speed for v_2 (point C).

The stable and invalid regions for MPPT can be obtained by stability analysis [4]. The driving torque of the wind turbine can be expressed as

$$T_m = \frac{\rho \pi R^5 C_P(\lambda)\,\omega_r^2}{2\lambda^3} \tag{7.19}$$

In addition, the unbalanced torque of the wind turbine with MPPT is calculated as

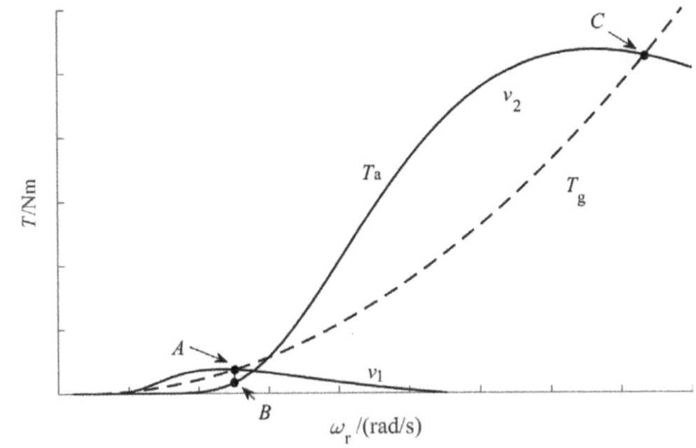

Figure 7.8 Example of Tracking Invalidity

$$\Delta T_{\mathrm{MPPT}}=\frac{1}{2}\rho\pi R^5\omega_{\mathrm{r}}^2\left(\frac{C_{\mathrm{P}}(\lambda)}{\lambda^3}-\frac{C_{\mathrm{P}}^{\max}}{\lambda_{\mathrm{opt}}^3}\right) \tag{7.20}$$

The $C_P(\lambda)/\lambda^3 - \lambda$ curve of a wind turbine is shown in Figure 7.9 [4]. Values λ_S and λ_U correspond to the balancing torque ($\Delta T_{MPPT} = 0$), where λ_S is equivalent to λ_{opt}, and λ_U is a very small TSR.

When either $\lambda = \lambda_S$ or $\lambda = \lambda_U$, $\Delta T_{MPPT} = 0$, and the rotor speed remains constant. When $\lambda_U < \lambda < \lambda_S$, $\Delta T_{MPPT} > 0$, and the rotor accelerates. When $\lambda > \lambda_S$ or $\lambda < \lambda_U$, $\Delta T_{MPPT} < 0$, and the rotor decelerates. Hence, there are two equilibrium points, ω_U and ω_S, corresponding to λ_U and λ_S under constant wind speed, respectively. Likewise, the acceleration and deceleration regions are $\omega_U < \omega_r < \omega_S$ and $\omega_r > \omega_S \cup \omega_r < \omega_U$, respectively.

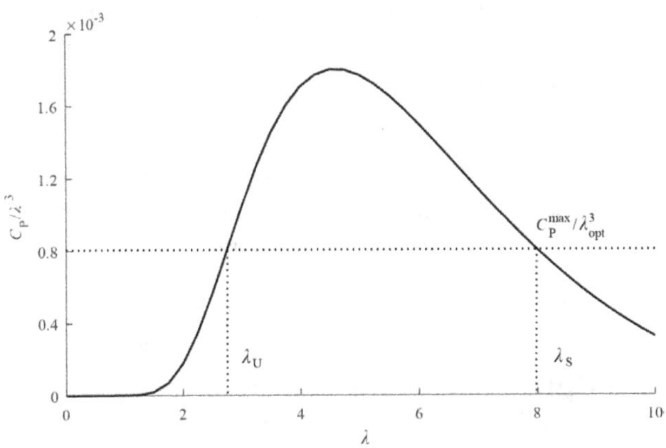

Figure 7.9 Determination of Points λ_U and λ_S [4]

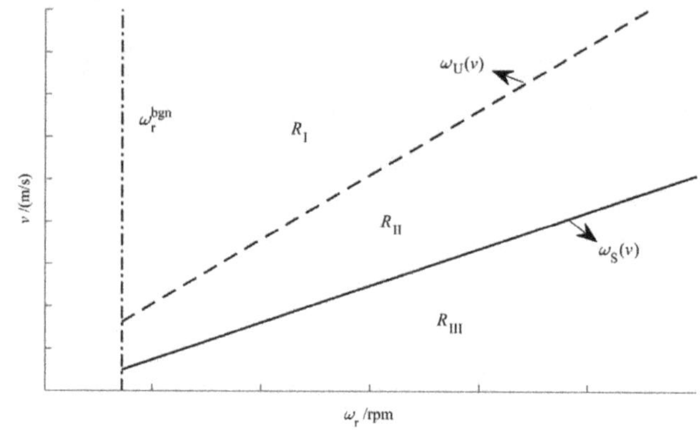

Figure 7.10 Distribution of Acceleration and Deceleration Regions on Rotor Speed–Wind Speed Plane [4]

If the wind turbine operates at the equilibrium point ω_S, accelerating turbulence decelerates the rotor, whereas decelerating turbulence accelerates the rotor. The rotor speed returns to ω_S after the turbulence. If the wind turbine operates at the equilibrium point ω_U, an accelerating turbulence makes the rotor accelerate continuously, whereas a decelerating turbulence makes the rotor decelerate to ω_r^{bgn} (i.e., $T_g = 0$). As the turbulent wind makes the rotor speed notably deviate from ω_U, ω_S is a stable equilibrium point, whereas ω_U is an unstable equilibrium point. Under constant wind speed, the domain of attraction for ω_S is $(\omega_U, \omega_S] \cup (\omega_S, \infty)$, and the rotor speed within this region converges to ω_S. Clearly, MPPT is valid in this region. However, in the region $(0, \omega_U)$, the rotor speed cannot converge to ω_S, thus leading to tracking invalidity.

Further considering the fluctuation of wind speed, the equilibrium points and distribution of the acceleration and deceleration regions of the wind turbine can be analyzed on the $\omega_r - v$ plane. The solid line in Figure 7.10 [4], representing $\omega_S(v) = \left\{(\omega_r, v)\,|\omega_r = \lambda_S v/R \cap \omega_r > \omega_r^{bgn}\right\}$ comprises a set of stable equilibrium points ω_S corresponding to different wind speeds. The points on $\omega_S(v)$ are related to λ_{opt}. The dashed line, representing $\omega_U(v) = \left\{(\omega_r, v)\,|\omega_r = \lambda_U v/R \cap \omega_r > \omega_r^{bgn}\right\}$, is a set of unstable equilibrium points ω_U corresponding to different wind speeds. The points on $\omega_U(v)$ are related to a very small TSR.

$R_{II} = \left\{(\omega_r, v)\mid \lambda_U v/R < \omega_r < \lambda_s v/R \cap \omega_r > \omega_r^{bgn}\right\}$ is an acceleration region of the wind turbine corresponding to $\lambda_U < \lambda < \lambda_S$, whereas $R_{III} = \left\{(\omega_r, v)|\omega_r > \lambda_s v/R \cap v > 0 \cap \omega_r > \omega_r^{bgn}\right\}$ is a deceleration region corresponding to $\lambda > \lambda_S$. The region $R_{II} \cup R_{III}$ is similar to the basin of the $\omega_S(v)$ region. If the wind turbine operates in the region $R_{II} \cup R_{III}$, its rotor speed approaches $\omega_S(v)$. Unless the wind speed fluctuates violently and causes wind turbines to enter the region R_I, the operation always occurs around $\omega_S(v)$ in the region $R_{II} \cup R_{III}$.

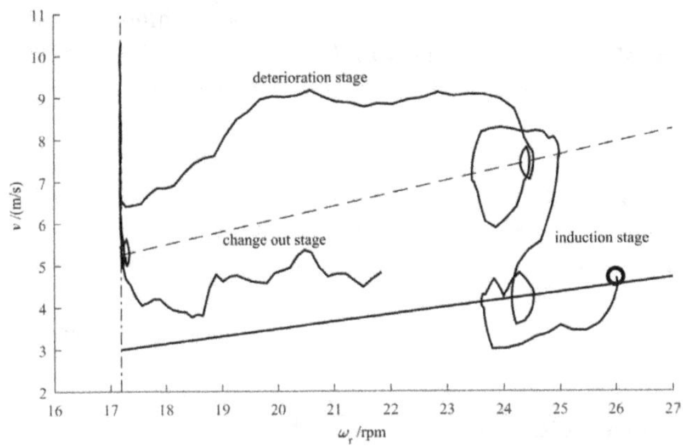

Figure 7.11 *Simulated Trajectory of Tracking in Validity on Rotor Speed–Wind Speed plane [4]*

$R_{\mathrm{I}} = \left\{(\omega_r, v)\,|\omega_r < \lambda_{\mathrm{U}} v/R \cap \omega_r > \omega_r^{\mathrm{bgn}}\right\}$ is a deceleration region of the wind turbine corresponding to $\lambda < \lambda_{\mathrm{U}}$. When operating in this region, the small TSR makes the wind turbine to enter a deep stall, which renders the aerodynamic torque smaller than the electromagnetic torque. The rotor decelerates, and its speed moves away from $\omega_{\mathrm{U}}(v)$, further deepening stall. Unless a decreasing wind speed makes the rotor speed return to the region R_{II}, the wind turbine cannot leave deep stall, and the rotor speed reduces to ω_r^{bgn}. As the wind energy extraction efficiency seriously deteriorates if the wind turbine operates in the region R_{I}, MPPT must occur in the region $R_{\mathrm{II}} \cup R_{\mathrm{III}}$. Otherwise, a stall may be continuous and deep, leading to tracking invalidity. However, the ability of the wind turbine rotor to accelerate and resume $\omega_{\mathrm{S}}(v)$ tracking depends on the wind speed, which is an uncontrollable factor.

To illustrate the entire process of tracking invalidity, the trajectory of wind turbine operation on the rotor speed–wind speed plane during a simulation is shown in Figure 7.11 [4], where the circle indicates the starting operation point. In this example, the trajectory corresponding to tracking invalidity can be divided into three stages:

1. Induction stage. Although the wind turbine accelerates in the region R_{II}, its large inertia weakens acceleration, and the increase in rotor speed cannot keep up with the wind speed. This causes the TSR of the wind turbine to gradually decrease during MPPT, eventually leading to operation in the region R_{I}.
2. Deterioration stage. After entering the region R_{I}, a rising or sustained wind speed impedes the wind turbine from accelerating to the optimal TSR. Instead, the rotor speed continuously drops to ω_{bgn} due to the cycle of deceleration and deep stall. Consequently, tracking invalidity persists, and the power coefficient is very low.

3. Recovery stage. Rising or sustained wind speeds hinder tracking invalidity recovery after entering the region R_{I}. Only a decrease in wind speed can allow the wind turbine to return to the region R_{II} and cease tracking invalidity.

During turbine tracking invalidity, the rotor speed cannot respond to rapidly increasing wind speeds given the slow turbine dynamics. The probability of tracking invalidity increases under strong turbulence, and the maximum loss of wind energy extraction can reach up to 10 percent [4]. Tracking invalidity can be prevented by reducing the deviation between the actual and optimal rotor speeds. Moreover, it can be mitigated with improved wind energy extraction and tracking strategies.

7.3 Effects of turbulence on MPPT performance

The mean wind speed, TI, and turbulence frequency are single-valued statistical indices reflecting three types of turbulence characteristics. The mean wind speed indicates the magnitude of a turbulent wind profile, whereas the TI describes the range of wind fluctuations, and the turbulence frequency reflects the rate of wind fluctuation.

In this section, the relations between wind energy extraction efficiency P_{favg} and these three indices are investigated via numerical simulations of a VSWT to analyze the influence of turbulence on the MPPT performance. The following considerations should be noted for this analysis:

1. When analyzing the relation between wind energy extraction efficiency and each index, the values of the other two indices are fixed.
2. As the turbulence frequency cannot be set before generating turbulent wind profiles, wind profiles with similar values of ω_{eff} are selected from several stochastically generated profiles, and the selected profiles are considered to have the same turbulence frequency.
3. Wind energy extraction efficiency P_{favg} and the ratio of time spent operating at maximum power point R_{h} are used to evaluate the MPPT performance. R_{h} is calculated as

$$R_{\mathrm{h}} = \frac{t_{\left(C_{\mathrm{p}}>90\%C_{\mathrm{p}}^{\max}\right)}}{t_{(total)}} \tag{7.21}$$

Given that the turbulence indices are statistical, P_{favg} and R_{h} are reported as their average values across 20 randomly selected wind profiles.

7.3.1 *Mean wind speed*

Figure 7.12 shows the positive correlation of P_{favg} and R_{h} with the mean wind speed. Hence, improving the dynamic response of VSWTs promotes operation near the maximum power point. Correspondingly, the wind energy extraction efficiency increases.

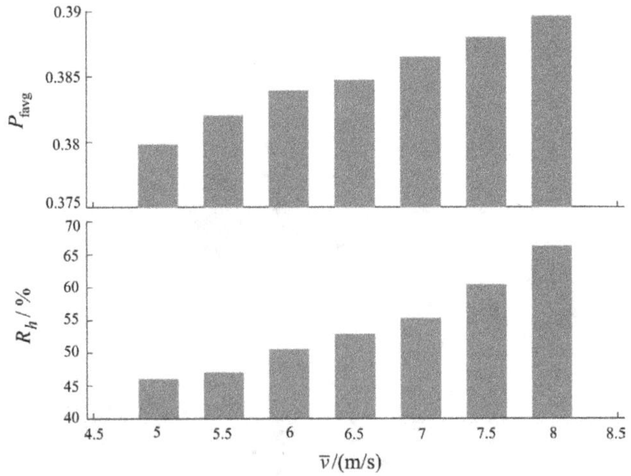

Figure 7.12 P_{favg} *and* R_h *according to mean wind speed*

Both P_{favg} and R_h improve with increasing mean wind speed because the driving torque on the wind turbine increases with the wind speed and enhances the wind turbine acceleration. This is conducive to timely tracking wind speed variations. Overall, increasing P_{favg} and R_h leads to high wind energy extraction.

7.3.2 Turbulence intensity

Figure 7.13 shows the negative correlation of P_{favg} and R_h with the TI.

As the TI measures the wind fluctuation, its increase reflects a wider variation of the wind speed and leads to more demanding tracking requirements for wind turbines. In fact, responding to rapid wind variations is difficult for VSWTs. Consequently, the divergence between the actual and optimal rotor speeds reduces both P_{favg} and R_h, undermining wind energy extraction.

7.3.3 Turbulence frequency

The effects of the turbulence frequency on wind energy extraction efficiency P_{favg} are more complex than those of the other factors on P_{favg}. In fact, the effects of turbulence frequency on P_{favg} also depend on the mean wind speed and TI. Therefore, univariate analysis is insufficient to determine the influence of turbulence frequency on the wind turbine characteristics.

The direct effect of the turbulence frequency on wind energy extraction efficiency P_{favg} and the ratio of time spent operating at the maximum power point R_h is first determined. Then, the combined effects of mean wind speed and TI for varying turbulence frequency on P_{favg} are analyzed. These indirect effects are verified through simulations considering periodic wind and turbulent wind.

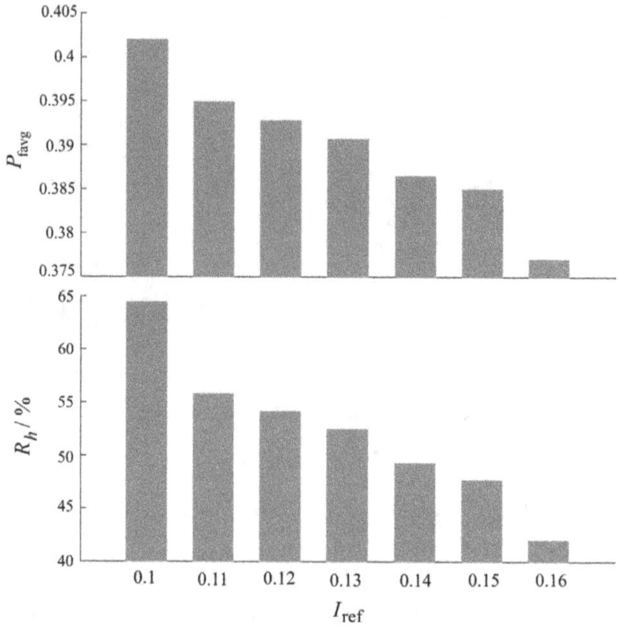

Figure 7.13 P_{favg} *and* R_h *According to TI*

7.3.3.1 Direct effects of turbulence frequency

Figure 7.14 shows the negative correlation of P_{favg} and R_h with the turbulence frequency. As higher turbulence frequency reflects more rapid wind fluctuation, the tracking difficulty increases. Hence, high turbulence frequency reduces both P_{favg} and R_h.

7.3.3.2 Indirect effects of turbulence frequency

The indirect effects of turbulence frequency on the MPPT performance reflect the sensitivity of wind energy extraction efficiency P_{favg} to the mean wind speed and TI. Hence, although the MPPT performance increases by high mean wind speed [45, 46] and low TI [4, 47, 48], it also depends on the turbulence frequency. Consider $S_{P\text{-}\bar{v}}$ to be the sensitivity of P_{favg} to the mean wind speed and $S_{P\text{-}TI}$ to be the sensitivity of P_{favg} to TI.

i. *Sensitivity of wind energy extraction efficiency to mean wind speed under varying turbulence frequency*

Sensitivity $S_{P\text{-}\bar{v}}$ can be defined as [6]:

$$S_{P\text{-}\bar{v}} = |\Delta P_{favg}/\Delta v_m| \tag{7.22}$$

where ΔP_{favg} and Δv_m are the variations of mean wind energy extraction efficiency and mean wind speed, respectively.

Figure 7.15 shows $S_{P\text{-}\bar{v}}$ according to wind frequency *f* of periodic wind. Although the VSWT acceleration improves at high mean wind speed, thus increasing efficiency

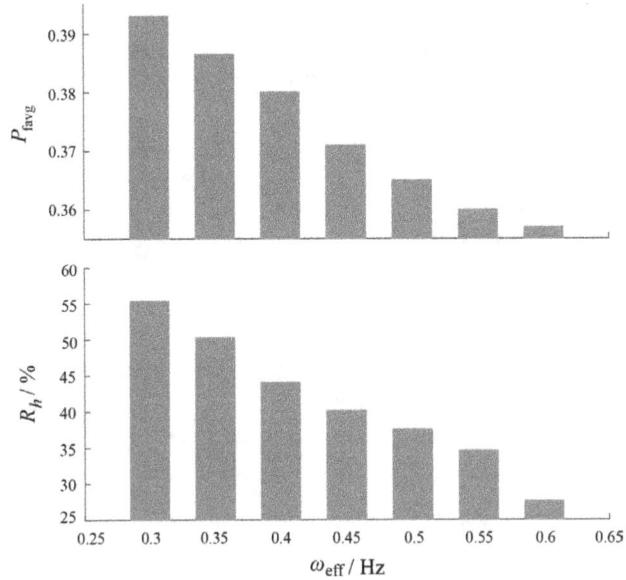

Figure 7.14 P_{favg} and R_h According to Turbulence Frequency

P_{favg} [47], the wind turbine cannot properly respond to the high-frequency variations in wind speed, as shown in Figure 7.16. In contrast, the wind turbine can suitably respond to low-frequency variations in wind speed. The unimodal-peak sensitivity curves in Figure 7.15 show this behavior.

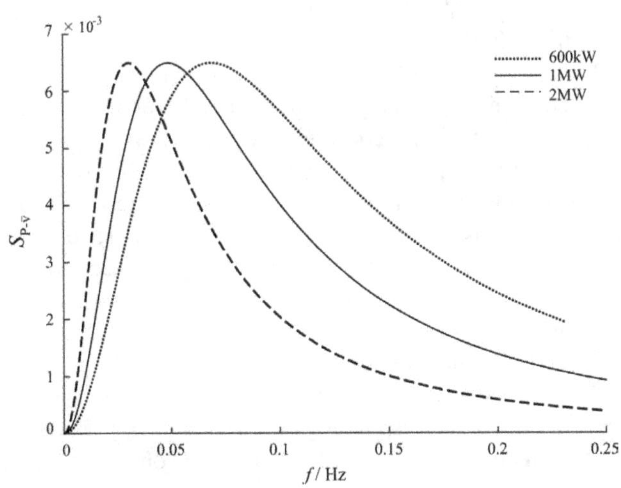

Figure 7.15 Sensitivity $S_{P\text{-}\bar{v}}$ According to Frequency f of Periodic Wind for Wind Turbines with Different Capacities

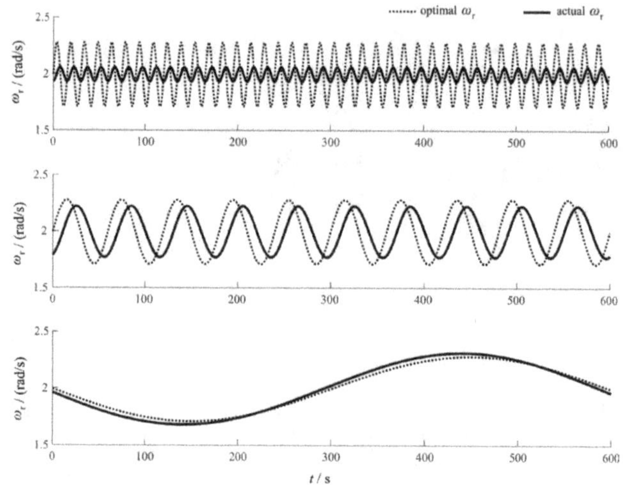

Figure 7.16 Wind Turbine Response to Wind Speed at Different Frequencies

A high mean wind speed improves the acceleration ability of a VSWT and thus the MPPT performance. As the wind frequency increases, the performance first increases until it reaches a peak, and then it decreases. In fact, the turbulent wind not only is the objective of MPPT but also provides excitation to the system. The VSWT is a system with high inertia that cannot properly respond to such excitation at very high frequencies. Hence, the wind turbine dynamics becomes insensitive to high-frequency excitation. The dynamic response becomes noticeable only as the excitation frequency approaches the frequency of the VSWT dynamic response. Therefore, the sensitivity peaks shift to lower frequencies as the wind turbine capacity increases, as shown in Figure 7.15.

To verify that the regularity of the indirect effects of turbulence frequency on the MPPT performance obtained by analyzing periodic wind applies in practice, a more realistic model of turbulent wind is considered to obtain the $S_{P\text{-}\bar{v}}$–ω_{eff} curve shown in Figure 7.17. These simulation results show that $S_{P\text{-}\bar{v}}$ intensifies with increasing ω_{eff}, consistent with the results obtained from the analysis considering the periodic wind.

To determine the reasons why the peak in Figure 7.15 does not occur in Figure 7.17, five wind conditions with $\bar{v}$ = 5 m/s and TI = 0.25 for L varying from 100 to 500 m in increments of 100 m are analyzed from 200 sequences of simulated turbulent wind profiles per condition. The mean wind energy extraction efficiency across the 200 sequences per condition is denoted as P_{favg}^{turb}, whose results are listed in Table 7.1. Sinusoidal wind speed profiles v_{sin}^{f} have frequency f and the same mean value and TI, and the corresponding mean wind energy extraction efficiency is denoted as $P_{\text{favg}}^{\text{sin}}$. Frequency f_{equ} is traversed to reach consensus between $P_{\text{favg}}^{\text{sin}}$ and $P_{\text{favg}}^{\text{turb}}$, and its results are also listed in Table 7.1.

For every integral length scale L, the energy extraction efficiency of the VSWT under turbulent wind is consistent with that under sinusoidal wind at frequencies of

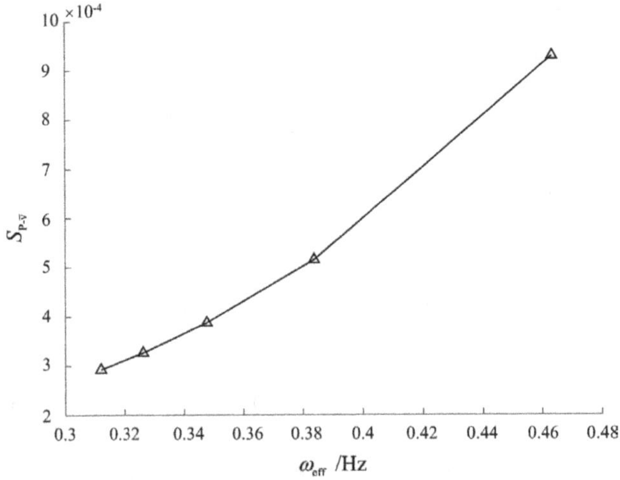

Figure 7.17 Sensitivity $S_{P\text{-}\bar{v}}$ According to Frequency ω_{eff} Under Turbulent Wind

0.008–0.0115 Hz, indicating that the frequency of turbulent wind is distributed in this range.

The $S_{P\text{-}\bar{v}}$–f curve obtained from the CART3 wind turbine model under sinusoidal wind is shown in Figure 7.18. Sensitivity $S_{P\text{-}\bar{v}}$ increases with the wind frequency in the range 0.008–0.0115 Hz, confirming the results in Figure 7.17.

ii. *Sensitivity of wind energy extraction efficiency to TI under varying turbulence frequency*

Sensitivity $S_{P\text{-}TI}$ can be defined as [6]:

$$S_{P\text{-}TI} = |\Delta P_{favg}/\Delta TI| \tag{7.23}$$

where ΔTI is the variation of TI.

Figure 7.19 shows $S_{P\text{-}TI}$ according to wind frequency f of periodic wind. Sensitivity $S_{P\text{-}TI}$ reduces with decreasing wind frequency. Hence, the negative effect of strong turbulence, which reduces the wind energy extraction efficiency, is mitigated at low wind frequency. In fact, the TI only represents the range of wind speed fluctuation, but MPPT depends on both the TI and the frequency. Therefore, the tracking difficulty caused by strong turbulence can be mitigated at low turbulence frequency.

Table 7.1 Frequency band distribution of turbulent wind

L (m)	**100**	**200**	**300**	**400**	**500**
P^{turb}_{favg}	0.4040	0.4092	0.4114	0.4172	0.4135
f_{equ}(Hz)	0.0115	0.0098	0.0088	0.0083	0.0080

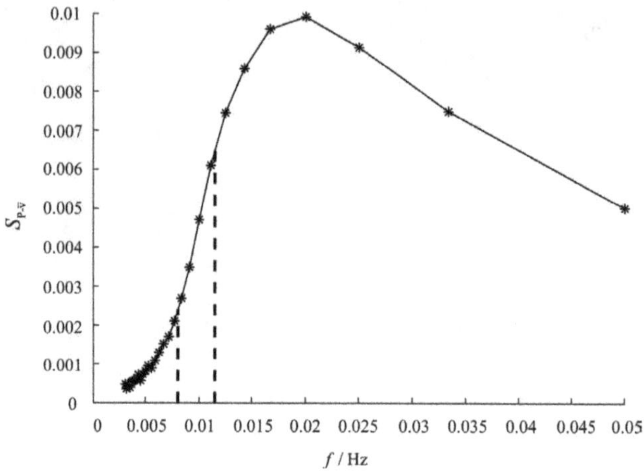

Figure 7.18 Sensitivity $S_{P\text{-}\bar{v}}$ According to Sinusoidal Wind at Frequency f

The $S_{P\text{-}TI}$–ω_{eff} curve to verify the trend in Figure 7.19 for the case of turbulent wind is shown in Figure 7.20. Sensitivity $S_{P\text{-}TI}$ increases for high turbulence frequency, being consistent with the results considering the periodic wind.

Under turbulent wind, the overall effects of mean wind speed and TI on MPPT become more noticeable as the turbulence frequency increases. Hence, the turbulence characteristics must be considered to design control strategies. Regardless of analyzing direct or indirect effects, the relation between the turbulence frequency and wind energy extraction efficiency should be considered for MPPT.

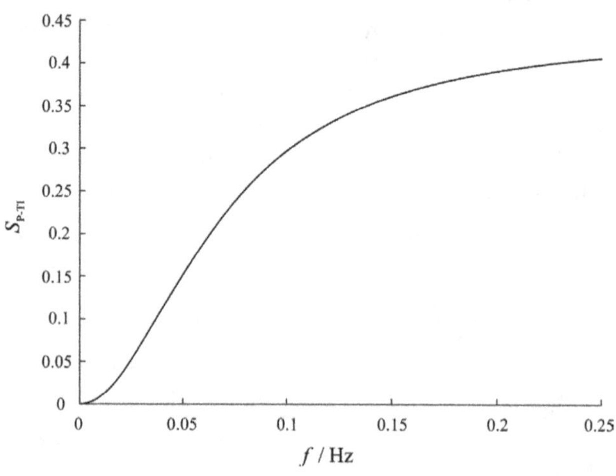

Figure 7.19 Sensitivity $S_{P\text{-}TI}$ According to Frequency f of Periodic Wind

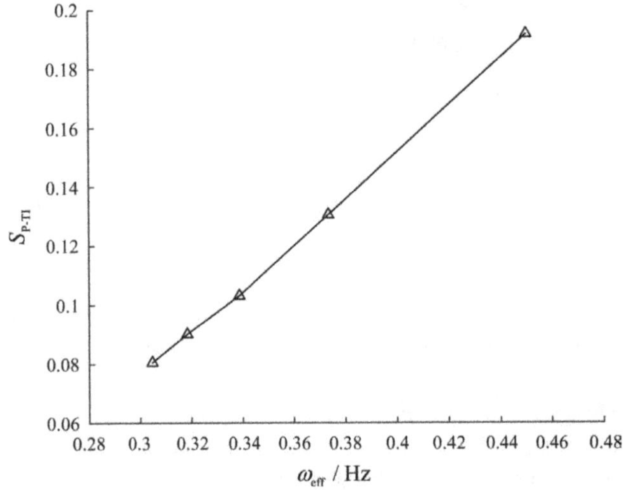

Figure 7.20 Sensitivity $S_{P\text{-}TI}$ According to Frequency ω_{eff} Under Turbulent Wind

7.4 MPPT reference input modification to increase wind energy extraction

Turbulent wind speed causing tracking loss has motivated studies on the improvement of control strategies. For instance, advanced control algorithms such as nonlinear control [14, 15], sliding mode control [49–52], and model predictive control [53–55] have been applied to MPPT. These methods enhance wind energy extraction by reducing the optimal speed tracking error by increasing the torque adjustment. However, a perfect C_P tracker would capture 3 percent more energy below the rated value by demanding huge power swings around three to four times the rated power, thus making it impractical and rendering these methods unsuitable for MPPT [56].

In this section, a novel method called reference input modification is proposed to improve the wind energy extraction efficiency. Unlike the minimization of optimal speed tracking error, the proposed method improves wind energy extraction by relaxing the tracking target (i.e., not tracking the optimal speed under certain amplitude or frequency characteristics of wind speed). As the modified tracking target is more appropriate for the slow dynamics of large-scale VSWTs, wind energy extraction can be improved. The main difference between the proposed and conventional methods is that the former aims to improve the tracking efficiency, while conventional methods aim to reduce the tracking error. If a VSWT rotor with large inertia fails to respond timely to rapid wind fluctuations, these objectives are not equivalent. Thus, MPPT should be focused on enhancing the tracking efficiency rather than adapting the rotor operation to the wind speed.

Decreased torque gain (DTG) control is similar to the proposed reference input modification, as it only modifies the speed–torque curve based on OT control, achieving higher tracking efficiency under wind gusts by partially omitting optimal

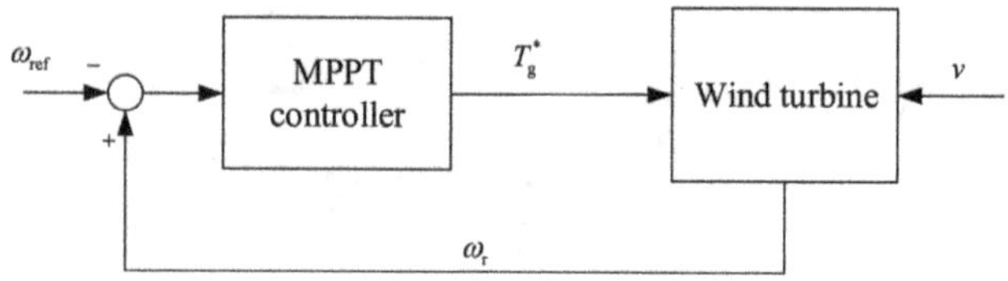

Figure 7.21 Block Diagram of Conventional MPPT Based on TSR Control

speed tracking during wind lulls. Therefore, DTG control mainly improves wind energy extraction by adjusting the tracking target, and a decrease in the torque gain can be considered as a form of reference input modification.

7.4.1 Reference input modification analysis

The dramatic increase in rotor inertia as the capacity of VSWTs increases and the highly fluctuating wind speeds render wind turbines with slow dynamics unable to timely track the optimal rotor speed, considerably deteriorating the wind energy conversion efficiency [57]. Although various studies have been devoted to aggressively regulate the rotor speed and speed up MPPT [14,50,51], the conversion efficiency is limited by the inertia and saturation of the generator torque. Consequently, maximizing the wind energy extraction based on MPPT is impractical for large-scale VSWTs under turbulence.

The proposed reference input modification aims to improve wind energy extraction by adopting a new concept that differs from existing controller improvement methods. For instance, the basic framework of conventional TSR control is shown in Figure 7.21, which aims to improve MPPT. In contrast, the proposed method modifies the reference input, as shown in Figure 7.22.

Simulation results from optimal rotor speed control excited by a step variation of wind speed (5 and 8 m/s) illustrate MPPT [58]. The corresponding trajectories of the rotor speed and power coefficient are shown in Figure 7.23. The VSWT focuses on reaching the optimal rotor speed while neglecting operation, leading to a considerable power coefficient reduction.

Compared with the optimal rotor speed, a conservative speed reference is easy to track. To demonstrate that a conservative speed reference can increase the harvested wind energy, the simulation results corresponding to a constant speed reference (optimal rotor speed of 6.6 m/s) are also shown in Figure 7.23. The conversion efficiency E/E_{max} for tracking the constant speed reference (84.75 percent) is higher

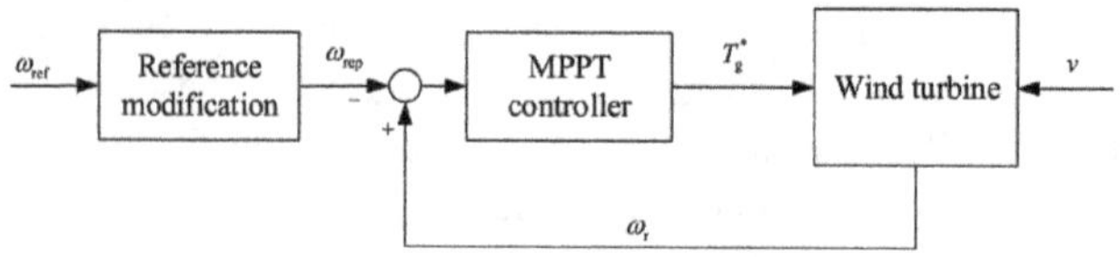

Figure 7.22 Block Diagram of Proposed MPPT Based on Reference Input Modification

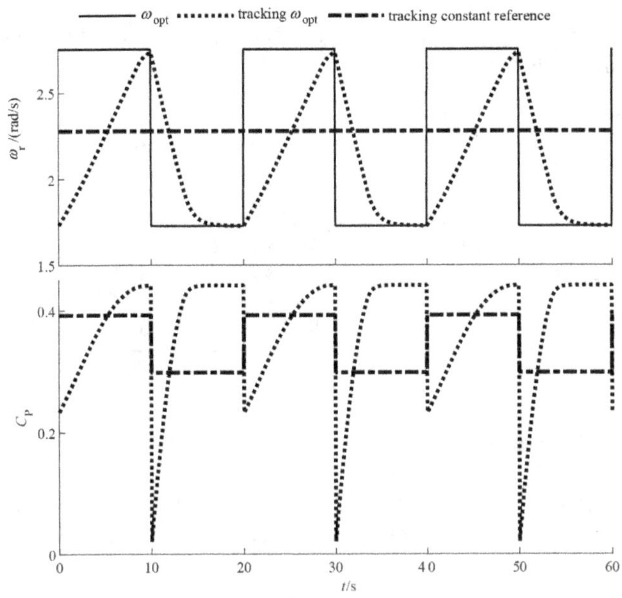

Figure 7.23 Rotor Speed and Power Coefficient Under Two Different Speed References

than that for tracking the optimal rotor speed (83.65 percent). This is because the C_P for tracking the optimal rotor speed significantly decreases once wind speed changes and reaches C_P^{max} for a long period. In contrast, a constant speed reference that can be suitably tracked prevents the serious decline in the power coefficient, and C_P is steadily maintained at a relatively high level. Hence, for large-scale VSWTs operating under turbulence, the wind energy extraction can be maximized using an appropriate speed reference, whose amplitude and frequency are consistent with the slow dynamics of such wind turbines.

To obtain an appropriate reference, optimization of the speed reference considering the VSWT characteristics is proposed. The rotor speed reference is determined by smoothing the optimal rotor speed through a first-order low-pass filter, whose coefficient is optimized according to the conversion efficiency using trial and error. Low-pass filters are also used in [14, 50] because a less turbulent rotor speed reference may prevent excessive transient loads during MPPT. To verify the speed reference optimization for maximizing wind energy extraction implemented on a second-order sliding mode controller [50], the wind energy extraction for optimal rotor speed $E^{opt\omega}$ and optimized rotor speed reference $E^{csv\omega}$ are compared using FAST (fatigue, aerodynamics, structures, and turbulence) simulations on the CART3 wind turbine model [59] under 10 min turbulence. The turbulent wind profiles with different mean wind speed v and TI are generated using the Kaimal spectrum [32]. As verified in Table 7.2, compared with the optimal rotor speed, the same controller

Table 7.2 Efficiency improvement using speed reference optimization under 10 min turbulence

Parameter		$E^{opt\omega}$ (MJ)	$E^{csv\omega}$ (MJ)	Improvement rate (%)
$\bar{v}$ (m/s)	TI			
5.17	0.20	29.10	30.78	5.77
5.78	0.23	43.66	45.58	4.40
6.52	0.21	62.52	64.74	3.55
7.53	0.16	92.36	95.24	3.12

with optimized rotor speed reference effectively increases the wind energy extraction for the CART3 wind turbine model.

For large-scale VSWTs, wind energy extraction can be further improved compared with MPPT using a speed reference that considers the slow dynamic response of large-scale VSWT rotors to turbulence. Previous studies on wind energy harvesting have been mostly focused on control algorithms for optimal rotor speed tracking, neglecting the contradiction between the requirements for fast MPPT and slow wind turbine dynamics. In contrast, the proposed method for maximum wind energy extraction applies a suitable speed reference considering the characteristics of large-scale VSWTs.

A conservative speed reference that conforms to the slow dynamic characteristics of large-scale VSWTs may improve energy extraction compared with conventional references that neglect such characteristics. Thus, a new method that differs from other types of control improvements has been established to mitigate tracking loss. The proposed method considers the conflict between VSWT slow dynamics and high-speed turbulence for improved wind energy extraction.

7.4.2 DTG

Reference input modification based on TSR control is discussed in Section 7.4.1. In OT control, existing improvement methods usually adjust the torque curves to consider the dynamics of wind turbines, being similar to the proposed reference input modification. DTG control [2] is analyzed in this section as a representative example of improved OT control.

Considering that the power available from the wind is proportional to the cube of the wind speed, maximizing the energy extraction from wind gusts is more beneficial than maximizing extraction from wind lulls [2, 3, 60]. Thus, DTG control reduces the torque curve gain K_d instead of K_{opt} [2]. The electromagnetic torque of DTG control can be expressed as

$$T_g^* = \frac{K_d}{n_g}\omega_r^2 - \left(n_g D_g + \frac{D_r}{n_g}\right)\omega_r \tag{7.24}$$

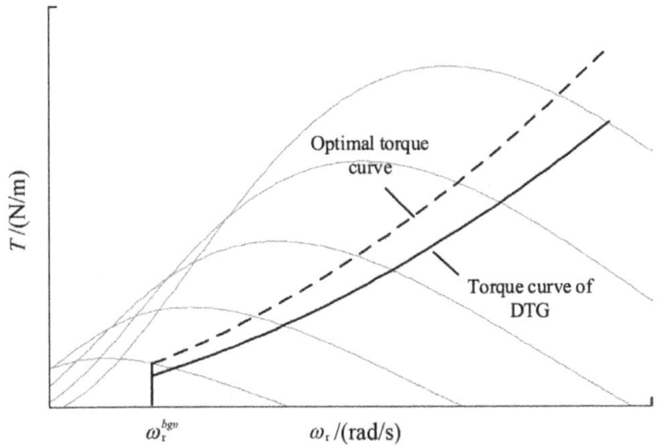

Figure 7.24 Operation Curves of OT and DTG Control

As given by (7.24) and shown in Figure 7.24, DTG control improves the tracking performance of a wind turbine under turbulence by decreasing the value of the electromagnetic torque and increasing the discrepancy of unbalanced torque. Compared with conventional OT control, DTG control focuses more on the wind energy extraction during wind gusts aiming for a fast response to increasing wind speed. The acceleration of the wind turbine increases as the electromagnetic torque decreases, thereby improving the rotor acceleration performance. On the other hand, the wind turbine decelerates slowly with smaller electromagnetic torque while slowing down. Therefore, considering that the wind energy extracted from wind at high speed is higher than that at low speed, tracking during wind gusts can be improved by reducing the extraction performance during wind lulls to improve the total wind energy extraction efficiency.

DTG control modifies the reference speed by adjusting the OT curve (Figure 7.24). Unlike TSR control that directly tracks the optimal reference speed calculated according to the wind speed by minimizing the error, the reference speed in OT control is determined by the intersection of the OT curve with the aerodynamic power curves (thin solid lines in Figure 7.24). The reference speed corresponding to the OT curve is equal to the optimal one calculated from the wind speed. In contrast, DTG control modifies the position of the equilibrium points by adjusting the torque curve, thereby modifying the tracking target of the wind turbine. In addition, DTG control improves the acceleration ability of the wind turbine at the expense of its deceleration ability. Hence, tracking during wind lulls is undermined to strengthen tracking during wind gusts. The total wind energy extraction is improved by adjusting the tracking target. Therefore, although DTG control is originally intended to improve the dynamic performance, it also conforms to the concept of reference input modification.

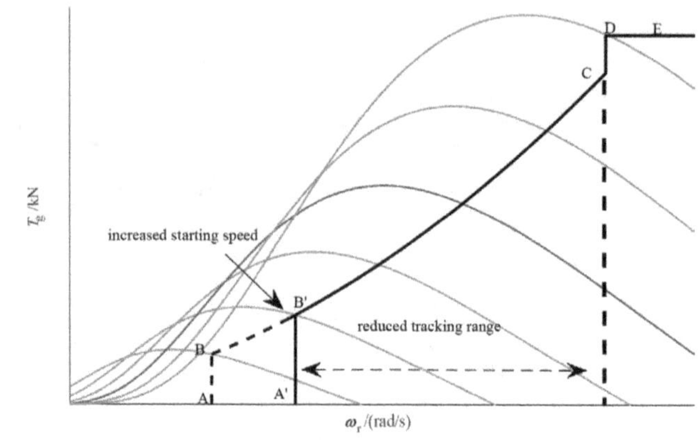

Figure 7.25 OT Curve With Increased Starting Speed for RTR

7.4.3 Reduction of tracking range

The reduction of tracking range (RTR) [8] method also corresponds to reference input modification. This method aims to track only the wind speed range with high wind energy concentration. The total wind energy extraction of the wind turbine is enhanced even for slow wind turbine dynamics. Like DTG control, the RTR method improves wind energy extraction by modifying the tracking target.

As the acceleration of turbines with increasing wind speed can be supported by shortening the tracking distance, improved OT control has been achieved using the RTR [48]. The OT curve is also employed, and the tracking range is reduced by simply increasing the starting speed, which is constant and usually set to the optimal rotor speed corresponding to the cut-in wind speed from conventional OT control [61]. As illustrated in Figure 7.25, the starting speed is increased by shifting segment A–B rightwards to A′–B′, while segment C–D remains unchanged.

A simulation illustrates the improvement due to the RTR considering a two-mass model of the CART3 wind turbine [59] excited by a step variation of wind speed from 3.5 to 6.0 m/s. Figure 7.26 shows the simulated trajectories of the rotor speed and aerodynamic power corresponding to the RTR and DTG curve. The improvement due to the DTG curve can also be obtained by reducing the tracking range. These simulation results provide the following insights:

(1) When the OT curve with increased starting speed is applied, the acceleration of the turbine during wind gusts is suitable given the shorter tracking distance for the rotor speed.

(2) As the rotor speed remains above the starting speed, the turbine cannot reach λ_{opt} at low wind speed. Hence, maximization of wind energy extraction at low wind speed is partially discarded to achieve higher energy extraction at high wind speed.

(3) A trade-off is determined by the starting speed between wind energy extraction at high and low wind speeds. The energy extracted at high and low wind speeds (6.0 and 3.5 m/s) for different starting speeds is listed in Table 7.3 to illustrate this trade-off. The optimal starting speed also varies with the wind conditions. However,

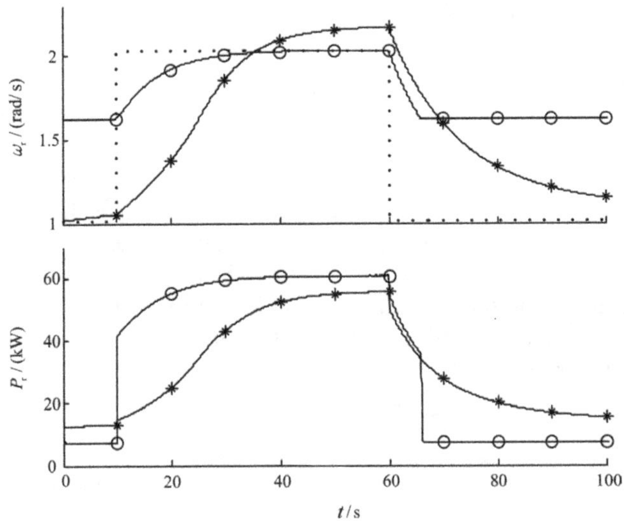

Figure 7.26 *Energy Harvesting Improvement Due to RTR (Line with Circles) Compared with DTG Control (Line with Asterisks) (Dotted Line: Optimal Rotor Speed)*

only the mean wind speed is considered in [48] when adjusting the starting speed, and an anemometer is required.

The RTR only tracks a conservative interval of wind speed in which the wind energy is concentrated. In contrast, conventional MPPT considers the complete range of wind speed variation. The total wind energy extraction improves by limiting the MPPT range at low wind speeds to reduce the distance when tracking wind gusts, leading to an overall improvement of the tracking effect.

Although this method mainly increases the starting speed of the MPPT, turbulence and dynamic performance of the wind turbine greatly influence the optimal tracking range. Thus, the dynamic optimization and adjustment of the optimal tracking range (i.e., starting speed) must be addressed considering the time-varying turbulent wind.

Table 7.3 *Energy extraction at high/low wind speeds for different rotor starting speeds*

Starting speed (rad/s)	**1.2**	**1.4**	**1.6**	**1.8**	**2.0**
Energy extraction at high wind speed of 6.0 m/s (MJ)	3.550	3.781	3.810	3.756	3.614
Energy extraction at low wind speed of 3.5 m/s (MJ)	0.720	0.687	0.625	0.537	0.429
Total energy extraction (MJ)	4.270	4.468	4.435	4.293	4.043

7.5 MPPT using ETR

Based on the RTR control with reference input modification proposed in Section 7.4, an optimization scheme for maximizing wind energy extraction is introduced in this section [9]. The scheme considers the correlation between the optimal tracking range and wind energy distribution. The ETR corresponding to the maximum wind energy distribution is derived. This range comprehensively considers turbulent wind characteristics for optimization by reducing the tracking range. Furthermore, the ETR estimation is presented. This MPPT method is easy to implement, as the reference input optimization for matching the tracking performance of the wind turbine rotor only requires the reference input of the tracking controller without modifying the system structure and controller design.

7.5.1 ETR

In the MPPT region, the available wind power for a wind turbine is proportional to the cube of the wind speed:

$$P_a(v) = 0.5\rho\pi R^2 C_P^{max} v^3 \tag{7.25}$$

In addition, the probability density function of the available wind power is defined as

$$f_{Pa}(v) = \frac{P_a(v) f_v(v)}{\bar{P}_a} = \frac{v^3 f_v(v)}{\int_0^\infty v^3 f_v(v)\,dv} \tag{7.26}$$

where $f_v(v)$ is the probability density function of the wind speed and $\bar{P}_a$ is the mean available wind power given by:

$$\bar{P}_a = \int_0^\infty P_a(v) f_v(v)\,dv = 0.5\pi\rho R^2 C_P^{max} \int_0^\infty v^3 f_v(v)\,dv \tag{7.27}$$

Clearly, $f_{Pa}(v)$ is very similar to the wind power density distribution [62, 63], except for the air density that can be assumed to be constant [63]. Function $f_{Pa}(v)$ describes the distribution of the maximum amount of wind power that can be extracted by a VSWT at different wind speeds. However, tracking loss limits the wind energy that can be effectively extracted.

Based on the wind power density distribution in [63], the occupation ratio of available wind power in wind speed interval $U_v = (v_l, v_u)$ can be calculated as

$$r = \int_{v_1}^{v_u} f_{Pa}(v)\,dv = \frac{\int_{v_1}^{v_u} P_a(v) f_v(v)\,dv}{\bar{P}_a} = \frac{\int_{v_1}^{v_u} v^3 f_v(v)\,dv}{\int_0^\infty v^3 f_v(v)\,dv} \tag{7.28}$$

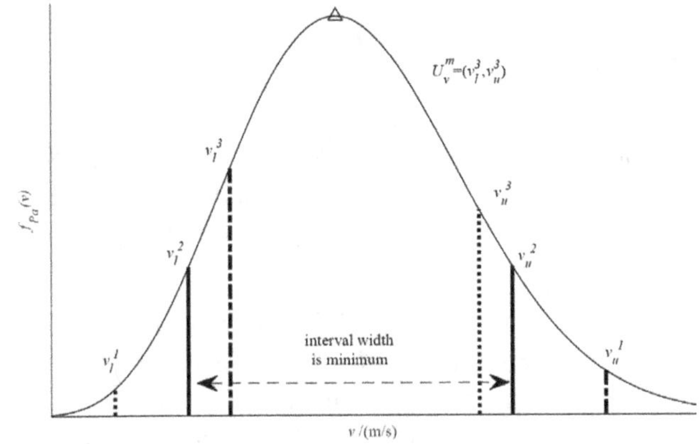

Figure 7.27 Wind Speed Interval $U_v^m = (v_l^m, v_u^m)$ of Maximum Energy Carrier (Indicated by Δ)

The available wind energy extracted by a wind turbine operating in the MPPT region during period t can be expressed as

$$\bar{E}_a = t\int_0^\infty P_a(v) f_v(v)\,dv = \bar{P}_a \cdot t \tag{7.29}$$

Occupation ratio r can also be computed using (7.30) by considering the available wind energy extraction:

$$r = \frac{E_a(U_v)}{\bar{E}_a} = \frac{\int_{v_1}^{v_u} P_a(v)\cdot t\cdot f_v(v)\,dv}{\bar{P}_a \cdot t} \tag{7.30}$$

where $E_a(U_v)$ is the available wind energy contained in wind speed interval U_v during period t.

Using the occupation ratio, the wind speed interval of the maximum energy carrier $U_v^m = (v_l^m, v_u^m)$ is defined as that which provides the same occupation ratio r with the minimum interval width. As illustrated in Figure 7.27, this definition can be interpreted as an extension of wind speed regarding the maximum energy carrier [64] given by the wind speed interval providing the highest wind energy concentration.

According to the definition of the wind speed interval corresponding to the maximum energy carrier, its boundaries can be demonstrated to correspond to the same value of f_{Pa}:

$$f_{Pa}(v_l^m) = f_{Pa}(v_u^m) \tag{7.31}$$

The proof is equivalent to solving the following problem of extreme values subject to constraints

$$\begin{aligned} &\text{Min } F(v_l^m, v_{\mathrm{u}}^m) = v_{\mathrm{u}}^m - v_l^m \\ &\text{s.t. } \int_{v_l^m}^{v_{\mathrm{u}}^m} f_{\mathrm{Pa}}(v)\,\mathrm{d}v = \text{const} \end{aligned} \tag{7.32}$$

which can be solved using Lagrange multipliers.

When both the distribution of wind energy and tracking performance of the wind turbine are considered, ETR U_ω^m for that wind turbine can be obtained by transforming the wind speed interval of the maximum energy carrier to the corresponding tracking range of rotor speed using (7.2) as follows

$$U_\omega^m = \{(\omega_l^m, \omega_u^m | \omega_l^m = \lambda_{opt} v_l^m(r_t)/R, \omega_u^m = \lambda_{opt} v_u^m(r_t)/R\} \tag{7.33}$$

where occupation ratio r_t depends on the tracking performance of the wind turbine. The tracking performance is approximated by the overall efficiency of the wind turbine operated using conventional OT control under turbulence and can be obtained from simulations or experiments as follows

$$r_{\mathrm{t}} \approx \frac{\sum_{i=1}^{N} P_{\mathrm{cap}}(i)}{\sum_{i=1}^{N} P_{\mathrm{a}}(i) \cos^3\psi(i)} \tag{7.34}$$

where $P_{\mathrm{a}}(i)$ is the available wind power at the time step i.

Equation 7.32 provides a direct relation between the ETR and wind conditions. Therefore, the ETR changes according to the characteristics of turbulent wind. The ETRs under different turbulence intensities and mean wind speeds are shown in Figures 7.28 and 7.29, respectively.

As shown in Figures 7.28 and 7.29, the ETR width is positively correlated with the TI at a given mean wind speed. Likewise, the ETR width is positively correlated with the mean wind speed at a given TI.

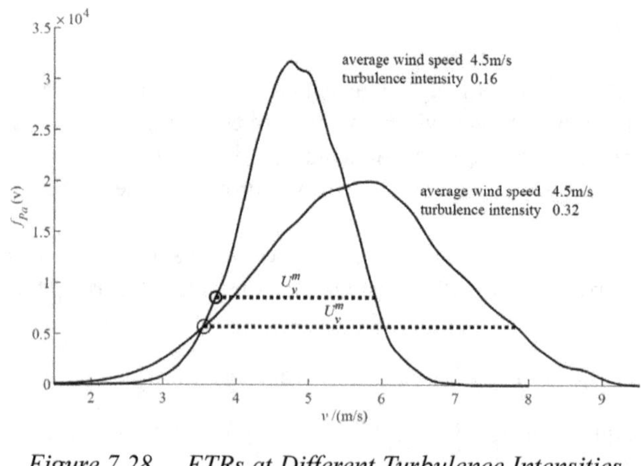

Figure 7.28 ETRs at Different Turbulence Intensities

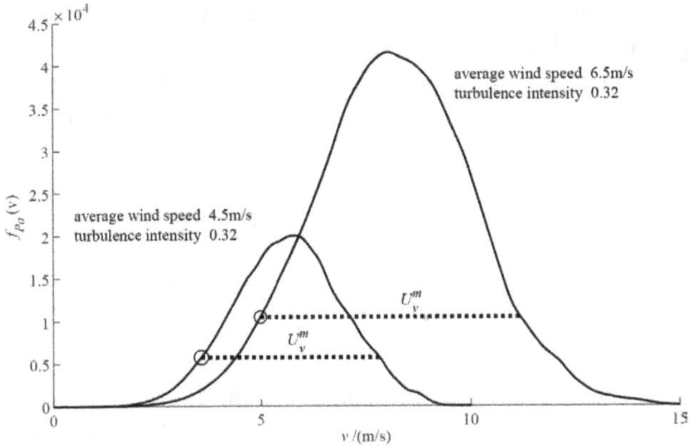

Figure 7.29 ETRs at Different Mean Wind Speeds

7.5.2 ETR estimation based on wind energy distribution

The ETR not only reflects the wind speed interval providing the highest energy concentration but also considers the wind turbine dynamics. Estimating the optimal starting speed from the optimal RTR can improve the wind energy extraction. This section details the optimization of the tracking range, from which the ETR is estimated based on the wind energy distribution.

For wind data observed by the wind speed estimator, $f_v(v)$ is first estimated by the kernel method [65]. Then, using (7.31), a series of auxiliary lines parallel to the v axis and intersecting the $f_{Pa}(v)$ curve on the $v - f_{Pa}$ plane are used to approximate U_v^m as illustrated in Figure 7.30. Given that $f_{Pa}(v)$ is usually unimodal because the

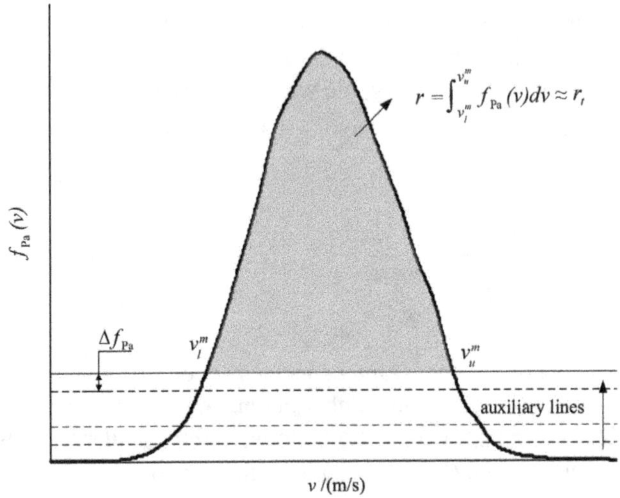

Figure 7.30 Search Procedure for U_v^m Based on $f_{Pa}(v)$

wind power density distribution often exhibits a nearly Gaussian shape [62], there should be two intersection points of an auxiliary line with the $f_{Pa}(v)$ curve. For a wind turbine with known r_t, searching for U_ω^m is described as follows.

(1) Consider the samples from observed wind data and estimate $f_v(v)$ using the kernel method as follows:

$$f_v(v) = \frac{1}{Nh}\sum_{i=1}^{N} K\left(\frac{v - v_i}{h}\right) \tag{7.35}$$

$$K(v) = \frac{1}{\sqrt{2\pi}}e^{-(v^2/2)}, h = 1.06\hat{\sigma}N^{-1/5} \tag{7.36}$$

where $K(v)$ is a Gaussian kernel function, N is the number of samples, σ is the standard deviation of the distribution, and h is the bandwidth of the kernel smoothing window.

(2) Calculate $f_{Pa}(v)$ according to (7.26). The intersection points of an auxiliary line with the $f_{Pa}(v)$ curve are numerically calculated using the bisection method.

(3) Initialize a candidate wind speed interval U_v by choosing the auxiliary line with zero f_{Pa}, that is, the v axis.

(4) Increase f_{Pa} by a step size of Δf_{Pa} (e.g., 0.01) to shift the auxiliary. Correspondingly, update candidate U_v using the intersection points of the auxiliary line with the $f_{Pa}(v)$ curve.

(5) Compute occupation ratio r associated with U_v according to (7.30) by calculating the area enclosed by the $f_{Pa}(v)$ curve.

(6) Check whether the following condition is satisfied: if $r < r_t$, define the wind speed interval of the maximum energy carrier U_v^m as U_v, and go to step 7. Otherwise, return to step 4.

(7) Define U_ω^m according to U_v^m using (7.33) and estimate $\omega_{\text{bgn}}^{\text{opt}}$ as the lower bound of U_ω^m (i.e., ω_l^m).

To avoid the computational burden of estimating $f_v(v)$, approximations to determine ETR U_ω^m and optimal starting speed $\omega_{\text{bgn}}^{\text{opt}}$ are adopted based on collecting wind data into intervals. Specifically, the wind speed range is divided into N_B equal intervals (known as bins), $U_v^j = \left(v_1^j, v_u^j\right), j = 1, 2, \ldots, N_B$, as shown in Figure 7.31. Occupation ratio r^j corresponding to each bin U_v^j is calculated as follows:

$$r^j = \frac{E_a\left(U_v^j\right)}{\bar{E}_a} \approx \frac{\sum_{v_1^j \le v_i \le v_u^j} P_a(v_i)}{\sum_{i=1}^{N} P_a(v_i)}, i = 1, \ldots, N,\ j = 1, \ldots, N_B \tag{7.37}$$

Then, U_v^m can be approximated by the connected set of bins that provide a specific occupation ratio with the minimum number of bins. Assuming r^j exhibits a unimodal distribution, U_v^m can be approximated by removing the boundary bins with small r^j from the wind speed range (i.e., $\cup_{j=1}^{N_B} U_v^j$) until the sum over r^j from the remaining bins approaches r_t. The estimation procedure of the effective tracking interval is described as follows.

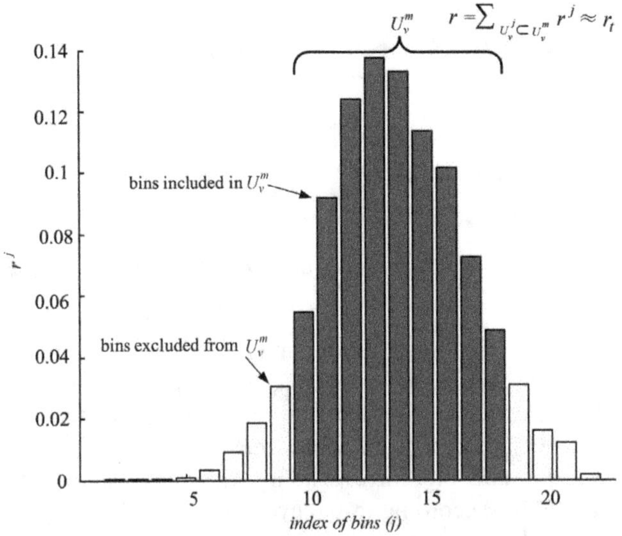

Figure 7.31 Search Procedure for U_v^m Based on Data Bins

(1) Divide wind speed range into N_B bins with the same width (e.g., 0.10 m/s). Distribute the wind data into each bin according to their values, and compute the occupation ratio r^j for each bin using (7.37). Initial candidate wind speed interval U_v is the total range of wind speed variation: $U_v = \cup_{j=1}^{N_B} U_v^j$.

(2) Determine the bin with the minimum r^j from the two boundary bins and update the candidate U_v by excluding this bin.

(3) Compute occupation ratio r associated with U_v by summing ratios r^j of the bins included in U_v.

(4) Check whether the following condition is satisfied: if $r < r_t$, define the wind speed interval with maximum energy carrier U_v^m as U_v and go to step 5. Otherwise, return to step 2.

(5) Define U_ω^m according to U_v^m using (7.33) and estimate $\omega_{\text{bgn}}^{\text{opt}}$ as the lower bound of U_ω^m, that is, ω_l^m.

7.5.3 Simulation studies and experimental validation

The improved MPPT based on ETR is described in Figure 7.32. A wind speed estimator based on the Newton–Raphson method [22] is employed to obtain real-time wind data. The estimator comprises aerodynamic torque estimation and wind speed estimation. At the end of each update cycle, U_ω^m is calculated using recent wind data collected during the current cycle, and $\omega_{\text{bgn}}^{\text{opt}}$ is updated by its lower bound ω_l^m. The update cycle is usually set from 20 min to several hours, depending on the time scale of wind variations.

The results from FAST simulations and wind turbine simulator (WTS) experiments on the CART3 wind turbine model verify the proposed MPPT using ETR.

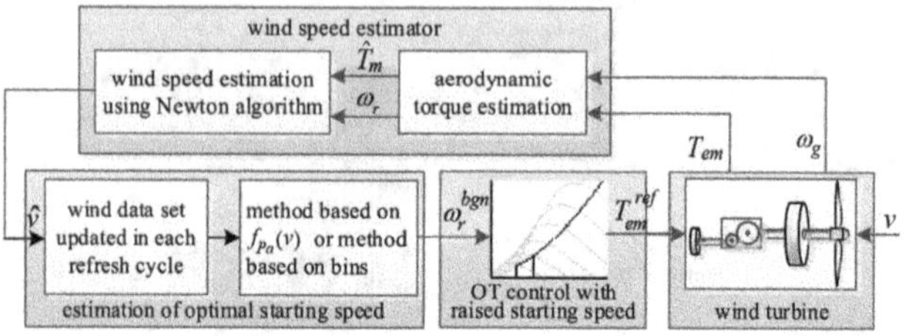

Figure 7.32 Block Diagram of Proposed MPPT Using ETR

The proposed method is compared with three MPPT methods: conventional OT control, inertia compensation control (ICC), and adaptive torque control (ATC).

Two wind profiles with a time horizon of 6 h are generated by two components: long-term evolution of wind speed chosen from the 10 min mean wind speed measurements and the corresponding turbulent wind described by the Kaimal spectrum [32]. According to the cycle used in ATC, which ranges from 10 min to 3 h [2, 60], the cycles in the ATC and MPPT using ETR are set to 20 min. According to (7.34), the occupation ratio r_t for the CART3 wind turbine model is set to approximately 0.8. The appropriate value selected for $\gamma_{\Delta M}$ is 0.002. A differential-based acceleration observation with a low-pass filter is used for ICC, and K_f is set to 10.0.

7.5.3.1 FAST simulation results

The performance of the evaluated methods is listed in Table 7.4 and depicted in Figure 7.33. For wind speed profile 1 with small wind fluctuation, although ATC, ICC, and the proposed MPPT using ETR can capture over 1.5 percent more energy from the available wind energy compared with the conventional OT control, the proposed method is more efficient. For wind speed profile 2 with large wind fluctuation, ICC and the proposed method also capture over 1.5 percent more wind energy than the conventional OT control. However, $\bar{C}_p$ over 6 h in ATC is considerably below

Table 7.4 Efficiency of evaluated methods using FAST simulation

Wind speed profile	MPPT method	$\bar{C}_p$ over 6 h	The increase compared with OT control (%)
1	Conventional OT control	0.4046	–
	ATC	0.4120	1.83
	ICC	0.4117	1.75
	Proposed method using ETR	0.4141	2.35
2	Conventional OT control	0.4060	–
	ATC	0.4026	−0.84
	ICC	0.4129	1.70
	Proposed method using ETR	0.4135	1.85

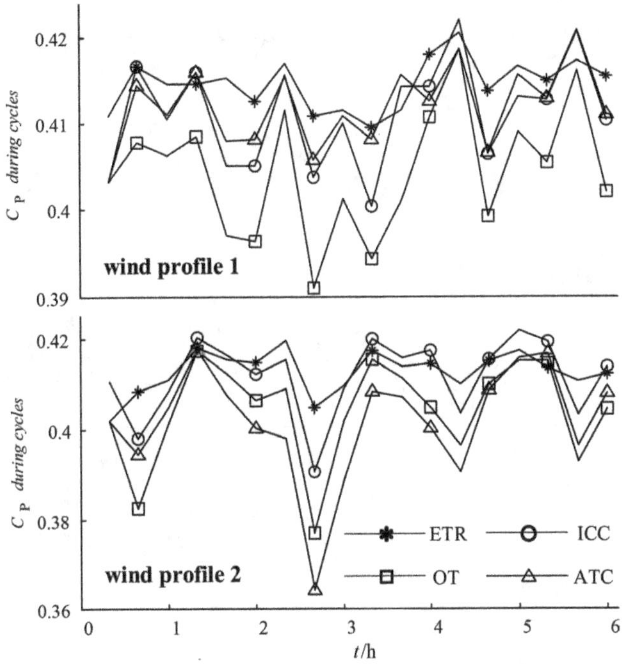

Figure 7.33 FAST Simulation Results of $\bar{C}_p$ Over 6 h

that in the conventional OT control, because the perturbation of adaptive gain M is affected by the large wind fluctuation.

To evaluate the drivetrain load and output power variation using the proposed method, the dynamic trajectories over a period of 5 min from wind speed profile 2 are shown in Figure 7.34. For the proposed MPPT using ETR, although small fluctuations of the generator torque and electric power result from constant speed control at ω_{bgn}^{opt}, they occur more readily owing to the increase in ω_{bgn}^{opt}, and the variations of generator torque and electric power above ω_{bgn}^{opt} are almost the same as those of the conventional OT control. In contrast, when applying ICC, a similar MPPT performance improvement is obtained by the high-amplitude and high-frequency control action. Therefore, K_f should be appropriately selected to balance the trade-off between wind energy extraction and either drivetrain load or output power variation.

7.5.3.2 WTS experimental results

To further verify the proposed MPPT using ETR, a 10 kW WTS VSWT test bench (Figure 7.35) that resembles the dynamics of the CART3 wind turbine model is established.

Figure 7.36 shows the test bench that consists of three main parts:

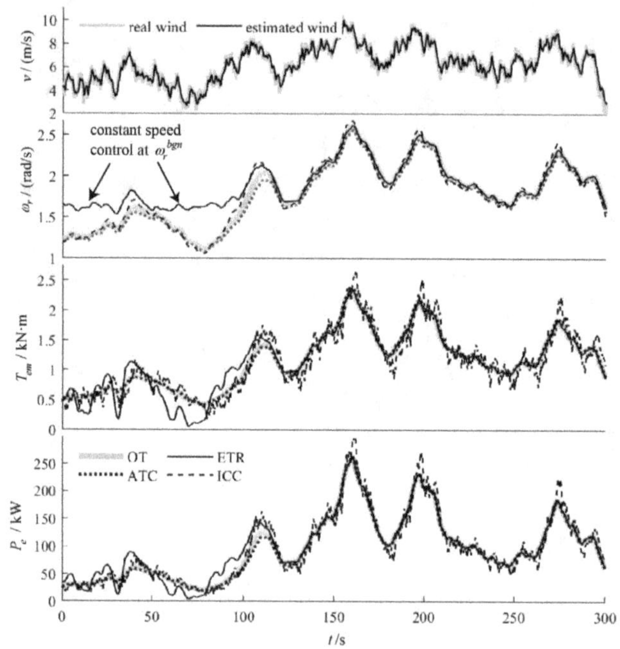

Figure 7.34 FAST Simulation Results of Dynamic Trajectories Over 5 min

1. The WTS includes a DC motor operated by a drive, a flywheel, and simulation software running in a Beckhoff programmable logic controller. The software mainly performs the transform to scale down the VSWT operational quantities to suit the motor characteristics [66], the aerodynamic simulation for calculating the aerodynamic torque, and the rotor inertia compensation for resembling the slow mechanical dynamics of real VSWT rotors [67]. By combining the deployment of the flywheel [68] and simulation software, the aerodynamic behavior

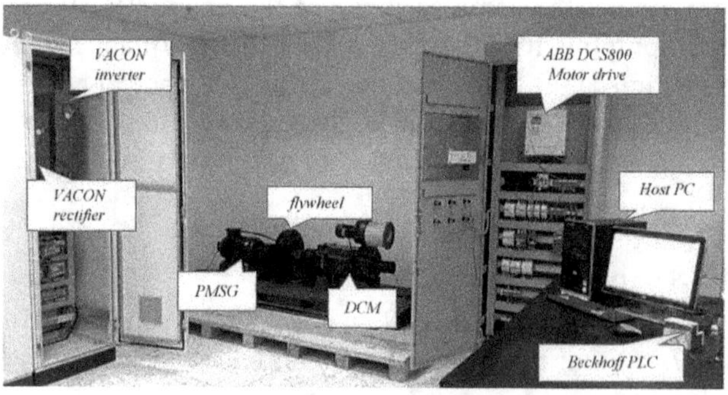

Figure 7.35 Laboratory Implementation for Experimental Evaluation

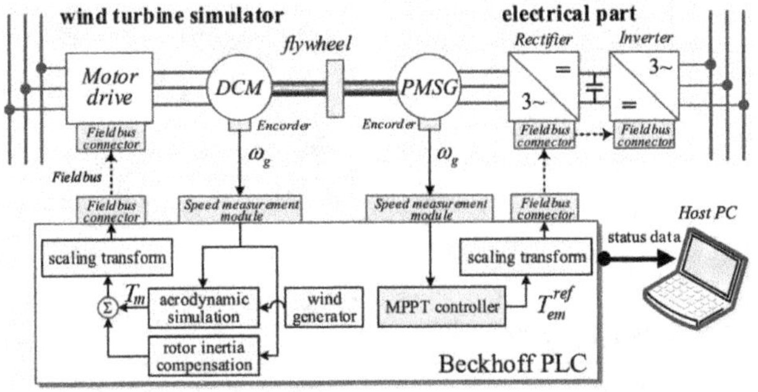

Figure 7.36 Schematic Diagram of WTS VSWT Test Bench

and rotor dynamics of the CART3 wind turbine model with large inertia can be simulated by the WTS.

2. The electrical part is the same as that for the VSWT and consists of a permanent-magnet synchronous generator and a grid-connected converter, which in turn comprises a generator-side rectifier and a grid-side inverter. The rectifier receives the electromagnetic torque reference and controls the generator torque.
3. MPPT is implemented on the programmable logic controller, which sends electromagnetic torque commands to the rectifier.

By evaluating the considered methods using the test bench, the test results in Table 7.5 are obtained. In addition, the $\bar{C}_p$ values obtained from the test bench are shown in Figure 7.37, and the dynamic trajectories are shown in Figure 7.38. The WTS experimental results are similar to the FAST simulation results.

From the WTS experimental results, the following can be noted:

1. The torque curve modification, including the DTG curve and the RTR, should be dynamically and appropriately adjusted. Otherwise, the improvement in MPPT efficiency cannot be maintained under varying wind conditions.

Table 7.5 Efficiency of evaluated methods using WTS test bench

Wind speed profile	MPPT method	$\bar{C}_p$ over 6 h	The increase compared with OT control (%)
1	Conventional OT control	0.4119	–
	ATC	0.4174	1.34
	ICC	0.4160	1.00
	Proposed method using ETR	0.4196	1.87
2	Conventional OT control	0.4115	–
	ATC	0.4052	−1.53
	ICC	0.4172	1.39
	Proposed method using ETR	0.4176	1.48

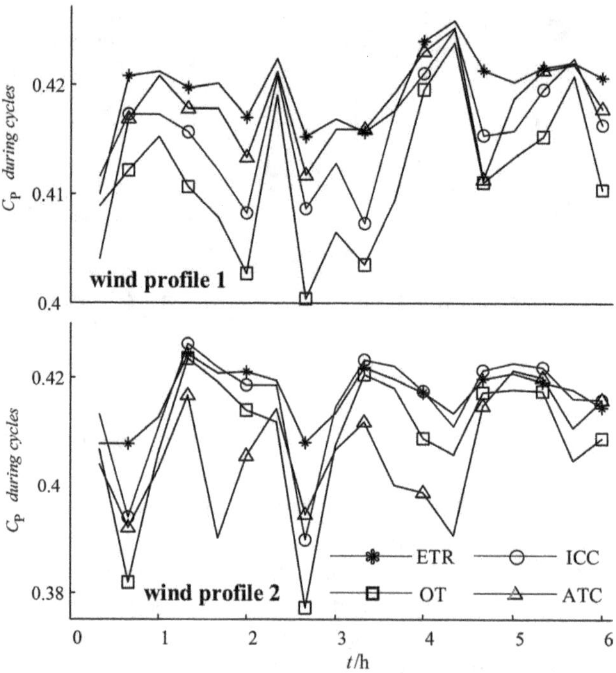

Figure 7.37 Experimental Results of $\bar{C}_p$ Over 6 h

2. The direct relation of the reduced tracking range with the mean wind speed and TI is established by the proposed MPPT using ETR, which can adjust the starting speed appropriately and achieve a relatively stable MPPT performance.
3. The MPPT efficiency can be improved using the ETR without notably affecting the generator torque/electric power fluctuation and drivetrain load.

The ETR corresponds to the wind speed interval, where the wind energy is mostly distributed. Moreover, a direct relation between the reduced tracking range and wind conditions can be established. By combining the conventional OT control and ETR estimation with a wind speed estimator, simple and efficient OT control can be achieved.

The FAST simulations and WTS experiments verify the high performance of the proposed MPPT using ETR. The torque curve (i.e., starting speed) can be rapidly and conveniently optimized to increase the MPPT efficiency and its stability under varying wind conditions. By reducing the tracking range without changing the torque curve gain, the MPPT performance is effectively improved.

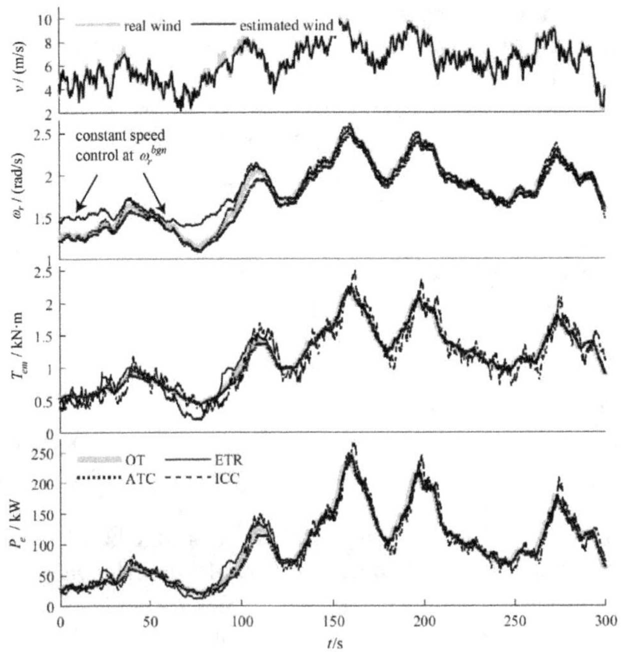

Figure 7.38 Experimental Results of Dynamic Trajectories Over 5 min

References

[1] Fingersh L.J., Carlin P.W. 'Results from the NREL variable-speed test bed'. *Proceedings of 17th ASME Wind Energy Symposium*. Nevada, USA: ASME/AEAA; 1998. pp. 103–13.

[2] Johnson K.E., Fingersh L.J., Balas M.J., *et al.* 'Methods for increasing region 2 power capture on a variable-speed wind turbine'. *Journal of solar energy engineering*. 2004;**126**(4):1092–100.

[3] Hand M.M., Johnson K.E., Fingersh L.J. *Advanced control design and field testing for wind turbines at the National Renewable Energy Laboratory*. No. NREL/CP-500-36118. National Renewable Energy Laboratory; 2004.

[4] Yin M., Kuai D., Li Q., *et al.* 'A phenomenon of maximum power point tracking invalidity of wind turbines'. *Proceedings of the CSEE*. 2011;**30**(18):40–7.

[5] Zhang X. *Optimizing Maximum Power Point Tracking of Wind Turbines with Consideration of Turbulence Effects*. Nanjing: Nanjing University of Science and Technology; 2014.

[6] Zhou L., Yin M., Zhang Z. 'Indirect effects of turbulence frequency on maximum power point tracking of wind turbine'. *10th International Conference on Advances in Power System Control, Operation & Management (APSCOM 2015)*. Hong Kong; 2015. pp. 1–6.

[7] Zhou L., Yin M., Zhang Z. 'Effects of turbulence frequency on MPPT of wind turbine'. *Proceedings of PCIM Europe 2015*. International Exhibition and Conference for Power Electronics, Intelligent Motion, Renewable Energy and Energy Management. VDE; 2015.

[8] Yin M., Zhang X., Ye X. 'Improved MPPT control based on the reduction of tracking range'. *Proceedings of the CSEE*. 2012;**32**(27):24–31.

[9] Yin M., Li W., Chung C.Y., Zhou L., Chen Z., Zou Y. 'Optimal torque control based on effective tracking range for maximum power point tracking of wind turbines under varying wind conditions'. *IET Renewable Power Generation*. 2017;**11**(4):501–10.

[10] Ekanayake J.B., Holdsworth L., XueGuang W., Jenkins N. 'Dynamic modeling of doubly fed induction generator wind turbines'. *IEEE Transactions on Power Systems*. 2003;**18**(2):803–9.

[11] Geng H., Yang G. 'Output power control for variable-speed variable-pitch wind generation systems'. *IEEE Transactions on Energy Conversion*. 2010;**25**(2):494–503.

[12] Ragheb M., Ragheb A.M. 'Wind turbines theory-the betz equation and optimal rotor tip speed ratio'. *Fundamental and Advanced Topics in Wind Power*. 2011;**1**(1):19–38.

[13] Miller A., Muljadi E., Zinger D.S. 'A variable speed wind turbine power control'. *IEEE Transactions on Energy Conversion*. 1997;**12**(2):181–6.

[14] Boukhezzar B., Siguerdidjane H. 'Nonlinear control of a variable-speed wind turbine using a two-mass model'. *IEEE Transactions on Energy Conversion*. 2011;**26**(1):149–62.

[15] Boukhezzar B., Siguerdidjane H., Hand M.M. 'Nonlinear control of variable-speed wind turbines for generator torque limiting and power optimization'. *Journal of Solar Energy Engineering*. 2006;**128**(4):516–30.

[16] Abdullah M.A., Yatim A.H.M., Tan C.W.,*et al*. 'A review of maximum power point tracking algorithms for wind energy systems'. *Renewable and Sustainable Energy Reviews*. 2012;**16**(5):3220–7.

[17] Bossanyi E.A. 'The design of closed loop controllers for wind turbines'. *Wind Energy*. 2000;**3**(3):149–63.

[18] Burton T., Jenkins N., Sharpe D., *et al*. *Wind Energy Handbook*. New York: John Wiley and Sons; 2011.

[19] Song D., Yang J., Cai Z., Dong M., Su M., Wang Y. 'Wind estimation with a non-standard extended Kalman filter and its application on maximum power extraction for variable speed wind turbines'. *Applied Energy*. 2017;**190**(4):670–85.

[20] Yang B., Yu T., Shu H.,*et al*. 'Robust sliding-mode control of wind energy conversion systems for optimal power extraction via nonlinear perturbation observers'. *Applied Energy*. 2018;**210**(1):711–23.

[21] González L.G., Figueres E., Garcerá G.,*et al*. 'Maximum-power-point tracking with reduced mechanical stress applied to wind-energy-conversion-systems'. *Applied Energy*. 2010;**87**(7):2304–12.

[22] Soltani M.N., Knudsen T., Svenstrup M., *et al.* 'Estimation of rotor effective wind speed: A comparison'. *IEEE Transactions on Control Systems Technology*. 2013;**21**(4):1155–67.

[23] Calif R. 'PDF models and synthetic model for the wind speed fluctuations based on the resolution of Langevin equation'. *Applied Energy*. 2012;**99**(May):173–82.

[24] Nichita C., Luca D., Dakyo B., *et al.* 'Large band simulation of the wind speed for real time wind turbine simulators'. *IEEE transactions on energy conversion*. 2002;**17**(4):523–9.

[25] Li D., Chen C. 'Wind speed model for dynamic simulation of wind power generation system'. *Proceedings of the CSEE*. 2005;**25**(21):41–4.

[26] Leithead W.E., de la Salle S., Reardon D. 'Role and objectives of control for wind turbines'. *IEE Proceedings C Generation, Transmission and Distribution*. 1991;**138**(2):135–48.

[27] Tang C., Soong W.L., Freere P., *et al.* 'Dynamic wind turbine output power reduction under varying wind speed conditions due to inertia'. *Wind energy*. 2013;**16**(4):561–73.

[28] Li Y., He Y., Yang Y. 'Application of ARMA model and identification of model's order to simulation of wind'. *East China Electric Power*. 2010;**38**(3):395–8.

[29] Welfonder E., Neifer R., Spanner M. 'Development and experimental identification of dynamic models for wind turbines'. *Control Engineering Practice*. 1997;**5**(1):63–73.

[30] Bossanyi E.A. *GH Bladed Version3.67 User Manual*. Garrad Hassan and Partners Ltd; 2005.

[31] Jonkman B.J., Kilcher L. 'TurbSim user's guide: version 1.06. 00'. *National Renewable Energy Laboratory*. 2012.

[32] TC88-MT I. E. C. 'IEC 61400-1: wind turbines–part 1: design requirements'. *International Electrotechnical Commission*. 2005;**64**.

[33] Wu Y. *Wind Engineering and Structural Wind Resistance Design*. Harbin: Harbin Institute of Technology Press; 2014.

[34] Simiu E., Scanlan R.H. *'Wind effects on structures: fundamentals and applications to design'*. Third Edition. New York: John-Wiley and Sons,Inc; 1996.

[35] Petersen E.L., Mortensen N.G., Landberg L., *et al.* 'Wind power meteorology. Part I: climate and turbulence'. *Wind energy*. 1998;**1**(S1):25–45.

[36] Wang Y., Sun C., Li X. 'Short-term wind speed simulation corrected with field measured wind speed'. *Proceedings of the CSEE*. 2008;**28**(11):94–00.

[37] von Kármán T. 'Progress in the statistical theory of turbulence'. *Proceedings of the National Academy of Sciences*. 1948;**34**(11):530–9.

[38] Kaimal J.C., Wyngaard J.C., Izumi Y., *et al.* 'Spectral characteristics of surface-layer turbulence'. *Quarterly journal of the Royal Meteorological Society*. 1972;**98**(417):563–89.

[39] Morfiadakis E.E., Glinou G.L., Koulouvari M.J. 'The suitability of the von Karman spectrum for the structure of turbulence in a complex terrain wind farm'. *Journal of Wind Engineering and Industrial Aerodynamics*. 1996;**62**(2-3):237–57.

[40] Kaimal J.C., Businger J.A. 'Preliminary results obtained with a sonic anemometer-thermometer'. *Journal of Applied Meteorology*. 1963;**2**(1):180–6.
[41] Clifford S.F., Kaimal J.C., Lataitis R.J., *et al.* 'Ground-based remote profiling in atmospheric studies: An overview'. *Proceedings of the IEEE*. 1994;**82**(3):313–55.
[42] Sørensen P., Hansen A.D., Rosas P.A.C. 'Wind models for simulation of power fluctuations from wind farms'. *Journal of Wind Engineering and Industrial Aerodynamics*. 2002;**90**(12-15):1381–402.
[43] Sim C., Basu S., Manuel L. 'On space-time resolution of inflow representations for wind turbine loads analysis'. *Energies*. 2012;**5**(7):2071–92.
[44] Johnson K.E., Pao L.Y., Balas M.J., *et al.* 'Control of variable-speed wind turbines: standard and adaptive techniques for maximizing energy capture'. *IEEE Control Systems Magazine*. 2006;**26**(3):70–81.
[45] Yin M., Zhang X., Ye X., *et al.* 'Improved MPPT control based on the reduction of tracking range'. *Proceedings of the CSEE*. 2012;**32**(27):24–31.
[46] Chen J., Chen J., Gong C. 'Constant-bandwidth maximum power point tracking strategy for variable-speed wind turbines and its design details'. *IEEE Transactions on Industrial Electronics*. 2012;**60**(11):5050–8.
[47] Zhang X., Yin M., Zhou L., *et al.* 'Analysis on factors affecting performance of MPPT control'. *Automation of Electric Power Systems*. 2013;**37**(22):15–21.
[48] Zhang X., Yin M., Zhou L.,*et al.* 'Research on the compensation coefficients of the improved MPPT control based on reduction of tracking range'. *Advanced Materials Research*. 2013;**724-725**:598–604.
[49] Beltran B., Ahmed-Ali T., El Hachemi Benbouzid M. 'Sliding mode power control of variable-speed wind energy conversion systems'. *IEEE Transactions on Energy Conversion*. 2008;**23**(2):551–8.
[50] Mérida J., Aguilar L.T., Dávila J. 'Analysis and synthesis of sliding mode control for large scale variable speed wind turbine for power optimization'. *Renewable Energy*. 2014;**71**(2):715–28.
[51] Saravanakumar R., Jena D. 'Validation of an integral sliding mode control for optimal control of a three blade variable speed variable pitch wind turbine'. *International Journal of Electrical Power & Energy Systems*. 2015;**69**(1):421–9.
[52] Golnary F., Moradi H. 'Design and comparison of quasi continuous sliding mode control with feedback linearization for a large scale wind turbine with wind speed estimation'. *Renewable Energy*. 2018;**127**:495–8.
[53] Kusiak A., Li W., Song Z. 'Dynamic control of wind turbines'. *Renewable Energy*. 2010;**35**(2):456–63.
[54] Soliman M., Malik O.P., Westwick D.T. 'Multiple model predictive control for wind turbines with doubly fed induction generators'. *IEEE Transactions on Sustainable Energy*. 2011;**2**(3):215–25.
[55] Camponogara E., Scherer H.F. 'Distributed optimization for model predictive control of linear dynamic networks with control-input and output constraints'. *IEEE Transactions on Automation Science and Engineering*. 2010;**8**(1):233–42.

[56] Bossanyi E.A. 'Wind turbine control for load reduction'. *Wind Energy*. 2003;**6**(3):229–44.

[57] Liu J., Meng H., Hu Y., *et al.* 'A novel MPPT method for enhancing energy conversion efficiency taking power smoothing into account'. *Energy conversion and management*. 2015;**101**(9):738–48.

[58] Chen Z., Yin M., Zou Y., *et al.* 'Maximum wind energy extraction for variable speed wind Turbines with slow dynamic behavior'. *IEEE Transactions on Power Systems*. 2017;**32**(4):3321–2.

[59] Darrow P., Wright A., Johnson K. 'Wind turbine control design to reduce capital costs'. *47th AIAA Aerospace Sciences Meeting including The New Horizons Forum and Aerospace Exposition*. 2010:1039.

[60] Johnson K.E. 'Adaptive torque control of variable speed wind turbines'. *National Renewable Energy Laboratory*. 2004.

[61] Bianchi F.D., De Battista H., Mantz R.J. *Wind Turbine Control Systems: Principles, Modelling and Gain Scheduling Design*. Springer Science & Business Media; 2006.

[62] Morrissey M.L., Cook W.E., Greene J.S. 'An improved method for estimating the wind power density distribution function'. *Journal of Atmospheric and Oceanic Technology*. 2010;**27**(7):1153–64.

[63] Carta J.A., Mentado D. 'A continuous bivariate model for wind power density and wind turbine energy output estimations'. *Energy Conversion and Management*. 2007;**48**(2):420–32.

[64] Jamil M., Parsa S., Majidi M. 'Wind power statistics and an evaluation of wind energy density'. *Renewable Energy*. 1995;**6**(5-6):623–8.

[65] Bowman A.W., Azzalini A. *Applied Smoothing Techniques for Data Analysis: The Kernel Approach with S-Plus Illustrations*. OUP Oxford; 1997.

[66] Munteanu I., Bratcu A.I., Bacha S., *et al.* 'Hardware-in-the-loop-based simulator for a class of variable-speed wind energy conversion systems: design and performance assessment'. *IEEE transactions on energy conversion*. 2010;**25**(2):564–76.

[67] Li W., Yin M., Chen Z. 'An improved inertia compensation scheme for wind turbine simulator based on deviation mitigation'. *Modern Power Systems and Clean Energy*. 2016.

[68] Park J.Y., Lee J.K., K. Y O., *et al.* 'Design of simulator for 3MW wind turbine and its condition monitoring system'. *Proceedings of the International MultiConference of Engineers and Computer Scientists*. 2010:930–3.

Chapter 8

Recent advances in vibration control for wind turbines under multiple hazards

Aly Mousaad Aly[1] *and Milad Rezaee*[1]

The increasing demand for affordable, reliable, and clean energy drives the research for more advanced wind turbines. Wind farms are growing, with larger and taller wind turbines. However, wind turbines may be installed or planned to be installed under natural and mechanical multiple hazard dynamic loadings. Excessive vibrations, due to these loads, can have detrimental effects on energy production, structural lifecycle, and the initial cost of wind turbines. To alleviate these issues, we developed a novel control theory that enables semi-active controller tuning under the complex structural behavior and inherent system nonlinearity. The theory applies to different types of structures. The proposed theory enables the evaluation of semi-active controllers' performance of multi-degrees-of-freedom systems, without the need for time-consuming simulations. With this analytical probabilistic control theory, a wide range of controllers can be tested in a fraction of a second, and their parameters can be tuned to achieve system-level performance for different optimization objectives. Vibration reduction can prevent malfunctioning of the mechanical acceleration-sensitive parts, with increased wind turbines' availability and reduced maintenance costs. This can also lead to reduced capital cost, and therefore less expensive and more competitive generated electricity.

8.1 Introduction

The increasing energy consumption and the need for dependable, secure, and clean energy have led to explorations in non-conventional sources, particularly in renewable energy sources. During the past decade, wind energy has been the fastest-growing source of electricity in the world. It plays a significant role in achieving sustainability and powering the future. Today, wind energy has already become one of the main electricity sources in some countries such as Denmark, Portugal, Spain, and Germany, where the contribution of wind energy to the total power is 21, 18,

[1]Windstorm Impact, Science and Engineering (WISE) Research Lab; Civil and Environmental Engineering, Louisiana State University, Baton Rouge, LA, USA

16, and 9 percent, respectively [1]. In 2015, the contribution of wind energy to the nation's electricity in the United States was about 1 percent, while potentially it could be increased up to 20 percent, without major changes in the distribution system [2].

Wind turbines can be deployed in coastal zones, inland, and offshore areas, wherever the wind is powerful and uniform, and close enough to the market. However, installing wind turbines in areas with powerful winds and at a short distance to the market can lead to significant design challenges. Modern wind turbines likely face various natural hazards. The population concentration in coastal areas, and the availability of more powerful and uniform wind resources, accelerated the development of coastal and offshore wind farms. However, environmental loads, such as wind and waves are harsher in coastal and offshore areas, which can cause severe damages to wind turbines, or make the produced power more expensive and less dependable.

It has been less than 15 years since wind farms started to be installed in seismic prone areas. Therefore, due to the lack of experience, there is no guaranty that modern wind turbines can survive extreme earthquakes [3]. The concerns over the possibility of exposure of more modern wind turbines to multi-hazard loads make the understanding of the structural behavior more important than before. These concerns motivated a global endeavor to study and evaluate the performance of modern wind turbines, exposed to various loading scenarios. The efforts are continuing, to provide a comprehensive perspective of the performance and functionality at a system-level, for wind turbines to survive multi-hazard loads.

According to the findings of other researchers, an important trait of modern wind turbines that needs consideration is the flexibility of the structure and the challenge of damping deficit. Consequently, the exposure of a wind turbine to the intense environmental dynamic loads can cause excessive vibrations in the system. The undesirable motion can impose a number of aftereffects on the wind turbine, such as malfunctioning of mechanical parts in the nacelle or the rotor, power production disorders, reduced wind turbine's availability, crack deployment through the machine, reduced fatigue life of the system, increased likelihood of failure, and the need for more frequent maintenance. The direct and indirect consequences of these problems can lead to costly generated electricity.

Vibration mitigation through increasing damping can be an effective approach, to remedy these problems. However, vibration mitigation is not a straightforward approach that can be achieved following specific steps for all wind turbines, regardless of their environmental conditions. Since the wind turbines may confront unequal natural hazards during their useful life, it should be taken into consideration that each type of natural hazard could have a specific impact on the structure, depending on the overall environmental conditions and structural characteristics. Therefore, it is of utmost importance to assess the structural performance and evaluate the mitigation approaches for various scenarios, to ascertain a proposed solution for one hazard does not make the structure more vulnerable to another hazard. Structural modeling and evaluating the performance of dynamic systems with multi-degree-of-freedom is generally a time-consuming process. Evaluating these systems with

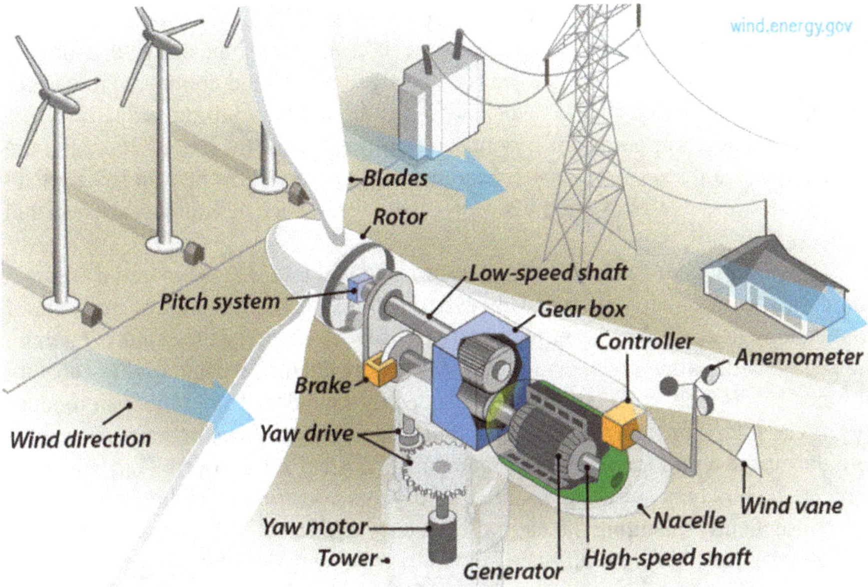

Figure 8.1 Anatomy of a wind turbine [4]

various control methods make the models even more tedious and time-consuming. To overcome these issues, an innovative analytical approach is proposed to evaluate the performance of the controlled wind turbines that are modeled as multi-degree-of-freedom systems. In the subsequent sections, an introduction to wind turbine systems, dynamic modeling, and control of wind turbines will be presented.

8.2 Wind turbines

Wind turbines are devices that harvest the power of the wind and convert it to a useful form of energy (electricity). Wind power makes the blades of a wind turbine turn around a rotor. The main shaft transfers this rotation to the generator via a gearbox unit and the generator converts the mechanical energy to electrical energy. A wind turbine consists of a hub, rotor, blades, nacelle, and tower. Figure 8.1 illustrates a detailed view of the inside of a wind turbine. The blades and the rotor are rotary parts, moving to bypass the wind through the rotor plane. The nacelle is a house for the mechanical parts and the tower is a foundation for the nacelle and the rotor to place them at an appropriate height. The materials that are used to build each component are continually changed and improved over the years, mainly because of significant changes in the height and size of the turbine, to harvest more wind energy by a single unit.

Developing technology yields increasing wind turbines' efficiency, rotor size, height, and more extracted energy. Developing technology also caused a continuous change in the material used in the construction of wind turbines. The trend is toward lighter weight and low-cost materials, especially for the blades and towers. The tower weight is typically 60 percent of the weight of the turbine above the foundation [5]. Therefore, the trend is toward using lighter steel and composite materials. An important issue that should be considered is the material fatigue properties. The majority of the components of a wind turbine, with a useful life of 30 years, are required to be able to sustain 4×10^8 fatigue stress cycles. This is even stricter than the fatigue stress cycles of other man-made structures, such as engines, aircraft, and bridges [5].

The tower can be made of various materials. Steel lattice is the material that was commonly used in early towers. However, because of the exposure of numbers of connections to corrosion, and the sensitivity of weak diagonals to wind excitation, the use of steel lattice towers is not safe to support the modern and higher capacity turbines [6]. Hence, steel lattice was replaced by cylindrical steel tubes in modern towers. Besides, for towers higher than 85 m, concrete can have better performance to balance the vibration excitation, compared to steel tube towers [7]. However, concrete towers may suffer the problem of thermal constraints, leading to cracking, and consequently changing the natural frequency and structural characteristics. The main material used in wind turbines from 2001 to 2010 in the US and worldwide was steel, particularly for large wind turbines [5]. Besides, in coastal areas, the part of fixed wind turbines that are placed in water is normally constructed from concrete.

According to the mentioned points, because the modern wind turbines are taller and lighter than the former generations, these structures are more flexible than before. This flexibility makes them sensitive to the environmental dynamic loads and causes excessive vibrations in the tower as well as the whole structure. Severe vibrations can cause functionality malfunctioning or permanent damages to acceleration-sensitive mechanical parts or may reduce the structure's fatigue life and increase the chance of systematic failure.

In a wind turbine, the damage can occur anywhere from mechanical and electrical parts to structural parts. However, from a civil engineering point of view, the damages can occur through the foundation, the tower steel tube, bolts, or blades. Structural failures are reported in various countries, such as Scotland, Spain, Germany, Japan, France, Wales, Denmark, and New Zealand [8]. The investigation of wind turbines' failures in Germany shows that all mechanical components have almost equal damage frequency as the primary structure [9].

Structural failure has been the third most common accidents in wind turbines, with 183 reported instances from the 1970s to 2017. Storm damage, tower collapse, poor control, lack of maintenance, and component failure were the main concerns in structural failures. Figure 8.2 shows several structural and blade failures between the 1970s and 2017. Figure 8.3 shows the collapse of two wind turbines, due to structural failure in Cape Breton, Canada, and Kansas, US. Failure of a wind turbine can have both societal and economic impacts. Modern wind energy harvesting is in a transient stage in most countries from just a potential source of energy to a dominant source of electricity. The failure of wind turbines can make doubt on the reliability of this energy and increase the power production costs. Considering the aforementioned points, and the consequences of the failure of wind turbines, it is crucial to

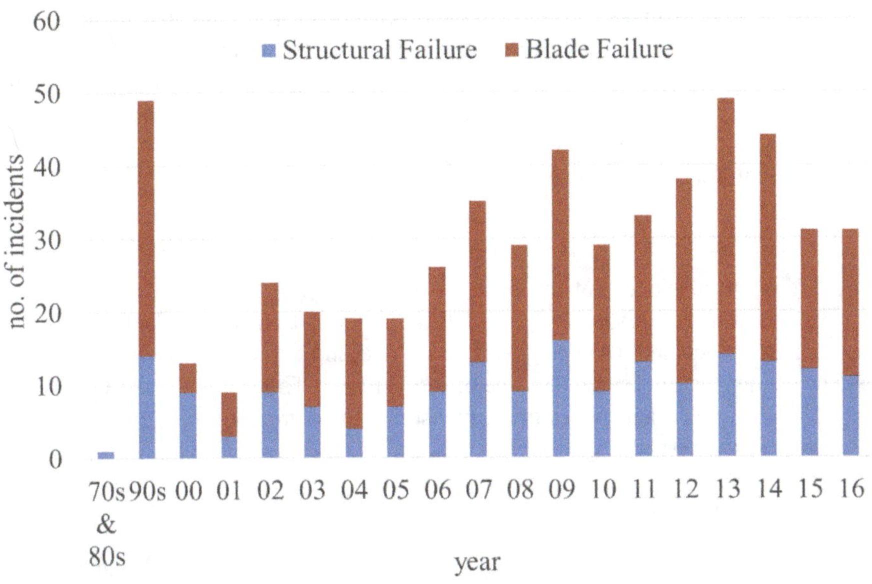

Figure 8.2 Structural and blade failures between 1970 and 2017

study the wind turbines from a structural point of view, subjected to dynamic loads, to advance their design.

8.3 Structural dynamics

Under the action of environmental loads, a wind turbine, as a tall and flexible structure, responds according to the load characteristics and the structure's properties. Like other tall structures, there are several studies on the structural dynamic

Figure 8.3 Wind turbine tower collapse: (a) Cape Breton, Canada [10] and (b) Kansas, USA [11]

behavior of wind turbines, particularly by increasing the size of these structures, the static analysis becomes insufficient to design the support structures. As a very basic assumption, a wind turbine can be considered as a spring-dashpot single-degree-of-freedom (SDOF) system excited with a harmonic or random force, or constrained mass at a top of a cantilevered beam (Figure 8.2) [12, 13]. Here, the governing equation of motion for the system can be obtained from Newton's second law as follows.

$$M\ddot{X} + C\dot{X} + KX = F(t) \tag{8.1}$$

where X is the displacement response, M, C, and K are the mass, damping coefficient, and stiffness, respectively, and F is the excitation force. The dots on the top of the displacement (X) designate the 1st and 2nd derivatives, representing the velocity and acceleration of the system, respectively. This equation can be rewritten in another form that relates the responses to the damping ratio and the natural frequency of the system:

$$\ddot{X} + 2\zeta\omega\dot{X} + \omega^2 X = \frac{1}{M}F(t) \tag{8.2}$$

here, ω and ζ are the natural frequency and the damping ratio of the system, respectively. As a very basic assumption, the natural frequency, f_{n1}, can be approximated as follows [13].

$$f_{n1}^2 = \frac{3.04}{4\pi^2}\frac{EI}{(M+0.227\mu L)L^3} \tag{8.3}$$

$\mu$is the tower mass per meter, M is the top mass, EI is the tower bending stiffness, and L is the tower height. Considering the nature of the environmental loads on wind turbines, the structure should be designed to have the first natural frequency between the two limits called 1P and 3P where 1P is a rotor loading frequency and 3P is the blade passing frequency for a three-bladed wind turbine. These frequency limits depend on the type of wind turbine, foundation, and the dominant environmental force. Designing the structure that its natural frequency places somewhere between 1P and 3P are called a soft-stiff design approach. Here, in this approach, the wind turbine is not that stiff to have a natural frequency higher than 3P (say 1 Hz) and is not as soft as a system with a natural frequency of about 0.17 Hz.

Simplifying a wind turbine as an SDOF system might cause the elimination of considering the real dynamic behavior of the system (Figure 8.4). Therefore, various studies are carried out on the dynamic of a wind turbine using a modal approach or finite element model that considers the structure as a multi-DOF (MDOF) system. Figure 8.5 shows an MDOF model of a fixed offshore wind turbine with lumped mass at various heights of the tower in the parked position eliminating the blades-tower coupled interaction. An example of this model is one of the first studies carried out on the effect of wave loads on offshore fixed wind turbines [14]. In this study, a 750 kW two-bladed wind turbine was modeled as a six-DOF system with concentrated masses in the nodes, and the masses were connected with a massless column having finite bending stiffness. Another method that is used in common simulation codes such as CAsT and FAST to find the response of parked or operating wind turbines

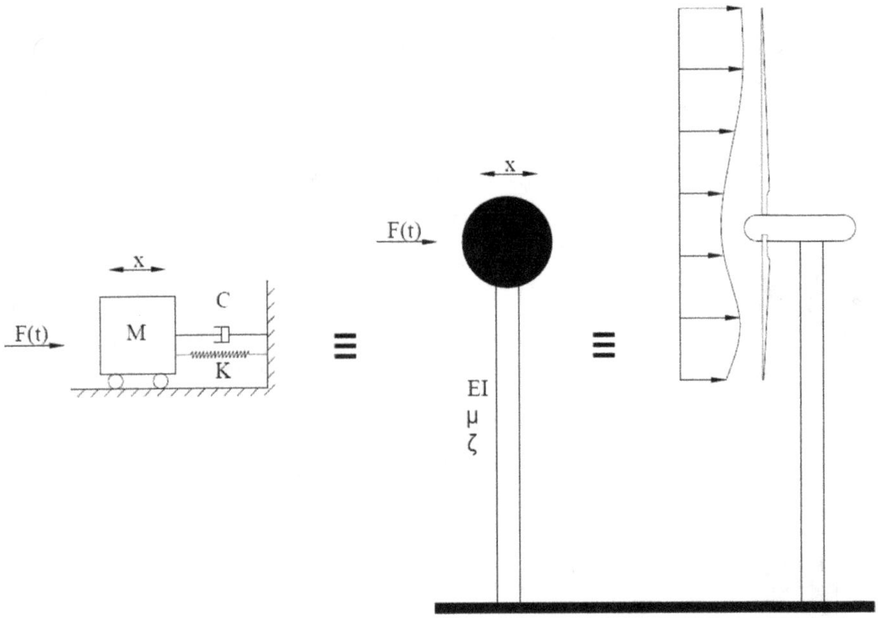

Figure 8.4 *Schematic representation of a wind turbine and its simplified equivalent SDOF structural and mechanical spring-dashpot model*

is modeling based on the solution of the equation of motion for an MDOF beam element for each time step [15–18].

In addition to the simplified models that mainly consider the wind turbine's tower or the wind turbine in the parked position, some models are more sophisticated which take the coupled effects of rotating blades, nacelle, and tower into account. Some of the examples of these models are used in the studies of [19–23]. In most of these studies, equations of motion for the system developed based on Euler–Lagrange formulation.

$$\frac{d}{dt}\left(\frac{\partial T}{\partial \dot{q}_i}\right) - \frac{\partial T}{\partial q_i} + \frac{\partial V}{\partial q_i} = Q_i \tag{8.4}$$

Here, T is the kinetic energy and V is the potential energy of the system, respectively; q_i is the displacement and Q_i is the generalized loading corresponding to the ith DOF. Therefore, the mass and stiffness matrices can be obtained by defining appropriate kinetic and potential energy equations. The potential energy term for a wind turbine, considering fore-aft and side-side direction of motion, can be written as follows [21]:

$$V_{\text{Total}} = \frac{1}{2}\sum_{j=1}^{3}\int_0^1 \begin{bmatrix} \text{EI}_{\text{in}}(x)\left(\frac{\partial \theta_{i,\text{in}}}{\partial x}\right)^2 \text{dx} + \text{EI}_{\text{out}}(x)\left(\frac{\partial \theta_{i,\text{out}}}{\partial x}\right)^2 \text{dx} \\ +2\text{EI}_{\text{inout}}(x)\left(\frac{\partial \theta_{i,\text{in}}}{\partial x}\right)\left(\frac{\partial \theta_{i,\text{out}}}{\partial x}\right)\text{dx} + N(x)\left(\frac{\partial u_{i,\text{in}}}{\partial x}\right)^2 \text{dx} \\ +N(x)\left(\frac{\partial u_{i,\text{out}}}{\partial x}\right)^2 \text{dx} + G(x)\left(\frac{\partial u_{i,\text{in}}}{\partial x}\right)^2 \text{dx} + G(x)\left(\frac{\partial u_{i,\text{out}}}{\partial x}\right)^2 \text{dx} \end{bmatrix} + \frac{1}{2}k_{n,\text{in}} \times q_{n,\text{in}}^2 + \frac{1}{2}k_{n,\text{out}} \times q_{n,\text{out}}^2 \tag{8.5}$$

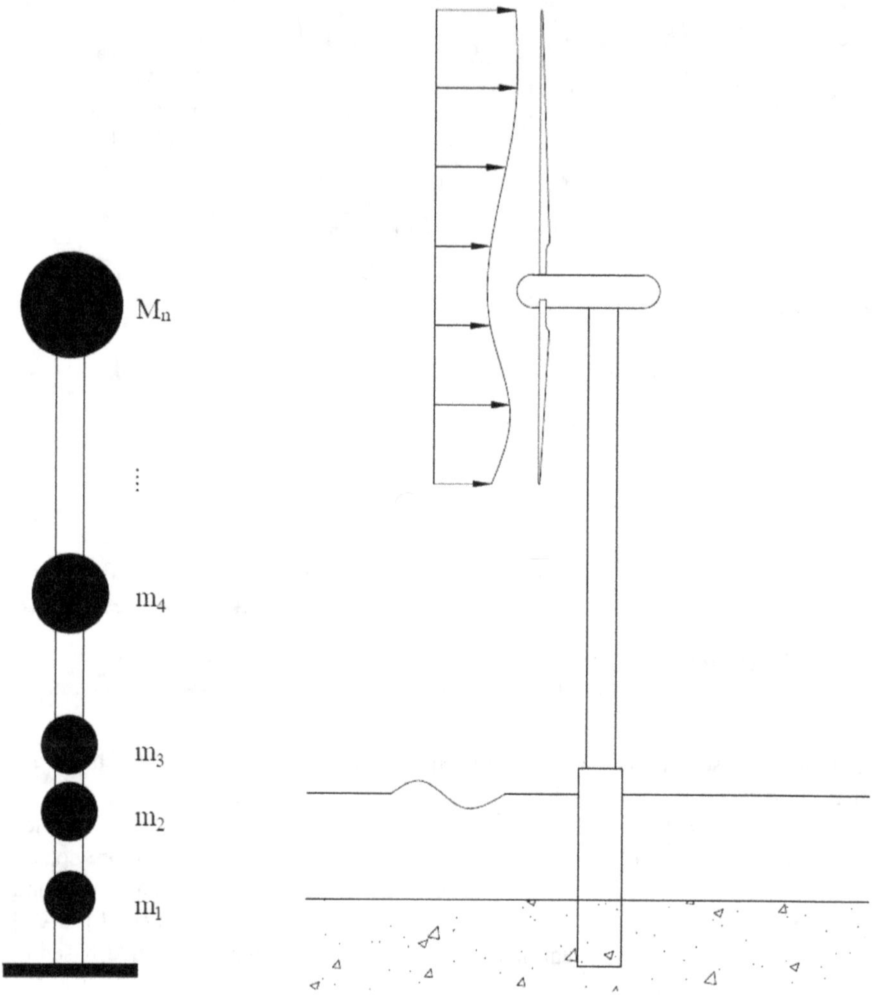

Figure 8.5 MDOF model of an offshore wind turbine on a mono-pile foundation

where L and E are the length and the modulus of elasticity of the blade, respectively. Ω is the rotor's rotational speed. The subscripts *in* and *out* are for in and out of rotor plane displacements and k is the stiffness of the tower/nacelle. For an operating wind turbine, the rotation of the blade leads to an excessive centrifugal stiffening effect. The gravity causes additional stiffness in the blades. Therefore, $N(x)$ and $G(x)$ which denote the centrifugal and gravitational forces, respectively, can be defined as:

$$N(x) = \Omega^2 \int_x^L \mu(\xi)\, \xi d\xi \tag{8.6}$$

$$G(x) = -\frac{1}{2} g \cos \Psi_i \int_x^L \mu(\xi)\, d\xi \tag{8.7}$$

where variable mass per unit length of the blade is denoted by $\mu(x)$, and ψ is the azimuthal angle of the ith blade because of the rotation. Blades have a 2π/3-phase difference.

$$\Psi_i = \Omega t + \frac{2\pi}{3}(i-1) \tag{8.8}$$

also, the rotations in in-plane and out-plane are:

$$\theta_{i,\mathrm{in}} = \frac{\partial u_{i,\mathrm{in}}}{\partial x} \tag{8.9}$$

$$\theta_{i,\mathrm{out}} = \frac{\partial u_{i,\mathrm{out}}}{\partial x} \tag{8.10}$$

the parameters $I_{in}(x)$, $I_{out}(x)$, and $I_{inout}(x)$ represent the in and out of plane second area moments of inertia and the second area products of inertia of the blade's cross-section, respectively. Additional information, including the cross-sectional *EI*, can be found in the National Renewable Energy Laboratory (NREL) documents [24]. Kinetic energy also can be calculated from the following equation:

$$T_{\mathrm{Nacelle}} = \frac{1}{2}\sum_{i=1}^{3}\int_0^L \mu(x)\, v_i^2 \mathrm{dx} + \frac{1}{2}M_n\dot{q}_{n,\mathrm{in}}^2 + \frac{1}{2}M_n\dot{q}_{n,\mathrm{out}}^2 \tag{8.11}$$

where, v_i is the absolute velocity of the ith blade and $\dot{q}$ is absolute in- and out-plane velocity of the nacelle and M_n is the mass of the tower [21]. Therefore, the wind turbine's mass, stiffness, and damping matrices can be derived.

Damping is assumed to be stiffness proportional damping for both the nacelle and tower and the blades. Therefore, with the damping ratios of ζ_n and ζ_b and natural frequencies of ω_n and ω_b for the nacelle/tower and blades, respectively, the damping can be obtained from the following relations:

$$\begin{aligned} c_n &= 2\frac{\zeta_n}{\omega_n}K_n \\ c_b &= 2\frac{\zeta_b}{\omega_b}K_b \end{aligned} \tag{8.12}$$

K_n and K_b are the stiffness matrices of the nacelle/tower and blades. For more details, refer to [20–23].

The presented models consider the structural properties of a wind turbine, regardless of the soil-structure interaction, controlled systems, and applied loads. More details on soil-structure interaction can be found in [21, 25–27]. Having an appropriate wind turbine model provides a tool to investigate the system's responses and its problems. Since the focus of current research is to reduce the excessive vibrations in wind turbines, one should go through the methods to control the structure.

8.4 Structural control

External dampers can be employed in structures to produce counter forces, reduce the responses, and therefore control structural vibrations. The first important step in the matureness of the application of structural control for civil engineering structures backs to 1994 when the first world conference on structural control was held in Los Angeles [28]. Since 1994, structural controllers are developed to a great extent and various types of them, including passive, semi-active, active, and hybrid controllers are investigated.

8.4.1 Passive control

Passive control employs tuned devices and there will be no continuous adjustment during the operation. Therefore, there is no need for an external power source and the design concept is much simpler than other types of controllers. Passive dampers can improve the damping ratio of civil engineering structures, which normally are less than 2 percent. Hence, passive dampers increase the dissipation capacity of the structures. Since the 1990s, impressive efforts have been done to implement the theory of applying external dampers into reality as a practical solution for civil engineering structure [29, 30].

Tuned mass dampers (TMDs), tuned liquid dampers (TLDs), and viscous fluid dampers are three examples of these controllers that have been used in full-scale civil engineering structures during the last decades. TMD conceptually presented by [31]. It consists of a mass connected to the primary mass with a tuned spring and an oil dashpot damping system. The use of these systems is investigated widely for different types of civil engineering structures such as bridges, tall buildings, and wind turbines [32–35]. Two of the tall buildings in the US that are equipped with TMDs are John Hancock Tower in Boston and Citicorp Center in New York [28].

TLDs are similar to TMDs in concept, where the motion of the fluid mass generates the counterforce to suppress vibrations of the system. Here, the dimension of the fluid tanker defines the damper's frequency that should be tuned to the natural frequency of the primary structure. Besides, the fluid viscosity and the opening of the barriers inside the tankers can adjust the damping ratio of the controller. Yokohama marine tower and Shin Yokohama are two examples of using TLDs in tall buildings [28, 36].

One commonly used class of dampers that employ fluids to dissipate energy is viscous damper (VD) that is widely used in mechanical systems and normally consists of a piston in the damper housing filled with a compound of silicone or oil [37]. Attractive parameters of this class of dampers that are interesting for civil engineering structures include the linear viscous response achieved over a broad frequency range, insensitivity to temperature, and compactness in comparison to stroke and output force [28]. Figure 8.6 illustrates a typical damper of this type. The movement of the piston through the viscous fluid dissipates energy and if the fluid is purely viscous (e.g., Newtonian), then the damper force is directly proportional to the velocity of the piston [28]. Therefore, it can be modeled as follows:

$$f_d = C_0\dot{x} \tag{8.13}$$

Figure 8.7 shows the implementation of viscous fluid dampers in civil engineering applications, via a bracing system, to control the inter-story drift.

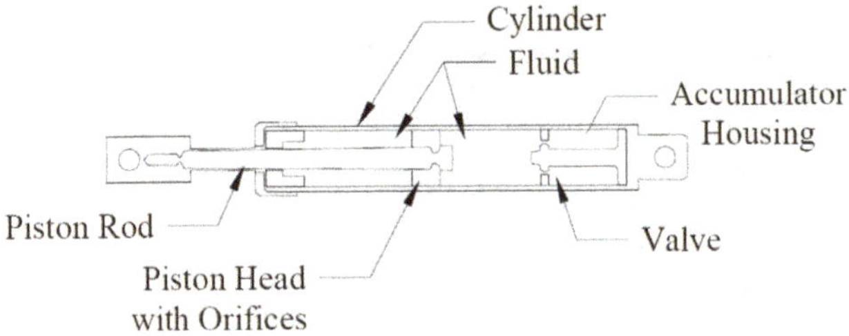

Figure 8.6 Fluid damper from Taylor devices [38]

8.4.2 Active control

In contrary to passive controller devices, active controllers inject energy into the system. These controllers apply force to the structure through actuators and in consequence, adjustments can be made. Due to the real-time changes that the active controllers make in the system, they may lead to instability with the consequences of structural/system failure [28]. Hence, the utmost endeavor should be done to design controllers to improve the dynamic responses of the system instead of making it worse. Considering this fact, different methods and algorithms are developed and various studies are carried out on active controllers. Some of these studies are sliding mode control [40, 41], saturation control [42, 43], reliability-based control [44–47], fuzzy control [48–50], and neural control [51]. A schematic block diagram

Figure 8.7 Viscous dampers in a building [39]

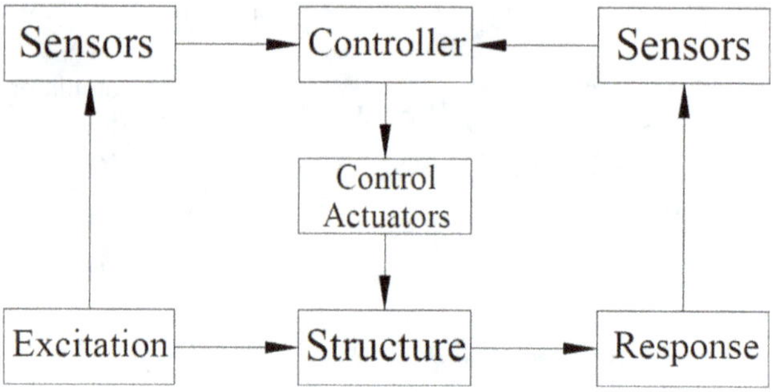

Figure 8.8 Block diagram of a generic structural control system

of the structural control problem is illustrated in Figure 8.8. As it is mentioned earlier, an important task is to specify an appropriate control strategy that receives the measured structural responses by sensors, calculate sufficient control signal, and send it to the actuator. So that the actuators improve the structure's performance by injecting enough forces to minimize the errors.

Figure 8.9 depicts the installation of an active mass damper at the top of a tall building. Here, to control the system, a small mass (about 1 percent of the total mass

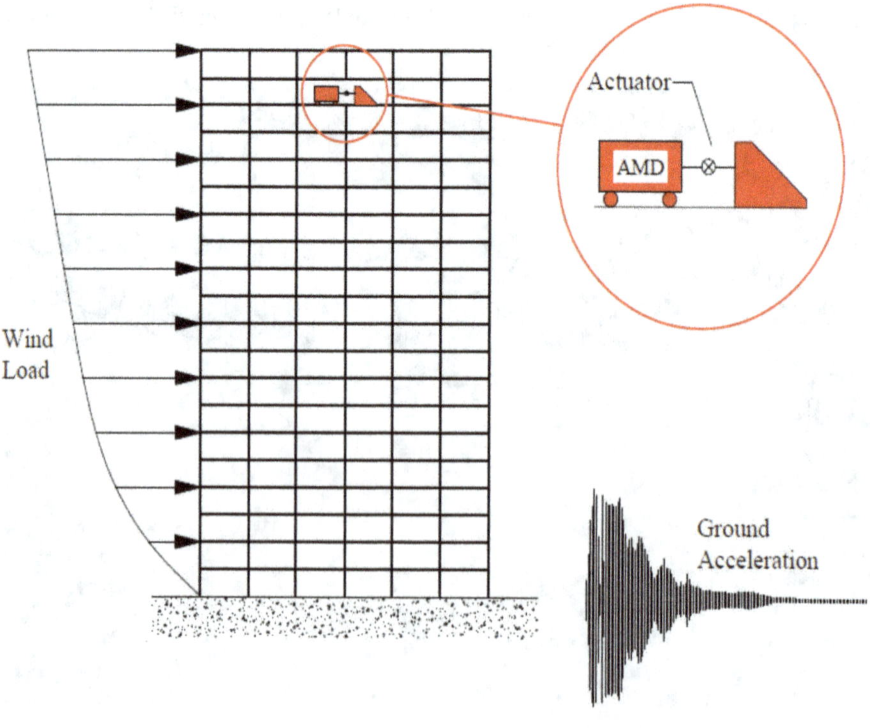

Figure 8.9 An active mass damper on top of a tall building [52]

of the system) is connected to the structure through an actuator. Sensors measure responses of the structure. A computer processes the data and according to the control algorithm, it sends appropriate signals to the amplifiers, which finally cause the reaction of the actuators to the excitation. The first full-scale installation of active control in a tall building was in the Kyobashi Center building in 1989. The installed control system consists of two active tuned mass dampers (AMDs), one to control transverse motion and the secondary AMD designed to reduce torsional motion [53]. For the list of civil engineering structures that employed external dampers refer to [53].

8.4.3 *Hybrid control*

Hybrid controllers are those that are a combination of active, semi-active, and passive dampers. An example of these systems is the structures that are equipped with various viscoelastic dampers and benefit an active base isolation module or active mass damper [18]. In a hybrid control scheme, an active control may be used to enhance the performance of a passive control scheme. Alternatively, a passive control system can be employed in an active control scheme to reduce the required power [28].

8.4.4 *Semi-active control*

Semi-active dampers are bounded with active and passive dampers. These types of systems do not inject mechanical or non-dissipative energy into the system. Therefore, they can be considered as passive controllers that can be turned on to change the properties of the damper whenever it is desired. Control strategies based on semi-active devices benefit the best features of both passive and active control systems and are receiving the attention of the engineers and scientists as a practical, optimal solution to protect civil engineering structural systems against earthquake and wind loading [28]. Semi-active dampers can mimic the active controller without requiring the associated large power sources. For instance, magnetorheological (MR) dampers require only 20–50 watts of power and can operate just on battery power. This is super critical and beneficial during seismic or hurricane events when the main power source to the structure may fail. Besides, for a wind turbine, which is a power generator plant, it would not be beneficial if the control system itself consumes a large amount of power that the turbine generates.

Moreover, because a semi-active controller cannot inject mechanical power into the system, and conceptually is a passive damper with adjustable properties, it does not suffer the instability problems that an active damper can have. Some of the examples of semi-active dampers are variable-orifice fluid dampers and controllable fluid dampers.

8.4.5 *Variable-orifice dampers*

In these types of dampers, the electromechanical, variable-orifice valve controls the opening of the fluid passage and so the hydraulic head loss in the cylindrical piston. This simply translates to altering the fluid resistance or the property of the damper

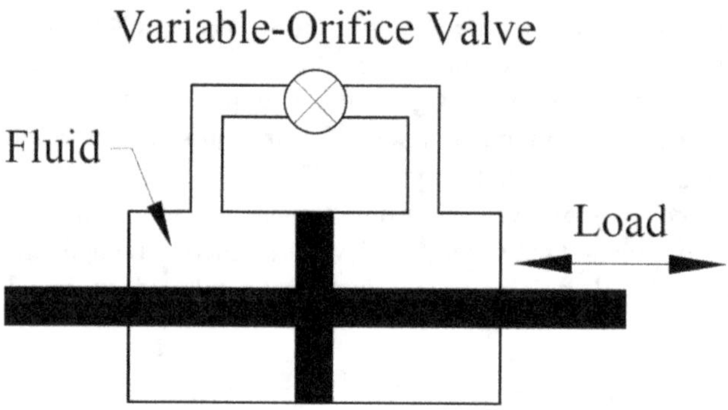

Figure 8.10 Variable-orifice damper [28]

with a very low amount of power. Figure 8.10 presents a schematic picture of such a device. Some studies implemented such a device to control different types of civil engineering structures [54, 55].

8.4.6 Controllable fluid damper

Since controllable valve dampers have various moving mechanical parts, they potentially have malfunctioning probability and maintenance issues, which can reduce their reliability. Controllable fluid dampers are those that the property of the fluid can be changed by electrical (for electrorheological or ER fluids) or magnetic field (for magnetorheological or MR fluids). Hence, since there are no moving parts other than the piston, ER and MR dampers are very reliable. Moreover, they can respond to the changes in milliseconds, so that they decrease the delay considerably, which is significantly important in high-frequency events.

MR/ER free-flowing, linear viscous fluid properties can be changed with controllable yield strength, to a semisolid by exposing to an electric/magnetic field, almost immediately. Typically, MR fluid dampers that have received more interest rather than ER fluid dampers are consist of magnetically polarizable micron-sized particles dispersed in a carrier medium such as mineral or silicone oil. Applying a magnetic field, then forms particle chains in the fluid. Hence, the fluid becomes semisolid with viscoplastic behavior [28]. See [56–61] for more information on the application of these types of dampers in civil engineering structures and wind turbines.

8.5 MR dampers

To predict and simulate the behavior of an MR damper, an appropriate model should be developed and validated experimentally. In the literature, several conceptual mechanical models are presented by different mathematical expressions. Effective control models of MR dampers are those that precisely can reproduce the nonlinear behavior of the system altered by a magnetic field. According to Sahin *et al.* (2012)

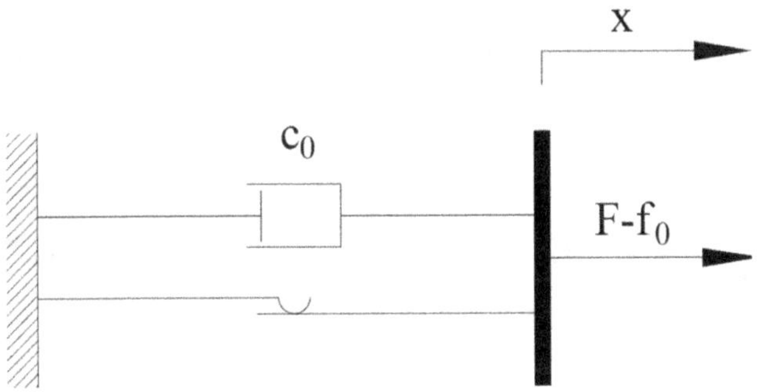

Figure 8.11 *Mechanical model representation of the Bingham model [68]*

[62], there are two main categories of models: (i) non-parametric and (ii) parametric models. In non-parametric models, the parameters in the model are not necessarily physical parameters such as neural network-based models [63, 64], neuro-fuzzy based models [65], and Chebyshev polynomials [66, 67]. Parametric models are those that consist of a set of mechanical elements such as springs, friction, and viscous dampers with various configurations. Therefore, the parameters of this category of models have the physical meaning which makes these types of models more desirable [62]. Brief information on some of the parametric models is presented as follows, including the formulations that are commonly used with these models.

8.5.1 Bingham model

A simplified mechanical scheme of this model is shown in Figure 8.11. In this model, a viscous damper is set parallel to a Coulomb friction element and the control force is governed as follows:

$$F_{MR} = f_c \text{sgn}(\dot{x}) + c_0\dot{x} + f_0 \tag{8.14}$$

where c_0, f_0, and f_c are damping coefficient, force offset, and frictional force related to the fluid yield stress, respectively, and a detailed information can be found in [68, 69].

8.5.2 Bouc-Wen model

As one of the earliest models, the Bouc-Wen model is used widely to model the dynamic behavior of MR dampers. A schematic simplified mechanical representation of this model is plotted in Figure 8.12. The control force of the MR damper, according to this model, can be calculated from the following equation:

$$F_{MR} = c_0\dot{x} + k_0(x - x_0) + \alpha z \tag{8.15}$$

here z is defined as:

$$\dot{z} = -\gamma\left|\dot{x}\right| z\,|z|^{n-1} - \beta\dot{x}\,|z|^{n} + A\dot{x} \tag{8.16}$$

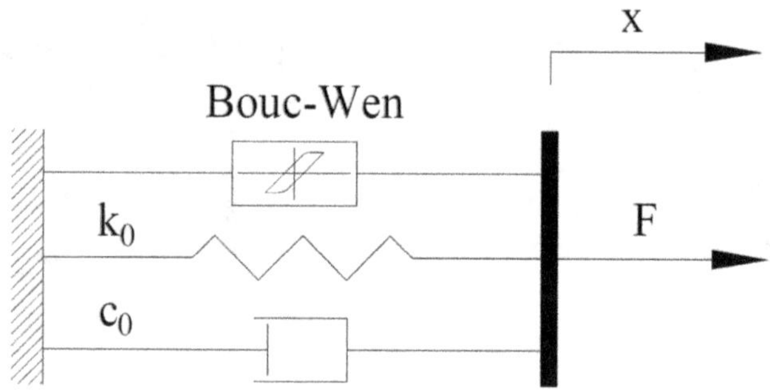

Figure 8.12 Mechanical model representation of Bouc-Wen for the MR damper [68]

To account for the unloading linearity and pre-yield to post-yield transmission, one can adjust the parameters γ, β, and A [8, 68, 70].

8.5.3 Modified Bouc-Wen model

In both Bingham and Bouc-Wen models, the force-velocity response does not roll-off when the velocities have small magnitude and the acceleration and velocity have opposite signs [68]. The modified version of the Bouc-Wen model can better predict the damper force in the region of small velocity with the opposite sign of acceleration and velocity. The mechanical scheme of the model is shown in Figure 8.13 and the force can be calculated with the following formulations:

$$F_{\mathrm{MR}} = c_1\dot{y} + k_1\left(x - x_0\right) \tag{8.17}$$

$$\dot{y} = \frac{1}{\left(c_0 + c_1\right)}\left\{\alpha z + c_0\dot{x} + k_0\left(x - y\right)\right\} \tag{8.18}$$

$$\dot{z} = -\gamma\left|\dot{x} - y\right| z\,|z|^{n-1} - \beta\left(\dot{x} - \dot{y}\right)|z|^{n} + A\left(\dot{x} - \dot{y}\right) \tag{8.19}$$

To find the parameters in this model for a 20-ton MR damper refer to [56] and for a 1.5-ton MR damper refer to [71]. There are some other models such as the Dahl

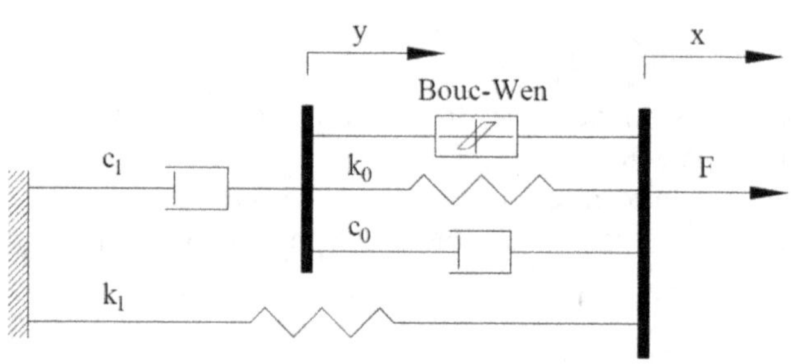

Figure 8.13 Schematic representation of modified Bouc-Wen for the MR damper [68]

model, Kwok model, and Gamota and Filisko (1991), and for more information you can refer to [62, 68].

8.5.4 Control algorithms

During the past decade, significant development has been achieved in the field of active and semi-active controllers in civil engineering structures and various control algorithms are developed. Some of these control algorithms and strategies are based on applying direct velocity feedback (DVFB) control, H_{∞} feedback control, wavelet function, fuzzy logic, genetic algorithms (GA), neural networks (NN), linear quadratic regulator (LQR), and linear Gaussian regulator (LGR) [50, 72–75].

One challenge in the use of semi-active technology is the nonlinearity nature of these systems. The nonlinearity requires control algorithms that are appropriate for implementation in full-scale structures. Various control algorithms are presented for the semi-active systems, including the skyhook damper control algorithm [76], the bang-bang controller [54], Lyapunov stability theory [77], modulated homogeneous friction algorithm [78], and clipped-optimal controllers [58, 79, 80].

The clipped-optimal controller is known as one of the most common algorithms. According to this algorithm, the linear optimal controller gains, $\mathbf{K}_c(s)$, should be designed to calculate the desired control force, f_c according to the measured structural responses, $\mathbf{y}$, and the measured force, f applied to the system [81]:

$$\mathrm{f}\_{c=L^{-1}}\left\{-\mathbf{K}_c(s)L\left\{\begin{matrix}\mathbf{y}\\ f\end{matrix}\right\}\right\} \tag{8.20}$$

and the control law is:

$$v = V_{max}H\left(\{f_c - f\}f\right) \tag{8.21}$$

A comparison study on the performance of different control algorithms for a three-story building excited with the south–north direction of the El-Centro earthquake showed that the largest reduction in inter-story displacement and the third-floor displacement was achieved with the clipped-optimal algorithm [81]. However, this study showed that the maximum reduction in peak acceleration obtained applying the bang-bang approach [81]. The same comparison has been done in a three-story building excited with El-Centro earthquake and controlled by various controller strategies, including Passive-off and on, Lyapunov, Decentralized bang-bang, Quasi-bang-bang, Maximum energy dissipation, Modulated homogenous friction, Clipped-optimal, and new quasi-bang-bang controllers give weight to the output in a way that is similar to a fuzzy logic controller [59]. The results revealed that the Lyapunov controller, quasi-bang-bang controller, and clipped-optimal control algorithm provided significant maximum displacement. Also, the clipped-optimal control algorithm gave a good reduction in the maximum absolute accelerations [59]. For more information on control algorithms refer to [57, 81, 82].

8.6 Controller design

In passive dampers, the system tuning includes setting the damping ratio and the frequency ratio once, without further adjustment during the operation. There are some studies on optimized passive damper parameters for wind and earthquake [35, 83–86]. The design procedure for semi-active controllers is more complicated. The performance of the controller depends on the selected algorithm that commands the damper. Moreover, the success of the controller to suppress excessive vibrations effectively depends on the desired control force as well as the control algorithm. Available methods to set the desired control force, and therefore design the optimal controller are based on the simulation. This makes existing methods to be very time consuming and requires significant computational resources. An alternative to the common simulation-based methods is an energy-based probabilistic approach [87] for SDOF systems. In this approach, the probability of dissipativity of the controller is calculated by solving the Lyapunov function and substituting the results into a stochastic dissipativity index presented by [72]. Then the probability of dissipativity can be applied in an energy-based method to obtain the desired control force without the simulation [72].

Although there are several studies on the control of wind turbines for structural performance improvement, there are just limited studies on the evaluation of the performance of wind turbines for multi-hazard loads and mitigation of excessive vibrations. Looking into the responses of the structure for various loading scenarios and control approaches is an important task because various excitation sources (e.g., wind, wave, earthquake, rotor imbalance) affect the structure differently and the controllers might improve a specific response by scarifying the other. To prevent that we need to investigate the performance of the system for multi-hazards.

In the literature, various control methods, including passive, semi-active, and active dampers installed in the nacelle, tower, or the foundation of the turbine are proposed and compared for wind and wave, or wind, and earthquake. For instance, the idea of installing vibration isolators for wind turbines is suggested in different studies. It is proposed to install vibration isolators at the top of the turbine's tower, or at the foundation using semi-active MR dampers to reduce the challenges regards the conventional base isolators for wind turbines [60, 88]. Both wind and seismic loading placed on the structure as disturbance forces and the responses quantified. The results showed that implementing an isolation system was beneficial for reducing certain key parameters of the turbine's structural response, particularly for the wind turbines in seismically active regions.

Offshore and coastal wind turbines' vibration mitigation is also investigated through different numerical studies considering a different type of floating and fixed foundations under the action of wind and wave. In a study that aimed to compare the effects of passive and active TMD on a floating wind turbine, a reduction in fatigue loads of the tower base was compared [17]. However, the stroke of the active damper was much greater than the passive damper. This can be a limitation as the available space in the nacelle or at the top of the wind turbine is limited and it is mainly occupied by mechanical parts.

The impacts of pendulum-type passive and semi-active dampers at the top of both floating and fixed offshore wind turbines are also investigated for parked and operating wind turbines [16]. The TMD's oscillations absorbed the energy of tower fore-aft and side-to-side motion, simultaneously. However, since the semi-active TMD (SATMD) was activated around 32 percent of the time in the fore-aft direction and 31 percent in the side-to-side direction, the power source of the system was increased meaningfully [16].

In a recent study on the control of fixed offshore wind turbines, multi-hazard loading on wind turbines including wind, wave, and earthquake is evaluated [89]. A SATMD is installed in the nacelle of a wind turbine to control the along-wind displacement of the nacelle/tower and it is shown that the SATMD was more effective than a passive damper to reduce the resonance responses. However, it is not shown that the controller was designed effectively for multi-hazards.

8.7 Advanced control theory to accelerate the optimal tuning of smart structures

The Windstorm Impact, Science and Engineering (WISE) research team at Louisiana State University (LSU) devoted efforts to developing advanced control theory to accelerate the optimal tuning of smart structures, to develop novel probabilistic analytical methods to address the complex behavior and inherent nonlinearity in semi-active control, for multiple hazards. The LSU WISE research group developed novel control methods that enabled semi-active controller tuning under complex behavior and inherent nonlinearity, without simulations [90]. The recently developed control theory enables the evaluation of semi-active controllers' performance of multi-degrees-of-freedom systems, without significant computational effort [91]. With this analytical probabilistic control theory, a wide range of controllers can be evaluated in a fraction of a second while optimum control parameters can be tuned to achieve different control objectives. The potential applications include semi-active vibration control in flexible structures under multiple hazard loads brought by wind, wave, and earthquakes.

In the dynamics of structures under multiple hazards, vibration control offers attractive means for protecting structures and inhabitants, as well as balancing between resilience and sustainability. A fundamental task related to the design of semi-active controllers to attenuate structural vibrations is the tuning of the parameters based on the physical properties (mass, stiffness, and damping), mitigation objectives, nature of excitation, etc. Besides, in smart damping, the performance of the system to mitigate excessive vibrations is governed by the corresponding ideal active control force and the algorithm alerting the damping characteristics of the device.

While there are several studies on the advantages of employing semi-active dampers to attenuate vibrations in flexible structures, the controller tuning process is mainly based on numerical simulations, hindering the consideration of different families of controllers, to achieve controller tuning-based performance under varying loading patterns, for instance, excitation due to multiple hazards. To address this challenge, the LSU WISE research team developed a novel control theory that enables the semi-active controller tuning, of multi-degrees-of-freedom systems, under complex behavior and inherent nonlinearity, without simulations [90]. According to the dissipative probability of the corresponding active control forces, a new concept of semi-active control gains was developed:

$$k_{s,i} = \tfrac{1}{2} G_{1,i} \left\{1 + \sin^2 \left[\pi \left(p - 0.5\right)\right]\right\} \quad \& \quad i = 1, 2, ..., m \tag{8.22}$$

$$\mathbf{k}_s = diag\left(k_{s,i}\right) \tag{8.23}$$

$$\omega_a = \sqrt{eig\left(M^{*-1}\left(\mathrm{K}_{m,a}\right)\right)}; \quad \mathrm{K}_{m,a} = \mathrm{K}^* + diag\left(\Phi G1\right) \tag{8.24}$$

$$\omega_s = \sqrt{eig\left(M^{*-1}\left(\mathrm{K}^* + \mathrm{k}_s\right)\right)} \tag{8.25}$$

$$c_{sl,i} = \frac{\left|G_{l,i}\right|}{\pi \omega_{s,i}} \left\{1 + \sin^2 \left[\pi \left(P - 0.5\right)\right]\right\} \quad \& \quad i = 1, 2, ..., m \tag{8.26}$$

$$C_{s2,i} = \frac{G_{2,i}\omega_{a,i}}{\pi \omega_{s,i}} \times \left\{\frac{\pi}{2} + \left[\sin[\pi(P-0.5)] \times \sqrt{1-\sin^2[\pi(P-0.5)]} + \tan^{-1}\left(\frac{\sin[\pi(P-0.5)]}{\sqrt{1-\sin^2[\pi(P-0.5)]}}\right)\right]\right\} \tag{8.27}$$

$$c_s = diag\left(c_{s1,i} + c_{s2,i}\right) \quad \& \quad i = 1, 2, ..., m \tag{8.28}$$

where P is the dissipative probability of the corresponding optimal active control forces, G_1 and G_2 are the active gains, and K^* is the generalized stiffness of the primary structure [90].

With this new theory, a family of controllers can be studied, and their performance can be evaluated in a fraction of a second. This enables controller tuning to achieve optimum parameters for different design objectives. The potential applications of this control theory include vibration control in flexible structures under multiple hazard loads brought by, for example, wind, wave, and seismic loadings. To show the potentiality of the new theory, we considered semi-active controller design in both onshore and offshore wind turbines, in parked and operating positions, under wind, wave, and earthquake external forces, in addition to mechanical unbalance [90].

The Lagrangian approach was employed to model the wind turbine considering the blade/tower coupling. Two types of control systems are investigated: (i) SATMD installed in the nacelle, and (ii) outer bracing with MR dampers (Figures 8.14 and 8.15) [90, 93]. The new control theory was employed to accelerate the tuning of the semi-active controllers subjected to multiple hazards (Figure 8.15).

The results show that for the operating wind turbine, earthquake excitations lead to excessive vibrations for a long period. However, the properly tuned semi-active damping system improved the performance of the wind turbine by reducing the magnitude and duration of excessive vibrations. For

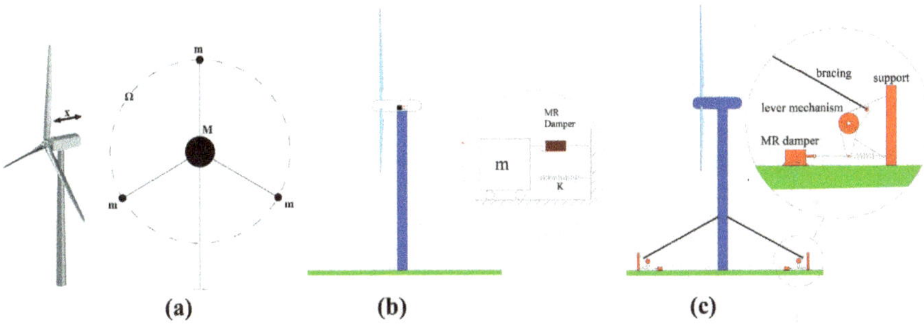

Figure 8.14 *Modeling and control of a 5 MW wind turbine: (a) lumped mass model, (b) SATMD, and (c) external bracing with magnetorheological (MR) dampers benefiting a lever mechanism to enhance the performance [92]*

vibration control under multiple hazards, including wind and earthquake, the control system is immediately available, compared to, for instance, the classical slow responding tuned mass damper.

This new control theory reduced the design and performance evaluation time significantly. For instance, the CPU-time required for a single performance evaluation case considering specific weighting matrices, using a 2.5 GHz Intel Core i5-2400S processor PC, was about 3 h. This amount of time increases significantly for aggressive primary active controllers, which means that several days/weeks are needed to evaluate the performance of a family of controllers under certain loading excisions. This amount of simulation time can increase exponentially when the objective is to design semi-active controllers for multiple hazards, for different optimization objectives—e.g., minimizing displacement, velocity, acceleration, or all responses. The new analytical control theory reduced the total time to a few seconds for a wide range of controllers and optimization objectives.

8.8 Summary

Structural design for multiple hazards is a challenge that requires the adoption and consideration of new control methods, theory, devices, and implementation techniques. Advanced control methods can yield more resilient and sustainable infrastructure, such as wind turbines, as well as other flexible structures, such as high-rise buildings, towers, and long-span bridges under the impact of wind, wave, and seismic excitations. The recently developed semi-active gain theory can be a benchmark for further investigations on structural control in wind turbines and other structures. Besides, the analytical approach has potential applications in various mechanical systems. The goal is to build new structures and retrofit existing ones in ways that can balance resilience with sustainability, enhance safety, and reduce the huge cost of rebuilding due to multiple hazards. This will broadly impact the wind/structural engineering research field and facilitate effective

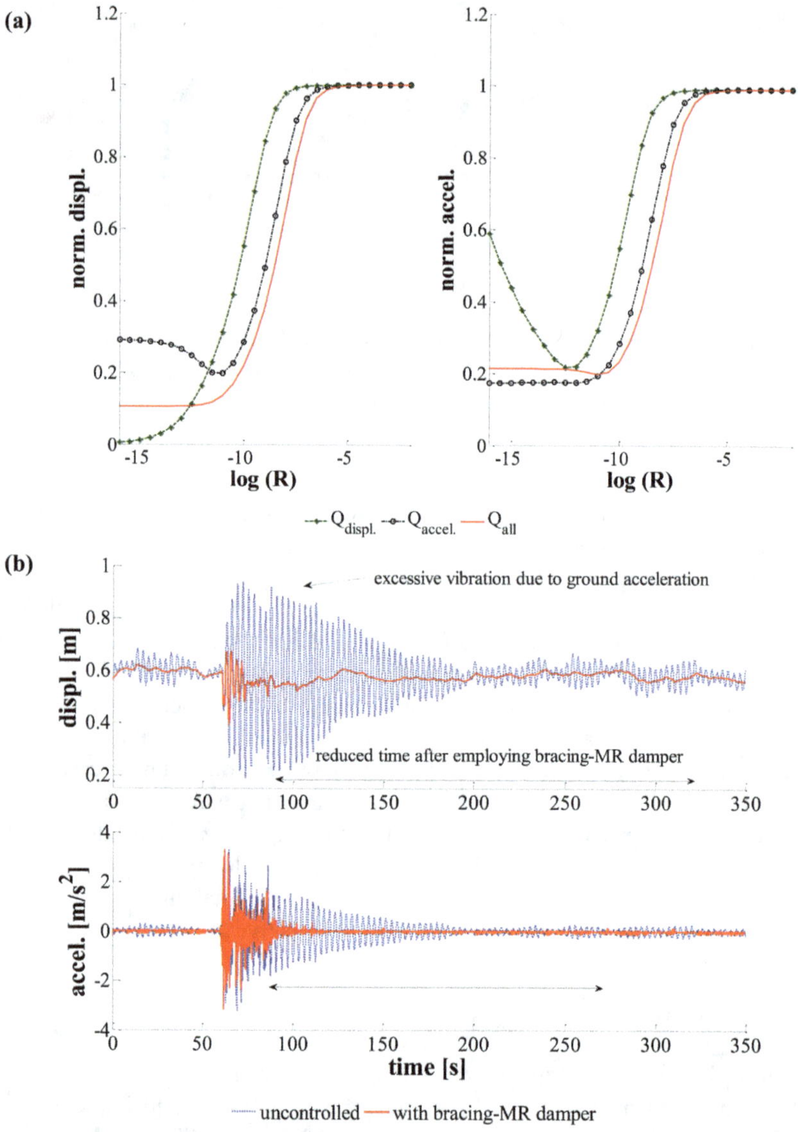

Figure 8.15 Results obtained by the proposed semi-active controller tuning theory: (a) effect of weighting the Q and R matrices to achieve desired control forces for certain optimization objectives—minimizing normalized root-mean-square (RMS) displacement and acceleration and (b) time histories of uncontrolled and controlled (employing MR dampers with an outer bracing system) displacement and acceleration responses of the nacelle of an operating wind turbine under earthquake loading (El Centro horizontal component of ground acceleration)

investments in the infrastructure industry that will result in more resilient and sustainable communities, improve the quality of life, and contribute to economic growth.

Acknowledgements

This research was partially supported by a grant from the Louisiana Board of Regents (award no. LEQSF (2016-19)-RD-A-02), as well as Faculty Grant from Louisiana State University (LSU). The results and findings are those of the authors and do not necessarily represent the official position of the funding agencies.

References

[1] WWEA. *World Wind Energy Report*. Bonn, Germany: World Wind Energy Association; 2011.

[2] Laks J.H., Pao L.Y., Wright A.D. 'Control of wind turbines: past, present, and future'. *American Control Conference*. 2009:2096–103.

[3] Katsanos E.I., Thöns S., Georgakis Christos T. 'Wind turbines and seismic hazard: a state-of-the-art review'. *Wind Energy*. 2016;**19**(11):2113–33.

[4] Office of Energy Efficiency & Renewable. *The Inside of a Wind Turbine*. U.S. Department of Energy; 2010.

[5] Ancona D., McVeigh J. *Wind Turbine-Materials and Manufacturing Fact Sheet*. Maryland, USA: Princeton Energy Resources International; 2001.

[6] Ciang C.C., Lee J.-R., Bang H.-J. 'Structural health monitoring for a wind turbine system: a review of damage detection methods'. *Measurement Science and Technology*. 2008;**19**(12):122001.

[7] Harte R., Van Zijl G.P.A.G. 'Structural stability of concrete wind turbines and solar chimney towers exposed to dynamic wind action'. *Journal of Wind Engineering and Industrial Aerodynamics*. 2007;**95**(9-11):1079–96.

[8] Rosenbloom E. *A Problem With Wind Power*. 2006. Available from http://www.aweo.org/problemwithwind.html [Accessed 5/18/2021].

[9] Hahn B., Durstewitz M., Rohrig K. 'Reliability of wind Turbines: experiences of 15 years with 1,500 Wts'. *Wind Energy. Proceedings of the Euromech Colloquium*. 2007:329–32.

[10] Ruskin B. *Wind Turbine Collapse Wreckage in Cape Breton Combed for Clues of Cause [online]*. Toronto: CBC News; 2016. Available from https://www.cbc.ca/news/canada/nova-scotia/wind-turbine-collapse-point-tupper-cape-breton-enercon-renewable-1.3734562.

[11] windAction. *Wind Turbine Collapse in Kansas*. 2014. Available from http://www.windaction.org/posts/41802-wind-turbine-collapse-in-kansas#.YKSSInmSmUk [Accessed 18 May 2021].

[12] Chen J., Liu Y., Bai X. 'Shaking table test and numerical analysis of offshore wind turbine tower systems controlled by TLCD'. *Earthquake Engineering and Engineering Vibration*. 2015;**14**(1):55–75.

[13] van der Tempel J., Molenaar D.-P. 'Wind turbine structural dynamics – a review of the principles for modern power generation, Onshore and offshore'. *Wind Engineering*. 2002;**26**(4):211–22.

[14] Rogers N., Still D. 'Structural dynamics of offshore wind Turbines subject to extreme wave loading N Rogers border wind limited, Hexham, Northumberland'. *EWEC-Conference*; 1999. pp. 369–74.

[15] Xu N., Ishihara T. 'Analytical formulae for wind turbine tower loading in the parked condition by using quasi-steady analysis'. *Wind Engineering*. 2014;**38**(3):291–309.

[16] Tsouroukdissian A.R. 'Smart novel semi-active tuned mass damper for fixed bottom and floating offshore wind'. *Offshore Technology Conference*; 2016. pp. 1–17.

[17] Lackner M.A., Rotea M.A. 'Structural control of floating wind turbines'. *Mechatronics*. 2011;**21**(4):704–19.

[18] Jonkman J.M. *Dynamics Modeling and Loads Analysis of an Offshore Floating Wind Turbine*. United States; 2007. Available from https://www.osti.gov/servlets/purl/921803.

[19] Dinh V.N., Basu B., Nagarajaiah S. 'Semi-active control of vibrations of SPAR type floating offshore wind turbines'. *Smart Structures and Systems*. 2016;**18**(4):683–705.

[20] Quilligan A., O'Connor A., Pakrashi V. 'Fragility analysis of steel and concrete wind turbine towers'. *Engineering Structures*. 2012;**36**(1):270–82.

[21] Fitzgerald B., Basu B. 'Structural control of wind turbines with soil structure interaction included'. *Engineering Structures*. 2016;**111**(1):131–51.

[22] Hansen M.H. 'Improved modal dynamics of wind turbines to avoid stall-induced vibrations'. *Wind Energy*. 2003;**6**(2):179–95.

[23] Arrigan J., Pakrashi V., Basu B., Nagarajaiah S. 'Control of flapwise vibrations in wind turbine blades using semi-active tuned mass dampers'. *Structural Control and Health Monitoring*. 2011;**18**(8):840–51.

[24] Jonkman J., Butterfield S., Musial W., Scott G. *Definition of a 5-MW Reference Wind Turbine for Offshore System Development [online]*. 2009. Available from https://www.nrel.gov/docs/fy09osti/38060.pdf [Accessed 18 May 2021].

[25] Abhinav K.A., Saha N. 'Dynamic analysis of an offshore wind turbine including soil effects'. *Procedia Engineering*. 2015;**116**:32–9.

[26] Harte M., Basu B., Nielsen S.R.K. 'Dynamic analysis of wind turbines including soil-structure interaction'. *Engineering Structures*. 2012;**45**(21/22):509–18.

[27] Lombardi D., Bhattacharya S., Muir Wood D., Muir D. 'Dynamic soil–structure interaction of monopile supported wind turbines in cohesive soil'. *Soil Dynamics and Earthquake Engineering*. 2013;**49**(5):165–80.

[28] Housner G.W., Bergman L.A., Caughey T.K., *et al.* 'Structural control: past, present, and future'. *Journal of Engineering Mechanics*. 1997;**123**(9):897–971.

[29] Soong T.T., Constantinou M.C. (eds.). *Passive and Active Structural Vibration Control*. Springer, Vienna; 1994.

[30] Soong T.T., Dargush G.F. *Passive Energy Dissipation Systems in Structural, Engineering*. Chichester, England: Wiley; 1997.

[31] Den Hartog J.P. *Mechanical Vibrations. New York, NY*: Crastre Press; 1947.

[32] Villaverde R. 'Seismic control of structures with damped resonant appendages'. *Proceedings of the First World Conference on Structural Control*; 1994. pp. 113–22.

[33] Lin Y.-Y., Cheng C.-M., Lee C.-H. 'A tuned mass damper for suppressing the coupled flexural and torsional buffeting response of long-span bridges'. *Engineering Structures*. 2000;**22**(9):1195–204.

[34] Murtagh P.J., Ghosh A., Basu B., Broderick B.M. 'Passive control of wind turbine vibrations including blade/tower interaction and rotationally sampled turbulence'. *Wind Energy*. 2008;**11**(4):305–17.

[35] Si Y., Karimi H.R., Gao H. 'Modelling and optimization of a passive structural control design for a spar-type floating wind turbine'. *Engineering Structures*. 2014;**69**(5):168–82.

[36] Tamura Y., Fujii K., Ohtsuki T., Wakahara T., Kohsaka R. 'Effectiveness of tuned liquid dampers under wind excitation'. *Engineering Structures*. 1995;**17**(9):609–21.

[37] Makris N., Constantinou M.C. *Viscous Dampers: Testing, Modeling and Application in Vibration and Seismic Isolation*. National Center for Earthquake Engineering Research. Technical Report NCEER-90-0028. Buffalo, NY; 1990.

[38] Constantinou M.C., Symans M.D., Tsopelas P., Taylor D.P. 'Fluid viscous dampers in applications of seismic energy dissipation and seismic isolation'. *Proceedings of the ATC 17-1 Seminar on Seismic Isolation, Passive Energy Dissipation, and Active Control*. 1993;**2**:581–92.

[39] Vijayaraja. Passive Energy Dissipation Devices in Earthquake Resistant Buildings. 2013. Available from https://buildcivil.wordpress.com/2013/11/25/passive-energy-dissipation-devices/ [Accessed 18 May 2021].

[40] Nonami K., Nishimura H., Tian H. *H∞/μ control-based frequency-shaped sliding mode control for flexible structures*. **289**. ASME-PUBLICATIONS-PVP; 1994. p. 19.

[41] Yang J.N., Wu J.C., Agrawal A.K. 'Sliding mode control for seismically excited linear structures'. *Journal of Engineering Mechanics*. 1995;**121**(12):1386–90.

[42] Chase J.G., Smith H.A. 'Robust H∞ control considering actuator saturation. I: theory'. *Journal of Engineering Mechanics*. 1996;**122**(10):976–83.

[43] Agrawal A.K., Yang J.N., Schmitendorf W.E., Jabbari F. 'Stability of actively controlled structures with actuator saturation'. *Journal of Structural Engineering*. 1997;**123**(4):505–12.

[44] Spencer B.F., Sain M.K., Kantor J.C., Montemagno C. 'Probabilistic stability measures for controlled structures subject to real parameter uncertainties'. *Smart Materials and Structures*. 1992;**1**(4):294–305.

[45] Spencer B.F., Sain M.K., Won C.-H., Kaspari D.C., Sain P.M. 'Reliability-based measures of structural control robustness'. *Structural Safety*. 1994;**15**(1-2):111–29.

[46] Field R.V., Voulgaris P.G., Bergman L.A. 'Methods to compute probabilistic measures of robustness for structural systems'. *Journal of Vibration and Control*. 1996;**2**(4):447–63.

[47] Field R.V., Voulgaris P.G., Bergman L.A. 'Probabilistic stability robustness of structural systems'. *Journal of Engineering Mechanics*. 1996;**122**(10):1012–21.

[48] Nagarajaiah S. 'Fuzzy controller for structures with hybrid isolation system'. *Proceedings of First World Conference on Structural Control, TA2*. 1994:67–76.

[49] Subramaniam R.S., Reinhorn A.M., Riley M.A., Nagarajaiah S. 'Hybrid control of structures using fuzzy logic'. *Computer-Aided Civil and Infrastructure Engineering*. 1996;**11**(1):1–17.

[50] Aly A.M. *Proposed Robust Tuned Mass Damper for Response Mitigation in Buildings Exposed to Multidirectional Wind*. **691**; 2012. p. 664.

[51] Ghaboussi J., Joghataie A. 'Active control of structures using neural networks'. *Journal of Engineering Mechanics*. 1995;**121**(4):555–67.

[52] Spencer B.F., Sain M.K. 'Controlling buildings: a new frontier in feedback'. *IEEE Control Systems Magazines*. 1997;**17**(6):19–35.

[53] Spencer B.F., Nagarajaiah S. 'State of the art of structural control'. *Journal of Structural Engineering*. 2003;**129**(7):845–56.

[54] Feng Q., Shinozuka M., Housner G. 'Use of a Variable Damper for Hybrid Control of Bridge Response Under Earthquake' in Housner G., Masri S. (eds.). *Proceedings of the U.S. National Workshop on Structural Control Research*. Los Angeles, CA: University of Southern California; 1990. pp. 107–12.

[55] Kamagata S., Kobori T. 'Autonomous adaptive control of active variable stiffness system for seismic ground motion'. *Proceedings of the First World Conference on Structural Control, TA4*. 1994:33–42.

[56] Spencer B.E Jr., Carlson J., Sain M.K., Yang G. 'On the current status of magnetorheological dampers: seismic protection of full-scale structures'. *Proceedings of the American Control Conference*; 1997. pp. 458–62.

[57] Spencer B.F., Dyke S.J., Sain M.K. 'Magnetorheological dampers: a new approach to seismic protection of structures'. *Proceedings of the 36th IEEE Conference on Decision and Control*; 1996. pp. 676–81.

[58] Dyke S.J., Spencer B.F Jr., Sain M.K., Carlson J.D. 'Seismic response reduction using magnetorheological dampers'. *Proceedings of the IFAC World Congress*; 1996. pp. 145–50.

[59] Aly A.M. 'Vibration control of buildings using magnetorheological damper: a new control algorithm'. *Journal of Engineering*. 2013;**2013**(9):1–10.

[60] Caterino N. 'Semi-active control of a wind turbine via magnetorheological dampers'. *Journal of Sound and Vibration*. 2015;**345**(8):1–17.

[61] Zasso A., Aly A.M., Resta F. 'Mr dampers with lever mechanism for response reduction in high-rise buildings under wind loads'. *5th European & African Conference on Wind Engineering; 2009*. 2009.

[62] Şahin İ., Çeşmeci Ş. 'Comparison of some existing parametric models for magnetorheological fluid dampers'. *Smart Materials and Structures*. 2010;**19**(3):035012.

[63] Chang C.-C., Roschke P. 'Neural network modeling of a magnetorheological damper'. *Journal of Intelligent Material Systems and Structures*. 1998;**9**(9):755–64.

[64] Chang C.-C., Zhou L. 'Neural network emulation of inverse dynamics for a magnetorheological damper'. *Journal of Structural Engineering*. 2002;**128**(2):231–9.

[65] Schurter K., Roschke P. 'Fuzzy modeling of a magnetorheological damper using ANFIS'. *EEE International Conference on Fuzzy Systems*; 2000. pp. 122–7.

[66] Ehrgott R.C., Masri S.F. 'Modeling the oscillatory dynamic behaviour of electrorheological materials in shear'. *Smart Materials and Structures*. 1992;**1**(4):275–85.

[67] Hanson R., Filisko F. 'Electrorheological dampers: Part II: Testing and modeling'. *Journal of Applied Mechanics*. 1996;**1050**:48109.

[68] Spencer B.F., Dyke S.J., Sain M.K., Carlson J.D. 'Phenomenological model for Magnetorheological Dampers'. *Journal of Engineering Mechanics*. 1997;**123**(3):230–8.

[69] Yang G., Spencer B.F., Carlson J.D., Sain M.K. 'Large-scale MR fluid dampers: Modeling and dynamic performance considerations'. *Engineering Structures*. 2002;**24**(3):309–23.

[70] Aly A.M., Zasso A., Resta F. 'On the dynamics of a very slender building under winds: response reduction using MR dampers with lever mechanism'. *The Structural Design of Tall and Special Buildings*. 2011;**20**(5):539–51.

[71] Dyke S.J., Spencer B.F., Sain M.K., Carlson J.D. 'Modeling and control of magnetorheological dampers for seismic response reduction'. *Smart Materials and Structures*. 1996;**5**(5):565–75.

[72] Inaudi J.A. 'Performance of variable-damping systems: theoretical analysis and simulation'. *3rd International Workshop on Structural Control for Civil and Infrastructure Engineering*; 2000. pp. 301–16.

[73] Fang J.Q., Li Q.S., Jeary A.P. 'Modified independent modal space control of m.d.o.f. systems'. *Journal of Sound and Vibration*. 2003;**261**(3):421–41.

[74] Yang J.N., Li Z., Liu S.C. 'Instantaneous optimal control with velocity and acceleration feedbacks'. *Journal of Probabilistic Engineering Mechanics*. 1991;**16**:204–21.

[75] Li Q.S., Liu D.K., Fang J.Q. 'Optimum design of actively controlled structures using genetic algorithms'. *Advances in Structural Engineering*. 1999;**2**(2):109–18.

[76] Karnopp D., Crosby M.J., Harwood R.A. 'Vibration control using semi-active force generators'. *Journal of Engineering for Industry*. 1974;**96**(2):619–26.

[77] Leitmann G. 'Semiactive control for vibration attenuation'. *Journal of Intelligent Material Systems and Structures*. 1994;**5**(6):841–6.

[78] Inaudi J.A. 'Modulated homogeneous friction: a semi-active damping strategy'. *Earthquake Engineering & Structural Dynamics*. 1997;**26**(3):361–76.

[79] Dyke S.J., Spencer B.F., Quast P., Sain M.K., Kaspari D.C., Soong T.T. 'Acceleration feedback control of MDOF structures'. *Journal of Engineering Mechanics*. 1996;**122**(9):907–18.

[80] Dyke S.J., Spencer B.F., Quast P., Kaspari D.C., Sain M.K. 'Implementation of an active mass driver using acceleration feedback control'. *Computer-Aided Civil and Infrastructure Engineering*. 1996;**11**(5):305–23.

[81] Dyke S.J., Spencer B.F Jr. 'A comparison of Semi-Active control strategies for the Mr Damper'. *IASTED International Congress on Intelligent Systems and Control*. 1997.

[82] Jansen L.M., Dyke S.J. 'Semiactive control strategies for MR dampers: Comparative study'. *Journal of Engineering Mechanics*. 2000;**126**(8):795–803.

[83] Feng M.Q., Mita A. 'Vibration control of tall buildings using mega sub configuration'. *Journal of Engineering Mechanics*. 1995;**121**(10):1082–8.

[84] Chang C.C., Qu W.L. 'Unified dynamic absorber design formulas for wind-induced vibration control of tall buildings'. *The Structural Design of Tall Buildings*. 1998;**7**(2):147–66.

[85] Min K.-W., Kim J., Lee H.-R. 'A design procedure of two-way liquid dampers for attenuation of wind-induced responses of tall buildings'. *Journal of Wind Engineering and Industrial Aerodynamics*. 2014;**129**(10):22–30.

[86] Pirner M., Urushadze S. 'Liquid damper for suppressing horizontal and vertical'. *WIT Transactions on Modelling and Simulation*. 2005;**41**:1329–49.

[87] Aly A.M., Christenson R.E. 'On the evaluation of the efficacy of a smart damper: a new equivalent energy-based probabilistic approach'. *Smart Materials and Structures*. 2008;**17**(4):045008.

[88] Van der Woude C., Narasimhan S. 'A study on vibration isolation for wind turbine structures'. *Engineering Structures*. 2014;**60**(5):223–34.

[89] Sun C. 'Semi-active control of monopile offshore wind turbines under multi-hazards'. *Mechanical Systems and Signal Processing*. 2018;**99**(9):285–305.

[90] Rezaee M., Aly A.M. 'Vibration control in wind turbines to achieve desired system-level performance under single and multiple hazard loadings'. *Structural Control and Health Monitoring*. 2018:p. e2261.

[91] Rezaee M., Aly A.M. 'Proposed theory of semiactive gains for smart dampers in MDOF systems'. *Journal of Structural Engineering*. 2019;**145**(12):04019155.

[92] Aly A.M. 'Control of wind-induced motion in high-rise buildings with hybrid TM/MR dampers'. *Wind and Structures*. 2015;**21**(5):565–95.

[93] Rezaee M., Aly A.M. 'Vibration control in wind turbines for performance enhancement: a comparative study'. *Wind and Structures*. 2016;**22**(1):107–31.

Part III

Environmental Concerns

Chapter 9

Wind farm noise propagation and viable noise reduction strategies

Aki Grönman[1]

This chapter aims to present the key mechanisms of wind turbine noise generation, how people are affected by it, and how noise-related problems can be reduced. Initially, the physical phenomena behind noise generation are presented. The second section gives an overview of the key annoyance-related issues, including the roles played by the noise level, noise character, and atmospheric conditions. In Section 9.1, the currently used methods to model both noise generation and propagation are reviewed. In the penultimate section, before the summary, different methods to reduce wind turbine noise are reviewed, including the possibilities emerging during wind farm- and turbine-design processes and wind farm operation.

9.1 Wind turbine noise sources

The noise from horizontal axis wind turbines (HAWTs) can be divided into two sources, namely mechanical and aerodynamic noise. Mechanical noise is related to the nacelle area and includes sources such as the gearbox, the generator, and the yaw and pitch mechanisms. For example, the yaw mechanism can have a tonal noise during its operation at a frequency of approximately 1 100 Hz, as presented by Liu [1]. In general, the A-weighted sound pressure level from the nacelle is lower compared to the noise caused by the airfoil, as was presented by Oerlemans *et al.* [2] for a GAMESA G58 wind turbine.

The aerodynamic noise can be divided into three sources: (i) airfoil self-noise, (ii) incoming turbulence noise, and (iii) discrete frequency noise due to the blade-passing frequency and its harmonics [3, 4]. The airfoil self-noise is caused by the interactions between the blade and the wake and the boundary layer, and it can be divided into five mechanisms, according to Brooks *et al.* [3]:

[1]LUT University, Finland

1. interaction between the turbulent boundary layer and the trailing edge
2. vortex shedding of the laminar boundary layer
3. separation stall
4. vortex shedding due to trailing edge bluntness
5. blade tip vortex formation.

The incoming turbulence noise is caused by the interaction between the turbulence in the wind and the blade, and it can have higher importance at high winds, as discussed by Lowson [4]. In this noise mechanism, the interaction between the turbulent eddies and the boundary layer causes blade surface pressure fluctuations that become audible.

The noise caused by the blade-tower interaction is of a low frequency, as discussed by Oerlemans *et al.* [2]. Compared to the blade noise, Oerlemans *et al.* [2] found that for their turbine, the noise increase due to blade-tower interaction was not significant, although a small increase in noise was found due to this phenomenon at a frequency of 1.6 kHz. A later numerical study by Zajamsek *et al.* [5] suggested that the tower could be responsible for most of the noise emitted; however, they lacked experimental proof. The mechanism behind the noise that comes when the blade passes the tower originates from the change of the angle of attack due to the flow-slowing effect caused by the tower. This change of the angle of attack can be in the order of 2 degrees, as discussed by van den Berg [6].

According to the latest understanding, for large three-blade HAWTs, the trailing edge noise (broadband-type noise) is the dominant noise source, as presented by Oerlemans *et al.* [2]. Their experimental study also showed that the blade noise production was allocated to the outer parts of the blade for their turbine, except in the tip area. This phenomenon is illustrated schematically in Figure 9.1. It is also worth mentioning that the noise is produced during the whole rotation, but due to the experimental setup (similar to what a person would experience in front of the turbine), only the noise during downward movement was recorded.

9.2 Human response to wind turbine noise and noise characteristics

There are several factors that can affect the acceptance of wind turbines in terms of noise. For example, it is known that involving local people during planning new wind farms can have a positive effect on the acceptability. Janhunen *et al.* [7] also highlighted the importance of people's involvement after the planning phase. Sometimes visual effects, such as flickering from the turbines, can also annoy people.

Several authors have reported that the time of the day affects the risk of becoming annoyed by turbine noise. The typical time at which people get annoyed is between evening/late evening and night [8–10]. This can be attributed to, for example, the lower background noise.

It is well known that the sound pressure level alone does not always explain why people are irritated by wind turbine noise. The noise characteristics can also have an

Figure 9.1 *Schematic presentation of the blade noise production location according to a study by Oerlemans* et al. *[2]*

effect. Typically, impulsive changes in the noise levels are described when the wind turbine noise-related annoyance is discussed. Zhu *et al.* [11] highlighted the roles of wind shear and turbulence on amplitude modulation for distances above 250 m from the turbine. Van de Berg [6] reported that impulsive thumping sound increased residents' annoyance close to a wind farm.

It should, however, be acknowledged that after a certain sound pressure level (dBA), the wind turbine sound becomes more easily audible, and it can annoy people. In general, the research suggests that above 40 dBA, the risk of sleep disturbance or other annoyance increases, as discussed by Kuwano *et al.* [10], Pedersen and Kaye [12], and Janhunen *et al.* [8]. The literature also indicates that above 45 dBA, the risk of annoyance and sleep disturbance may increase even more rapidly [8, 13].

These limits and noise characteristics are also considered in national legislation. In Finland [14], the average declared or measured sound pressure level should not exceed 45 dBA in residential areas during the daytime (between 7 AM and 10 PM) and 40 dBA during the night. In national parks, the limit is set to 40 dBA during both day and night. If the sound is impulsive or has a narrow-band character, 5 dBA is added to the measured values during control; therefore, the maximum allowed sound pressure level is 5 dBA lower in these cases. The approach whereby the maximum allowable noise level varies according to the time of day, location, and wind conditions is relatively common, according to Koppen and Fowler [15]. The turbine noise levels can also be connected to the background noise level, as is the case in

France, by allowing a 5 dBA sound pressure level increase during the daytime and a 3 dBA increase at night (relative to the background noise level) [15].

9.3 Atmospheric conditions and wind turbine noise

The atmospheric conditions can affect wind turbine noise. This section discusses two possible scenarios leading to neutral and stable atmospheric conditions. On a cloudy and/or windy night, the atmosphere is neutral. The wind profile is relatively uniform and follows the logarithmic shape. Hence, the wind speed close to the ground does not differ much from that at the nacelle height [9]. On the other hand, the atmosphere becomes stable during the night when the winds are low and the sky is clear. Under stable condition, the wind speed is relatively low close to the ground and high at the nacelle height. Therefore, the noise level can become higher than expected, leading to increased annoyance, as described by van den Berg [9]. This variation is illustrated in Figure 9.2, where the German air quality guideline profile is used to model the wind shear profiles. This model was shown by van den Berg [9] to provide a good estimation of the wind shear also at stable atmospheric conditions during the night. The shear profile is defined as

$$v = v_{ref}\left(\frac{h}{h_{ref}}\right)^{m} \tag{9.1}$$

where v_{ref} and h_{ref} are the reference wind speed and height, respectively. The exponent m depends on the atmospheric conditions and in Figure 9.2, $m = 0.22$ describes the neutral, $m = 0.20$ moderately unstable, and $m = 0.41$ the stable conditions. It should also be mentioned that already relatively small wind turbines can have noticeable

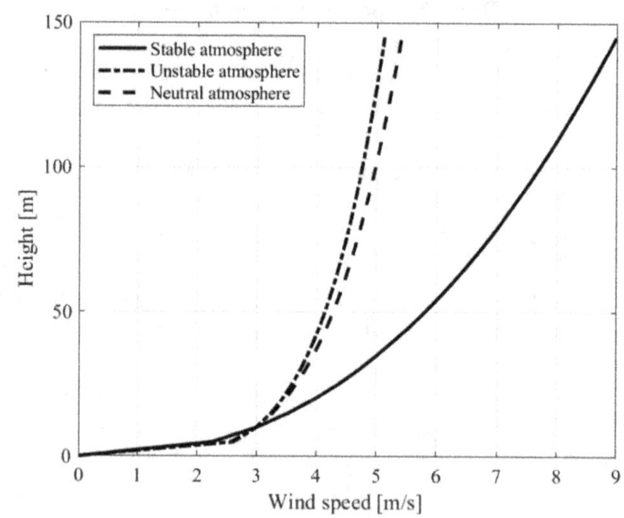

Figure 9.2 Effect of atmospheric conditions on wind shear profile. Reference height 10 m and reference wind speed 3 m/s

sound emission effects due to wind shear profiles, as discussed by Grönman *et al.* [16].

9.4 Wind turbine noise prediction

This chapter discusses the noise generation and propagation prediction methods. The noise generation prediction approaches are divided into semiempirical and computational aero-acoustics (CAA) methods. In addition, a simple sound immission level calculation method and a more complex parabolic equation method are used to introduce different sound propagation modeling approaches.

9.4.1 Semiempirical noise prediction methods

A commonly used semiempirical method to predict wind turbine noise was developed by Fuglsang and Madsen [17] based on earlier work by Brooks *et al.* [3] for airfoil self-noise as well as on the turbulent inflow model of Lowson [4]. The method has an accuracy of ±1.5 dBA for a single airfoil. This method predicts the airfoil self-noise sources presented in Section 9.1 and for turbulent inflow separately. The most significant drawback of this approach is that it does not consider the discrete frequency noise due to the blade-passing frequency and its harmonics. For a full turbine, a Blade Element Method (BEM) solution is required at each blade section to provide the required aerodynamic information. In a simplified BEM-based approach, Perfilev [18] modeled only the noise due to the interaction between the turbulent boundary layer and the trailing edge and the incoming turbulence-related noise as the primary noise sources. The overall sound power level L_{tot} was calculated with the interference law as:

$$L_{\mathrm{tot}} = 10\log\left(10^{L_{TE}} + 10^{L_{TI}}\right) \tag{9.2}$$

where the L_{TE} and L_{TI} represent the sound power levels for the trailing edge and incoming turbulence noise, respectively. The semiempirical methods and the BEM are even today widely used in the industry due to their low computational cost [11].

9.4.2 Computational aeroacoustics

Another approach is to use numerical methods that combine computational fluid dynamics (CFD) and different extensions of Lighthill's acoustic analogy from the 1950s. The general idea is that the turbine flow field is calculated first, after which the acoustic analogy is applied. As a result, the sound pressure level at a certain position relative to the turbine can be calculated.

The biggest drawback of computational aeroacoustics is that relatively large computational power is required to accurately solve the aerodynamics using large eddy simulation (LES) or direct numerical simulation (DNS). Some studies use the LES approach to simulate the aerodynamic performance of wind turbines and combine it with the acoustic analogy. However, it was highlighted by Zhu *et al.* [11]

in 2018 that the use of LES in full turbine simulations is still difficult with current computational powers. Some studies also use the less accurate Unsteady Reynolds Averaged Navier-Stokes approach (RANS) to overcome these challenges, which makes it easier to model larger turbines. As a trade-off between the LES's accuracy and RANS's computational efficiency, it is also possible to combine them into a hybrid method to predict turbine aerodynamics, as was done by Kaviani and Nejat [19] for a 1.5 MW turbine. From a practical point of view, the use of DNS is limited by the high Reynolds numbers and the associated need for extensive computational power. It is also worth mentioning that a hybrid method that combines both the semiempirical and CFD approaches is also available [11].

9.4.3 Noise propagation and immission modeling

There are several methods available to model noise propagation. As was discussed by Zhu *et al.* [11], the variables that affect wind farm noise propagation are the wake interaction with other turbines and the ground, turbulence levels, and the terrain.

For example, the sound immission level at a certain distance from a wind farm can be calculated with the approach that has been used in both Germany and the Netherlands at the national level, and the input data can be either measured or modeled. With this approach, the following equation was presented by van den Berg [9] to calculate the sound immission level L_{imm}:

$$L_{\text{imm}} = 10 \log \left[\sum_J \sum_K 10^{(L_j - D_{j,k})/10} \right] \tag{9.3}$$

where the sound power level L is calculated over k turbines at j sound power octave band levels. Attenuation $D_{j,k}$ takes into account the effects of ground and air absorption and geometrical spreading.

One option to model noise propagation is to use the parabolic equation model, as chosen by the Technical University of Denmark for their software [11], which is used here as a state-of-the-art example. It is fundamentally based on solving the wave equation. As a result, a steady solution at different frequencies is provided. This method is, however, also capable of yielding transient sound pressure level predictions.

The parabolic equation methods can be divided into scalar and vector approaches. The scalar approach is discussed by Zhu *et al.* [11] to be suitable in cases where both the noise source and the receiver are close to the ground. The following input information is required to perform the simulations according to Zhu *et al.* [11]:

1. terrain geometry
2. starter function
3. background flow field
4. ground impedance.

The modeling of the undulating terrain can be done using either terrain-following or domain decomposition approaches. The starter function requires the noise source information, for example, derived from the previously discussed noise

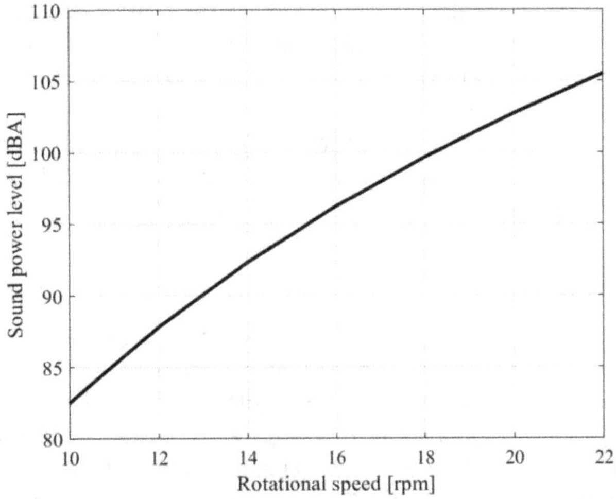

Figure 9.3 *Correlation between the wind turbine rotational speed and sound power level*

generation models. It is also important to acknowledge that the noise propagation modeling is not accurate if the wake is not modeled. In the most advanced approach, the background flow field can be predicted by combining the actuator line technique and LES. Also, experimental wind measurement data can be used as an input. The modeling of the ground impedance is done either with an empirical or a theoretical model [11].

9.5 Methods to reduce noise-related problems

9.5.1 Wind farm operation

There are different methods that a wind farm operator can utilize when the noise exceeds the allowed level. The first presented method is lowering the turbine rotational speed since there is a relationship between the emitted sound pressure level and the rotational speed (*N*), as shown by, for example, Wagner *et al.* [20] and van den Berg [9]. Figure 9.3 presents this dependency based on the logarithmic fit by van den Berg [9]:

$$L_W = 67.1 log\left(N\right) + 15.4 dBA \tag{9.4}$$

Another possible method is changing the blade pitch angle, but this also affects turbine power and may not be suitable from a cost-effectiveness point of view. van den Berg also discussed continuous pitch angle control (like in helicopters) and its positive influence on the sound emission [6] as a solution for fluctuating noise due to atmospheric stability.

Rotor tilt was also discussed by van den Berg [6] as a possible strategy to decrease the fluctuating sound from the wind turbine. However, it was concluded that the required change of the tilt angle would be substantial, and this would cause other challenges, such as a change of the rotor's normal area toward the wind, which is not considered an efficient approach [6]. It was also discussed by van den Berg [6] that when the turbines in a wind farm are synchronized and strengthen the sound emission, a wind turbine desynchronization using small random changes in pitch angles could be used. This condition was also described in [9] while discussing the varying impulsive sound detection from a wind farm.

9.5.2 Wind farm design

As previously discussed in Section 9.5.1, turbine operation can affect the noise emission from the turbines. In addition, the design of the farm can have a marked influence on the audibility of the turbines. The simplest way to reduce the risk of noise annoyance is to have enough distance between a wind farm and the closest residents. A study by van den Berg [9] noted that even people living 1 900 m from a wind farm had reported noise-related problems. National legislation is important in these situations as it stipulates the limits that should not be exceeded, as discussed earlier in this chapter.

From the wind farm design point of view, the location of the farm relative to residents and to the most common wind direction may also be used as a design criterion to limit the risk of noise-related annoyance. A schematic presentation of this effect is presented in Figure 9.4. In this figure, different areas of interest are presented following the literature. The turbine is at the center of the figure, and the wind direction is always from top to bottom (zero degrees). It was suggested by Janhunen *et al.* [8] after examining two wind farms in Finland that wind turbine noise becomes more audible when people are located approximately behind the turbine (with a small decline) (area 1). In an earlier study by Bockstael *et al.* [22], it was found that the highest risk of annoyance existed if the wind blew from the north or north-east when the studied people lived approximately between northwest and west of the turbines (area 4). In their work, there was not enough data for easterly winds to draw a conclusion, although indications that these were causing a higher annoyance risk were visible in their data. This potential finding would have led to areas overlapping area 1 from Janhunen *et al.* [8]. In the study of van den Berg [6–9], it was found that the highest wind turbine sound immission was recorded when the turbines were located approximately between the coming wind and the measurement location (area 2). This is also qualitatively in accordance with the findings of Janhunen *et al.* [8], although a high sound pressure level does not necessarily always lead to annoyance even if the risk increases, as previously discussed. In addition, Oerlemans [21] found that the highest measured sound pressure levels were detected upwind and downwind from a turbine. However, the swish amplitude was found to be 5 dB higher in a crosswind situation (area 3); this was further discussed as a possible reason for the reported amplitude modulation. From a physical point of view, the reasons behind this phenomenon were the convective amplification and the directivity of the trailing edge noise. In addition, the possible role of changing atmospheric conditions on the noise fluctuation several turbine diameters away was highlighted as an affecting

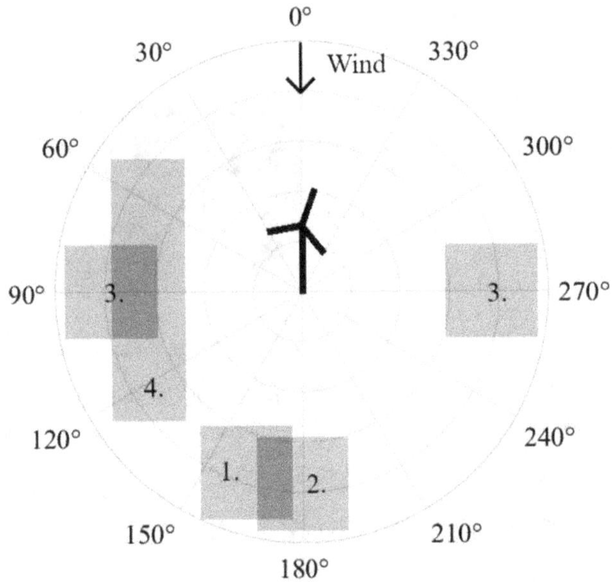

Figure 9.4 *Schematic presentation of the effect of wind direction on turbine noise audibility. Area 1 is the location of increased noise and audibility as per Janhunen* et al. *[8]. Area 2 is the location of increased noise, according to van den Berg [9]. Both incidences of area 3 are the locations of increased sound pressure amplitude in Oerlemans [21], and area 4 is the location of an increased risk of noise-related annoyance.*

factor. These findings may be interpreted in such a way that if a wind farm has a location in which the wind comes mostly from a certain direction, the risk of annoyance due to turbine noise decreases if the farm is built upwind (regarding the main wind direction) of a residential area.

9.5.3 Wind turbine aerodynamics

This section discusses noise reduction methods for three noise sources: airfoil self-noise, incoming turbulence noise, and the noise due to the interaction between the tower and the blades.

9.5.3.1 Airfoil self-noise

Different solutions, including passive and active flow control methods, have been studied to lower the airfoil self-noise, either at the airfoils themselves or for full-scale turbines. When discussing self-noise, it is also worth remembering that the trailing edge-related noise production is the main source, as previously discussed.

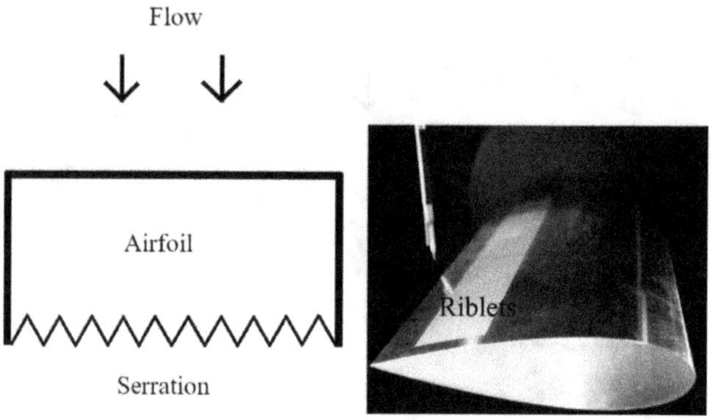

Figure 9.5 *Examples of passive airfoil self-noise reduction techniques. Left: serration; right: laser manufactured riblets on a NACA0024 airfoil*

Figure 9.5 presents a schematic overview of some of the noise reduction methods available.

Chong *et al.* [23] found up to 7 dB reductions in trailing edge noise using trailing edge serrations. In their study with a NACA0012 airfoil, a noise increase due to a bluntness of the serration root was found; thus, only one of the studied serrations was able to reduce the overall noise by a total of 3 dB. When considering the full turbine level, in a study by Oerlemans *et al.* [24], it was found that an optimized airfoil was able to reduce the noise by 0.5 dB, whereas a serrated blade reduced it by 3.2 dB. The principle behind the airfoil optimization was to reduce the boundary layer turbulence structure.

It is also possible to use porous media airfoils, either along the whole blade chord length or along smaller parts of the airfoils. However, Geyer [25] pointed out that this may even increase the sound pressure level. In addition, it was concluded that porous media airfoils are characterized by reduced aerodynamic performance. It is also worth mentioning that trailing edge brushes are also considered as a potential way to decrease turbine noise.

Regarding the active flow control methods, Wolf *et al.* [26] found that steady wall-normal suction led to reduced trailing edge noise. In their approach, the suction was realized by using a porous surface close to the trailing edge. Other active methods are also being studied, such as trailing edge blowing and plasma actuators (with a 3.8 dB noise reduction reported [27]), which aim to affect the blade wall boundary layers.

Another possible option to reduce the trailing edge noise is to use riblets, which are small flow-wise aligned protrusions that can reduce the airfoil drag in turbulent boundary layers. The height of the riblets is of similar magnitude to the thickness of the viscous sub-layer. For example, the European Union-funded Riblet4Wind project, which studied the effectiveness of riblets between 2015 and 2019, found that although full-scale turbine performance improved, the noise levels increased contrary to expectations. Besides, the European Union has funded a new project on riblets' noise reduction capability lasting from 2020 until 2022.

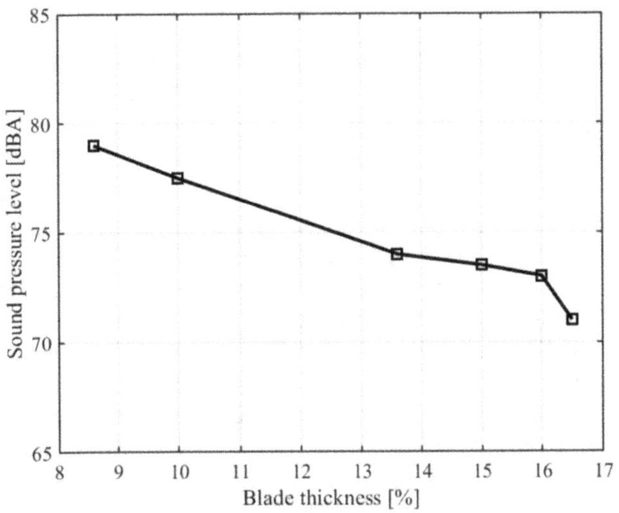

Figure 9.6 *Presentation of the role of the leading-edge shape on incoming turbulence noise. The data is presented at a zero-degree angle of attack. Data reproduced from [27].*

In conclusion, it seems that there are several investigated approaches that can reduce aerodynamic noise. Out of the discussed techniques, riblets may have more potential than the current tests have shown; in addition, serrations have already been introduced into turbines by several manufacturers [11].

9.5.3.2 Incoming turbulence noise

The noise due to the interaction between the blade and the flow turbulence was studied by, for example, Migliore and Oerlemans [28], who examined the effects of different blade shapes on the sound pressure level. Although their study concentrated on airfoils for small turbines, the general findings may also be used for a broader view. They did find that the leading-edge turbulent inflow noise increased when the leading-edge thickness decreased (see Figure 9.6), and the difference between the most silent airfoil and the noisiest airfoil was reported to be 8 dB. However, it was not clear whether the phenomenon was due to the thickness or sharpness changes at the leading edge. Overall, their analysis gives an overview of the importance and possibilities of leading-edge design on incoming turbulence noise reduction.

9.5.3.3 Noise due to the interaction between the tower and the blades

The interaction noise between the blades and the tower can be modified by increasing the distance between them. However, Doolan *et al.* [29] discussed that this change might, for example, lead to major changes in the turbines' mechanical lifespan. Another possible option discussed by Zajamsek *et al.* [5] is to modify the shape of the tower by modifying the fairings. It should also be acknowledged that the change of the fairing shape may lead to differences in the blade-tower interaction at different wind directions.

9.6 Summary

This chapter presented the fundamentals of the wind turbine noise production mechanisms by dividing the sources into airfoil self-noise, incoming turbulence noise, and noise due to the airfoil passing the tower. It further discussed how people respond to this specific noise source. It was found that the noise level does not alone explain why people get annoyed by wind turbine noise. Several reasons can explain these findings, such as the wind shear profile, turbulence, and wind direction. Several noise-reduction methods were reviewed to provide viable solutions to these issues. These methods can be used at the following stages of a wind farm project:

- aerodynamic design
- wind farm design
- wind farm operation.

In addition, the available methods to model wind turbine noise generation and propagation were reviewed to give a good overview. In general, noise generation methods can be divided into semiempirical and computational aeroacoustics methods or hybrid methods that combine both approaches. Despite the higher accuracy of computational aeroacoustics, their computational cost is still relatively high for industrial use. Therefore, semiempirical methods are currently widely used.

References

[1] Liu W.Y. 'A review on wind turbine noise mechanism and de-noising techniques'. *Renewable Energy*. 2017;**108**(C):311–20.

[2] Oerlemans S., Sijtsma P., Méndez López B. 'Location and quantification of noise sources on a wind turbine'. *Journal of Sound and Vibration*. 2007;**299**(4-5):869–83.

[3] Brooks T.F., Pope D.S., Marcolini M.A. *Airfoil self-noise and prediction*. Technical Report, NASA RP-1218. NASA; 1989.

[4] Lowson M. V. 'A new prediction model for wind turbine noise'. *Wind Engineering*. 1994;**18**(2):51–61.

[5] Zajamsek B., Yauwenas Y., Doolan C.J., *et al.* 'Experimental and numerical investigation of blade–tower interaction noise'. *Journal of Sound and Vibration*. 2019;**443**(5):362–75.

[6] van den Berg G.P. 'Wind turbine power and sound in relation to atmospheric stability'. *Wind Energy*. 2008;**11**(2):151–69.

[7] Janhunen S., Hujala M., Pätäri S. 'The acceptability of wind farms: the impact of public participation'. *Journal of Environmental Policy & Planning*. 2018;**20**(2):214–35.

[8] Janhunen S., Grönman A., Hynynen K., Hujala M., Kuisma M., Härkönen P. 'Audibility of wind turbine noise indoors – evidence from mixed method data'. 6th IEEE International Conference on Renewable Energy Research and Applications; San Diego, CA, USA, 5-8 Nov. 2017; 2017.

[9] van den Berg G.P. 'Effects of the wind profile at night on wind turbine sound'. *Journal of Sound and Vibration*. 2004;**277**(4-5):955–70.

[10] Kuwano S., Yano T., Kageyama T., Sueoka S., Tachibana H. 'Social survey on wind turbine noise in Japan'. *Noise Control Engineering Journal*. 2014;**62**(6):503–20.

[11] Zhu W.J., Shen W.Z., Barlas E., Bertagnolio F., Sørensen J.N. 'Wind turbine noise generation and propagation modeling at DTU wind energy: a review'. *Renewable and Sustainable Energy Reviews*. 2018;**88**(10):133–50.

[12] Pedersen E., Persson Waye K., Waye K.P. 'Wind turbine noise, annoyance and self-reported health and well-being in different living environments'. *Occupational and Environmental Medicine*. 2007;**64**(7):480–6.

[13] Bakker R.H., Pedersen E., van den Berg G.P., Stewart R.E., Lok W., Bouma J. 'Impact of wind turbine sound on annoyance, self-reported sleep disturbance and psychological distress'. *Science of the Total Environment*. 2012;**425**(4-6):42–51.

[14] Section 142 of the Environmental Protection Act (527/2014). *Government decree on guide values for the outdoor noise level of wind turbines in Finland*. Finland; 1107/2015.

[15] Koppen E., Fowler K. *International Legislation for Wind Turbine Noise*. Euronoise May 31-June 3. Maastricht, The Netherlands; 2015.

[16] Grönman A., Backman J., Avramenko A. 'Wind profile effect on small wind turbine noise generation'. Proceedings of ASME Turbo Expo; Düsseldorf, Germany, June 16-20'; 2014.

[17] Fuglsang P., Madsen H.A. *Implementation and Verification of an Aeroacoustic Noise Prediction Model for Wind Turbines*. RISO Report. Roskilde: RISO National Laboratory. RISO-R-867(EN); 1996. p. 54.

[18] Perfilev D. *Methodology for Wind Turbine Blade Geometry Optimization*. Dissertation. Finland: Lappeenranta University of Technology; 2013.

[19] Kaviani H.R., Nejat A. 'Aerodynamic noise prediction of a MW-class HAWT using shear wind profile'. *Journal of Wind Engineering and Industrial Aerodynamics*. 2017;**168**(Part A):164–76.

[20] Wagner S., Bareiß R., Guidati G. *Wind Turbine Noise*. Berlin: Springer; 1996.

[21] Oerlemans S., Schepers J.G. 'Prediction of wind turbine noise and validation against experiment'. *International Journal of Aeroacoustics*. 2009;**8**(6):555–84.

[22] Bockstael A., Dekoninck L., Can A., Oldoni D., De Coensel B., Botteldooren D. 'Reduction of wind turbine noise annoyance: an operational approach'. *Acta Acustica United with Acustica*. 2012;**98**(3):392–401.

[23] Chong T.P., Joseph P.F., Gruber M. 'Airfoil self noise reduction by non-flat plate type trailing edge serrations'. *Applied Acoustics*. 2013;**74**(4):607–13.

[24] Oerlemans S., Fisher M., Maeder T., Kögler K. 'Reduction of wind turbine noise using optimized airfoils and trailing-edge serrations'. *AIAA Journal.* 2009;**47**(6):1470–81.

[25] Geyer T.F. *Trailing Edge Noise Generation of Porous Airfoils*. Dissertation. Germany: Technical University of Cottbus; 2011.

[26] Wolf A., Lutz T., Würz W., Krämer E., Stalnov O., Seifert A. 'Trailing edge noise reduction of wind turbine blades by active flow control'. *Wind Energy.* 2015;**18**(5):909–23.

[27] Li Y., Zhang X., Huang X. 'The use of plasma actuators for bluff body broadband noise control'. *Experiments in Fluids.* 2010;**49**(2):367–77.

[28] Migliore P., Oerlemans S. 'Wind tunnel aeroacoustic tests of six airfoils for use on small wind turbines'. *Journal of Solar Energy Engineering.* 2004;**126**(4):974–85.

[29] Doolan C.J., Moreau D.J., Brooks L.A. 'Wind turbine noise mechanisms and some concepts for its control'. *Acoustics Australia.* 2012;**40**(1):7–13.

Chapter 10
Bird and bat collisions at wind farms

Golrokh Mirzaei[1] *and Mohsin M Jamali*[2]

Wind energy is considered green as wind turbines do not produce pollutants during their operation. One drawback of wind energy is wind turbines' interaction with birds and bats, especially with those listed as endangered species or those protected by the Migratory Bird Treaty Act (MBTA). One way to live in harmony with nature is to study collision rates and develop mitigation techniques. This work describes various electronic monitoring techniques, collision models, and several mitigation techniques. All these approaches may not work 100 percent of the time, but at least they will reduce bird/bat mortalities with the wind turbines. This would be a win-win approach and living in harmony.

10.1 Introduction

Wind energy is one of the forms of renewable energy that has received attention in the past decades and it is one of the rapidly growing energy sources. It is sustainable, abundant, clean, and cost-effective in terms of space and operational costs. Wind power installations have been increasing steadily for the last 25 years. According to one estimate, installed wind energy in the US at the end of 2018 is about 96 000 MW (Mega Watts). Wind turbines can produce power in the range of 8.0–9.5 MW. The turbine tower heights are around 220 m and rotor diameters are 174 m giving a rotor swept area of 23 779.0 m^2 with a blade tip speed of 324 km/hour [1]. A wind turbine facility may involve thousands of hectares.

Wind energy is considered green energy, carbon-neutral, and reduces greenhouse gasses. Wind turbines and wind farms are required to extract wind energy, which makes it a challenge due to its impact on wildlife. Wind turbines can threaten the life of habitants living within or passing the wind farm area. There is a direct relationship between the number of wind turbines and the mortality rate of wildlife in the area. Among all the inhabitants, birds and bats are the most vulnerable creatures as their lives are at a high risk in the vicinity of the wind farms. A wind

[1]Department of Computer Science and Engineering, The Ohio State University, Marion, OH, USA
[2]Department of Electrical Engineering, The University of Texas Permian Basin, Odessa, TX, USA

turbine has a stationary tower and rotating blades. Unintended bird collision with these stationary and rotating elements has been a concern to various stakeholders. It is estimated that bird and bat fatalities are from 3 to 6 birds per MW and 4 to 7 bats per MW per year, respectively [1]. Bird/bats can be attracted by turbine blades and get struck with the blades and eventually killed. Three migratory bats namely hoary, eastern red, and silver-haired bats, contribute to most bat fatalities [1]. Mexican free-tailed bats also contribute to some 41 to 86 percent bat fatalities mostly in the southern half of the United States [1]. Bat fatalities vary from regions to regions and seasons to seasons. Some bats may be attracted to wind turbines. This may be due to feeding on small insects underneath the wind turbine. Also, bats are subjects to Barotrauma, which is a physical damage to their lungs due to the sudden drop in the air pressure close to the wind turbines, which eventually cause death. In addition to the direct impacts, wind farms indirectly threaten the avian's life. Building wind farms cause displacement from feeding or nesting areas. Wind turbines affect movement and result in habitat degradation or loss. Building wind farms will be harmful to birds/bats especially at migratory hubs and close to breeding colonies.

Bats and birds play an important role in a healthy ecosystem. One of the main advantages of bats is consuming insects. It is reported that a large percentage of all bat species are insectivores and consume large number of insects [2]. For instance, Brazilian free-tailed bat (*Tadarida brasiliensis*) colonies consume 8.4 metric tons of insects overnight in the lactation time [3]. Bats provide other services to the environment, such as dispersing seeds and pollinating plants. Bats are also beneficial for farming and the economy. According to the US Fish and Wildlife Services, bats save billions of dollars in the United States in controlling pests by hunting insects that damage crops [4].

Bird fatalities have been recorded on all wind turbines, mainly by searching carcasses. Small passerines have the majority of fatalities. Bird mortality is due to various causes and sources. Bird fatalities vary from regions to regions. Fixed structures such as wind turbines and their rotating blades are one of the sources of mortality. Bird mortality also occurs due to cats, building structures, power lines, transmission towers, predators, and traffic. There are some negative effects of increased birds and bats mortality near wind turbines due to predator activities.

Nocturnally migrating birds and bats are considered at high risks. Bird species perceive wind turbines differently due to their vision, hearing, age, body conditions, breeding status, and foraging need. Avian species interact with wind turbines in different ways, and there are various causes of their collision with wind turbines. They may react to a wind turbine with perceiving risk or may not react at all. There are videos on the internet showing endangered species getting hit by moving blades. People have found dead bats under the wind turbines. It was discovered during the autopsy of bats that blades did not strike them but their lungs collapsed due to negative pressure under the wind blades. Complains from neighbors range from being eye soars and have constant whishing noise. Permitting is a long process due to the above negative effects and requiring environmental assessment. Wind power also requires negotiations with the landowners, dealing with neighbors during the permit process.

Wind turbines are only placed historically in high wind areas. Wind farms and associated infrastructures result in habitat loss, disturbances during the construction and post-construction process. If birds do not leave the site, then the probability of collision increases and loss of life will result. Ideally, a wind turbine should be placed in high wind areas and away from bird habitats or migrating pathways.

Section 10.2 describes monitoring activities, and Section 10.3 discusses the collision of birds with wind turbines and various collision models. Section 10.4 offers various mitigation techniques that have been experimented by wildlife biologists. Finally, Section 10.5 concludes this chapter.

10.2 Monitoring activities

Wind energy can be obtained from two different types of wind farms: onshore and offshore. In onshore wind farms, wind turbines are installed in a vast area on land, while offshore wind farms are installing wind turbines in bodies of water. In both cases, monitoring birds and bats is an essential task for understanding the movement of these creatures to preserve their population and prevent their collision with turbines. Monitoring activities can be performed using traditional or advanced approaches. The traditional monitoring techniques use manual observations of birds/bats such as mist netting and moon watching. These techniques are neither feasible nor efficient for a long-term basis, as they are weather dependent, labor-intensive, or may not be accurate. Automated techniques are more effective and accurate and can be run on long-term basis. In automated techniques, one or multiple sensors are used. Each sensor collects data separately, and data will be processed to provide useful information about the targets. According to Parise and Walker [5], radar, infrared (IR) imaging, and acoustics develop a strong understanding of the migration and activity patterns of bird/bats. In the subsequent sections, we describe monitoring activities in onshore and offshore wind farms.

10.2.1 Onshore monitoring

10.2.1.1 Acoustic monitoring

Acoustic monitoring has received considerable attention in the last decades for monitoring birds/bats. This is a powerful tool to identify birds and bats at the species level. Bird/bat identification is an important environmental practice as an invalid assessment of bat/bird distribution, which leads to making wrong decisions with serious conservation consequences [6]. Many studies have been conducted on acoustic monitoring for the identification of birds and bats [6–13]. In many studies, bat echolocation calls and birds nocturnal flight calls over the migration period are subjects of analysis.

Most of the birds/bats are detectable by their sound, rather than the vision. Acoustic monitoring uses birds/bats vocalization and provides information about the distribution at the species level. The birds/bats vocalizations are recorded by specific acoustic detectors that can record in different frequency ranges for bird and

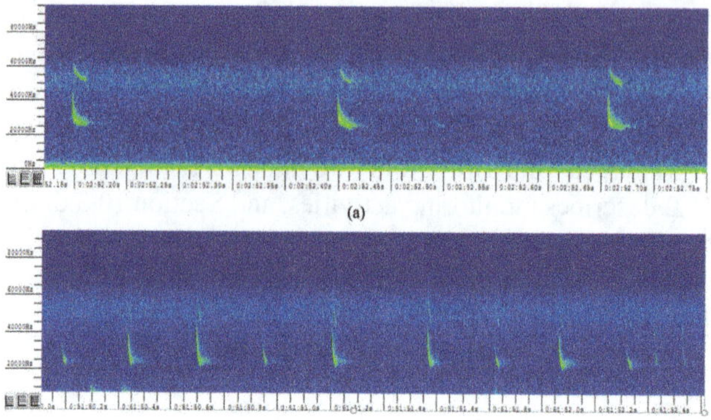

(a)

Figure 10.1 *Spectrogram of echolocation calls for two bat species (a) silvered-hair bat (*L. Noctivagans*) and (b) big brown bat (*E. Fuscusbat*)*

bat calls. Figure 10.1 shows the spectrogram of echolocation call for two bat species (*Lasionycteris Noctivagans* and *Eptesicus Fuscusbat*), and Figure 10.2 shows the spectrogram of bird flight calls for two bird species (*Passerculus sandwichensis, Catharus ustulatus*). Bat echolocation calls and bird nocturnal calls contain acoustic features and characteristics that can be used to differentiate the species. In acoustic processing, the significant features of the calls are extracted and used for identification. In the subsequent sections, we describe the main steps for acoustic data detection and processing.

10.2.1.1.1 Acoustic detectors

In acoustic monitoring, bird/bat calls are captured by specific detectors that convert the calls into electrical analog signals. In the case of the bat detector, the ultrasound is converted to an audible range. The more recent advanced detectors can store data for further processing. Generally, bat and bird calls are recorded using different detectors. The main difference between the bird and bat detectors is the sampling

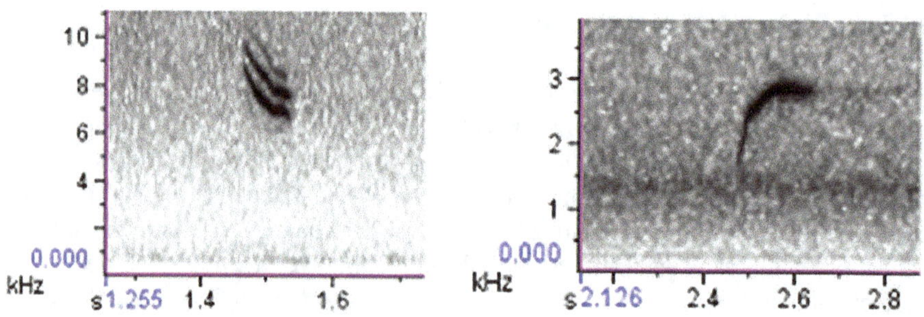

(a) (b)

Figure 10.2 *Spectrogram of flight calls for two bird species (a) Savannah sparrows (*P. sandwichensis*) (b) Swainson's thrush (*C. ustulatus*)* frequency of recording. According to the Nyquist-Shannon sampling theorem, the sampling frequency should be at least twice that of the highest frequency component in the signal. The birds' nocturnal frequency range is 5 Hz–20 kHz, and the frequency range of bat echolocation calls is 5–200 kHz. The commercially available detectors have sampling rates of 192, 256, 384, and 500 kHz [14] to detect the maximum frequency of bat calls. To assess the birds' and bats' movement in wind farms or potential areas for construction of wind farms, the detectors are set to record the calls over specific periods of time, mostly during migration time in the vicinity of the wind farms. Depending upon on the detector's type and its memory size, the recorded data are transferred to a processing system to free memory and allow collecting new data. The recently advanced detectors have a large size of memory (more than one terabyte in the case of SM4bat [14]) and can store data over a long period.

There are different types of bat detectors: Heterodyne (HET), Frequency Division (FD), Time Expansion (TE), and Full Spectrum (FS). Heterodyning detectors convert the call to an audible signal by filtering the ultrasonic portion of the signal, which means only a narrow bandwidth of the call is transformed. Thus, the main characteristics of the call may be lost. It is inexpensive, easy to use, and works in real-time. FD detectors distinguish all frequency ranges of the bat call and divide the incoming frequency using a user-defined division ratio of 4, 8, 10, 16, or 32 to convert the ultrasonic call to an audible signal. Similar to HET, FD detectors are suitable for real-time analysis. TE detectors record the full spectrum call and expand the time at a slower rate for post-processing. Thus, the entire structure of the call can be heard at a lower frequency. However, during the time that sound is playing at a lower speed, it will not detect other calls. FS detectors record the ultrasonic call and do not convert it to an audible signal. Thus, the entire structure of the call is preserved. Unlike the TE recorders, FS detectors are appropriate for real-time detection. So, they are superior to other detectors, as they detect the full spectrum, preserve the entire structure of the call, and work in real-time. However, the files produced by this kind of detectors are very large size, so they need a triggering mechanism for the recording system. They are also more expensive than other types. There are several commercially available bat detectors that are different in types and specifications. Figure 10.3 shows two bat detectors from Wildlife Acoustics [14] and Binary Acoustics Technology [15]. Bird song recorders are simpler than bat detectors as the frequency range of birds is in the hearing range, and there is no need to audible conversion. Thus, a recorder equipped with a built-in microphone to record the audio signals can be used for recording bird songs. However, cut-off filtering can be applied to remove the background noise while retaining the narrow calls.

10.2.1.1.2 Feature extraction

The birds/bats calls have specific characteristics (features) that can be used for identification at the species level. Figure 10.4 shows the spectrogram of the echolocation call for Myso bat. Some of the frequency-based features are shown in the

(a) (b)

Figure 10.3 Bat detectors with ultrasonic microphone (a) SM4BAT [14] and (b) AR125 [15]

spectrogram such as f_max (maximum frequency), f_maxp (frequency of maximum power), and f_k (frequency of knee which is the frequency at which the initial slope of the call most abruptly transitions to the slope of the body of the call). Different techniques have been used to extract call features of birds and bats. The most common techniques are mathematical transformation techniques such as Fourier Transform (FT), Discrete Wavelet Transform (DWT), and Mel-Frequency Cepstral Coefficients (MFCC). In the next section, we review these techniques briefly.

Short-Term Fourier Transform (STFT)—FT is the most popular mathematical transformation technique that decomposes a continuous signal into its frequency components. Thus, this technique represents the signal in the frequency domain. The bird/bat vocalizations are non-stationary, which means the properties of the signal, change with time. Thus, a time-dependent FT preserves the properties of the calls better. STFT represents the signal in the time-frequency domain and works better in birds/bat vocalization recognition. STFT partitions the signal $f(t)$ into several segments where each of them can be assumed stationary and then applies FT to each segment. Partitioning the signal is performed using a sliding window $w(t-u)$ where the window size is fixed. As STFT depends on the size of the window, the frequency components are given for a time interval, not for a time instance, a problem known as "uncertainty principle." One of the challenging tasks in STFT is finding

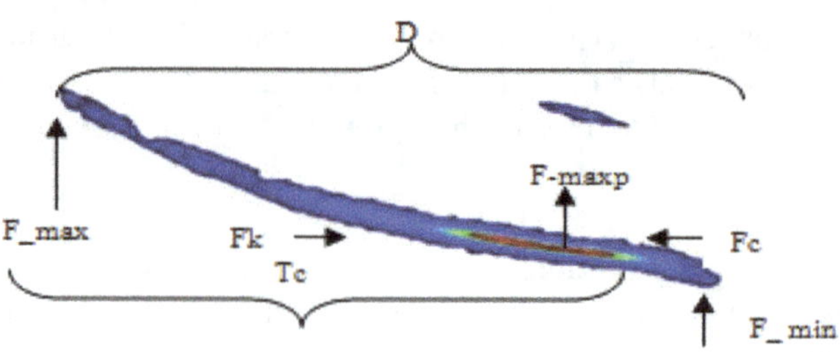

Figure 10.4 Bat echolocation call spectrogram (Myso bat) [16]

an appropriate length for the sliding window that preserves both time and frequency resolution. A narrow window provides poor frequency and good time resolution, while a wide window provides a good frequency and poor time resolution. The continuous-time STFT is formulated as [17]:

$$STFT\ f(u,w) = \int_{-\infty}^{+\infty} f(t)\, w(t-u) \exp(-jwt)\, dt \tag{10.1}$$

where $w(u)$ is the window function. The coefficients of the STFT are represented as the features of the signal, so features of the calls.

Discrete Wavelet Transform (DWT)—Wavelet Transform (WT) is another mathematical transformation technique for analyzing signals and has been used in many bird/bat studies. WT overcomes the limitation of STFT to find an appropriate length for the sliding window by introducing a *mother wavelet* that has a dynamic size, in contrast to the fixed size of the sliding window in STFT. WT is similar to STFT in the sense that the signal is multiplied to a function called wavelet, same as STFT that multiplies the signal to the sliding window. However, in WT, the Fourier transform of the segments is not taken, and the length of the window is changing. The continuous WT is defined as [18]

$$WTf(u,s) = \int_{-\infty}^{+\infty} f(t)\, \frac{1}{\sqrt{s}} \varphi_{u,s}(t)\, dt \tag{10.2}$$

The transformed signal is a function of s (scale) and u (translation) parameters. The scale is inversely proportional to the frequency. Scaling is a mathematical process that compresses or dilates the signal. In continuous WT, the decomposition is performed by scaling and shifting the mother wavelet $\varphi(t)$. However, in discrete WT, the decomposition is performed by passing the signal through a series of low- and high-pass filters to analyze signals in different scales and then followed by subsampling. High-pass filters are used to analyze the high frequencies of the signal, and low-pass filters are used to analyze the low frequencies. Subsampling is applied by eliminating half of the samples (subsampling by 2) due to Nyquist's rule. The following formulas show the filtering and sub-sampling operations in discrete WT:

$$y_H[k] = \sum_n x[n].g[-n+2k] \tag{10.3}$$

$$y_L[k] = \sum_n x[n].h[-n+2k] \tag{10.4}$$

where $g[n]$ and $h[n]$ are high- and low-pass filters, respectively, and $y_H[k]$ and $y_L[k]$ are the outputs of high- and low-pass filters, respectively. In most studies, the features are selected based on the statistical measures (e.g., maximum, mean, standard deviation) of wavelet detail coefficients from high-pass filters. Hsu *et al.* [11] proposed a technique that uses not only DWT coefficients in each subband independently but also uses the inter-subband correlation to extract features for bird song recognition.

Other Feature Extraction (FE) Techniques—Another popular FE technique is Mel Frequency Cepstrum (MFC) [19], which is commonly used in human speech recognition and extended to animal vocalizations. In this technique, the magnitudes of the spectrum obtained from FT are mapped into a Mel-frequency scale. Coefficients of the MFC are considered the sound features, which are obtained by taking discrete cosine transform of the logs of the magnitudes at each Mel frequencies. Other variations of this technique are Linear Predictive Coding (LPC), Linear Prediction Cepstrum Coefficients (LPCC), and their variations [19]. In a study by Qian *et al.* [10], features of bird sounds are extracted based on MFCC and statistical functions, including the zero-crossing-rate of the time signal, the root mean square frame energy, the pitch frequency, and the harmonics-to-noise ratio by autocorrelation function. Bastas *et al.* [8] proposed Spectrogram-based Image Frequency Statistics (SIFS) for extracting features of bird flight calls. In this technique, the spectrograms are constructed based on the sequence of Fast Fourier Transform (FFT) and are converted to binary spectrograms. Features of lowest, highest, mean, and median frequencies are computed for the whole image using morphological image processing. Pryadarshani *et al.* [20] provide a review of different FE techniques in automated bird song recognition.

FE techniques may provide many features that are not effective in determining the species and reduce the performance of the technique by increasing the computational time. Thus, dimension reduction techniques are usually applied to remove the unnecessary features while retaining the information of the signal. One of the popular techniques to reduce the dimensionality of the features is Principal Component Analysis (PCA) [11].

10.2.1.1.3 Classification

Classification is a supervised learning approach that defines the categorical labels of the instances based on previously annotated data. In bird/bat call classification, the feature vectors are the inputs of the classifier, and the bird/bat species is the output, as shown in Figure 10.5. The quality of the classifier depends on several factors. First, it depends on the extracted features. These features need to be significant enough to be able to differentiate between species. The more significant features are extracted, the higher accuracy in classification will be obtained.

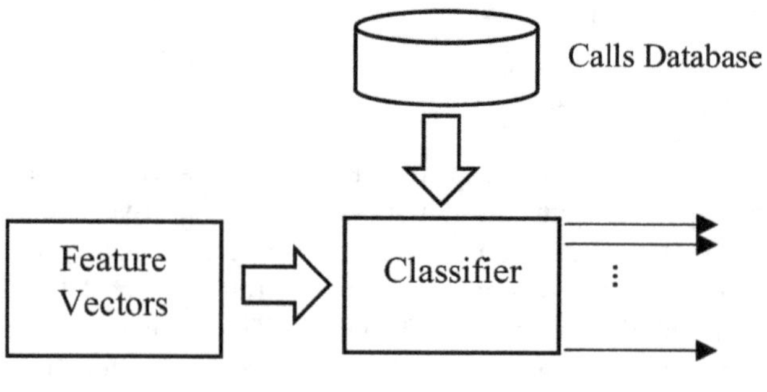

Figure 10.5 Classification

Second, the quality of the classifier depends on the quality of the database used in training [21]. The reference calls in the database have to be evaluated through other approaches and annotated with certainty. The database should also contain intra-species call variations related to habitat structure and geographical differences [6, 21]. Suppose the database does not have the variation of the calls and covers only a limited set of behaviors, habitats, and other factors. In that case, it will not represent the calls and therefore cause the classifier to be trained poorly.

There are many classification techniques proposed in the literature and used in birds/bat studies, such as Neural Networks, Support Vector Machine (SVM), K-Nearest Neighbor (KNN), Hidden Markov Model (HMM), etc. KNN is a simple and powerful classification technique that measures the similarity between instances based on a distance metric. The new instance is assigned to the class in which it has the minimum distance to its members. Different distance metrics can be used in this technique, such as Euclidean distance, Mahalanobis distance, Cosine similarity, and Pearson correlation. HMM is a statistical technique based on the Markov process representing probability distribution over sequences of observations. In this technique, it is assumed that the observations at each time instance are generated by some *hidden* states. The state of the hidden process satisfies the Markov property. The hidden state variable is discrete. HMM defines the probability distribution over a sequence of observations using probability distribution over the initial state, transition matrix, and the output model.

There are other classification techniques proposed and applied in the bird/bat studies. Mirzaei *et al.* [16] developed an Evolutionary Neural Network (ENN), which uses a genetic algorithm and neural network to classify bird and bat echolocation calls. The inputs of the classifier are feature vectors extracted by MFCC, STFT, and DWT. They showed DWT and ENN provide superior accuracy compared with the other techniques. Qian *et al.* [10] used an extreme learning machine [22] to classify bird songs. The proposed classifier is a feedforward neural network with one hidden layer that randomly chooses hidden nodes and analytically determines the neural network's output weights. The learning process is very fast, and it does not have the local minima issue as gradient-based algorithms.

Deep learning has been proven to be very effective in classification problems and has received considerable attention in the last decade. One of the most common deep learning-based classification techniques that has been widely used in bird/bat studies is Convolutional Neural Network (CNN). CNN is a specific type of feedforward neural network. The layers consist of convolutional layers followed by an activation function, pooling layers, and fully connected layers, as shown in Figure 10.6. CNN uses spatial information, and therefore it is well suited for object detection and classification in imaging. In CNN, FE is performed by a series of convolution and pooling layers, while classification is performed by mapping the extracted features into final output using a fully connected layer. The larger number of the convolution process enables to extract more details of the images. Most of the deep learning-based acoustic classifications in bird/bat studies are performed by applying CNN over the spectrograms of the calls. Toth and Czeba [24] proposed a CNN for bird song classification. The inputs of the network are the spectrograms of the songs

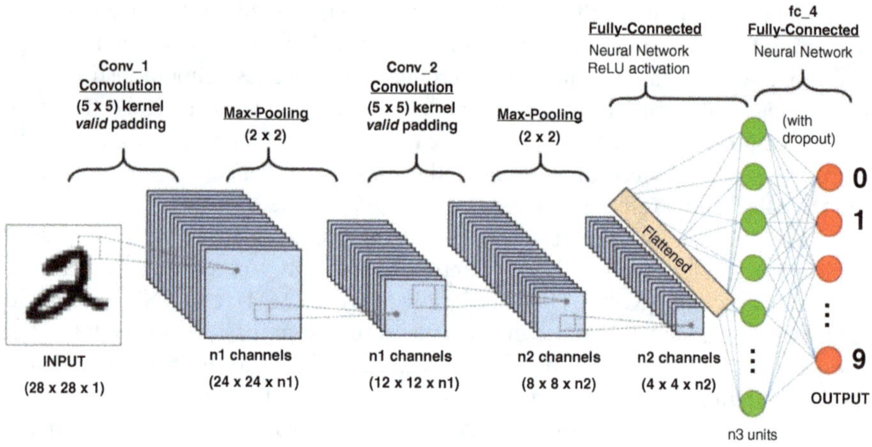

Figure 10.6 CNN [23]

generated from STFT. Aodha *et al.* [12] developed a CNN for classification of bat echolocation calls. They recorded the signals with a sampling rate of 44.1 kHz to retain calls in frequency ranges of 5–135 kHz. The audio signals are converted to a spectrogram using FFT. They used the spectrograms of the calls and predicted the presence of a search-phase bat call at every instance of time in the spectrograms. Their proposed CNN is developed in three convolution and max-pooling layers, followed by one fully connected layer.

10.2.1.2 Thermal imaging

Thermal imaging is a vision-based monitoring tool for monitoring birds/bats in wind farms. This approach can estimate the number of passes and the mortality rate, but it cannot identify the species. This technology works by detecting the IR energy (heat) from objects and converting it to electronic signals that are then digitized and used to produce an image. IR cameras are different in their capabilities and specifications, such as thermal sensitivity, thermal accuracy, pixel pitch, horizontal and vertical field of view, lens focal length, resolution, etc. One of the important characteristics of the IR camera is the frame rate, which defines how fast the camera detector creates the images. For targets moving with higher speed, the higher frame rates provide more accurate results. On average, birds' speed is between 9 and 22 ms (10 and 50 mph) [25, 26]. Larger birds fly faster on average than smaller birds, and flight speed varies depending on the bird's behavior.

IR monitoring also provides information about straightness index, heat, and direction. This tool, however, is not capable of distinguishing birds from bats and insects. However, unsupervised learning can be used to cluster the targets into different avian groups. Mirzaei *et al.* [27] used the Ant Clustering algorithm to group the targets captured by IR into different clusters of birds/bats/insects. In another study [28], they proposed a classification technique, regularized logistic regression-based,

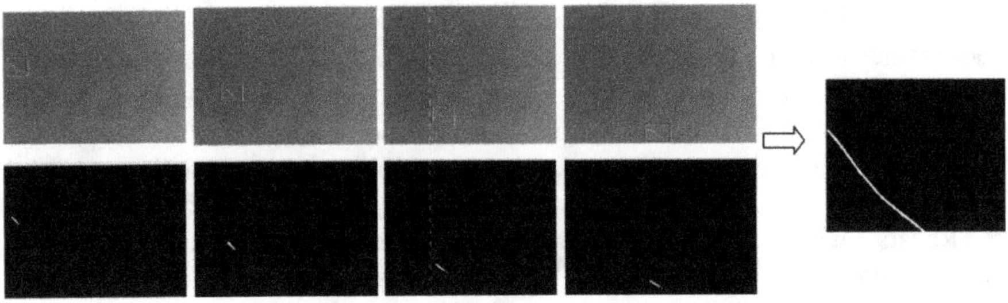

Figure 10.7 Target detection and tracking in IR imaging [31]

to classify targets using a pre-created database. The main processing of IR imaging is detection and tracking, as described in the next section.

10.2.1.2.1 Detection and tracking

Target detection is described as the process of identifying an object of interest on an image, and it is the initial task for tracking. The object of interest is called *foreground,* and the rest of the entities on the image are called *background.* Object detection is generally the process of extracting the foreground from the background through the motion change. The most commonly used detection techniques using motion change are frame differencing, background subtraction, motion segmentation, and matrix decomposition [29]. Tracking constructs the trajectory by finding the corresponding objects in the following frames. There are different techniques and approaches proposed for object detection and tracking. The approaches are different based on object representation, image features, motion, appearance, and shape. The most common tracking algorithms are Template Matching, Kalman filtering, Particle filtering, and Mean shift-based target tracking [30]. Mirzaei *et al.* [31] implemented a technique for bird/bat detection and tracking in wind farms using several steps: background subtraction using running average, adaptive thresholding using an enhanced Otsu thresholding, noise suppression using morphology, and tracking using labeling-based connectivity. Figure 10.7 shows the detection of a bird/bat in a sequence of images and the trajectory of the track. Wei *et al.* [32] developed a detection and tracking technology to track birds/bats in a wind farm. They initialized the background model by computing the average value of each pixel in the first few frames. Then they subtracted the adjacent frames and updated the background. They used region-labeling to assign pixels to different groups of objects. They also used break-point detection to avoid breaking one object into two or three. Mitchell and Clarke [33] used thermal imaging for detecting birds to assess the efficiency of IR cameras in monitoring rooster birds. They also compared the result of the thermal imaging monitoring with traditional monitoring technique. According to their results, the IR camera detection was higher during days for fourteen species compared to traditional techniques, and it did not differ between diurnal and nocturnal surveys for eight other species.

10.2.1.3 Radar technology

Radar stands for radio detection and ranging, and it is a useful tool for detection and tracking targets, including birds/bats. Radar works by transmitting radio signals in the directivity of the antenna and receiving the echo back. If there is an object in that direction, the radio wave is scattered, and a small portion of it will be received by radar.

Radars have been used in many applications such as military, aviation, meteorology, geology, traffic control, etc. The use of radars in avian monitoring is relatively novel, and there has been a great interest in this technology over the last decades. Radar can be used with different types of antennas: vertical, horizontal, and Doppler. Each of these antennas provides a different type of information about the birds and bats movement. A horizontal antenna scans across the landscape and provides information about the direction of the flight. A vertical antenna provides information about the number of the targets as well as the height of the flight. However, radars cannot differentiate birds and bats and cannot identify the species.

10.2.1.3.1 Radars in avian studies

Different types of radars have been used in bird/bat studies in wind farms. MERLIN avian and bat radars manufactured by DeTect [34] use dual vertical and horizontal radars to provide bird/bat passage range and the altitude in wind farm sites. The radar provides information for birds/bats for the area out to 2–4 nm and up to 3 048 m in altitude. In a project conducted by US Fish and Wildlife Service [35], two avian radars (MERLIN) are stationed around Great lakes since 2011 for bird and bat tracking. Komatineni *et al.* [36] used Doppler weather surveillance radar for birds' observation. Figure 10.8 shows two different types of radar for monitoring birds and bats.

Mirzaei *et al.* [39] used a marine radar with a T-bar antenna, which was operating in horizontal and vertical mode, and a parabolic dish to assess bird/bat movement in the western basin of Lake Erie. In a radar study conducted by New Jersey Audubon Society [40], birds/bats movement pattern, passage rate, altitude information, and flight direction were estimated at The Maple Ridge Wind Power Facility in

(a) (b)

Figure 10.8 Radars for monitoring birds and bats (a) MERLIN radar [34] and (b) Furuno Marine radar [37, 38]

New York. Radar provided altitude information to estimate the number of collisions with wind turbines in the altitudes at risk. Radars can provide information about altitude, range, direction, but they cannot identify the targets at species level.

Although different sensors such as radar, IR, acoustics, and other sensors could be used to collect data, however, each of these sensors provides some specific information and not complete information about the targets. For example, acoustics provide species information; however, direction and flight altitude cannot be obtained by acoustics. IR can be used to estimate the number of passes and straightness index, flight pattern, etc. However, it is not helpful for altitude information. Radar is a useful tool for altitude information, but not for species recognition. So, data fusion of these sensors could provide more robust and reliable results about birds/bats activities and movement patterns. Mirzaei *et al.* [41] used data fusion of acoustics, IR, and radar to assess birds'/bats' activities in Ottawa National Wildlife Refuge, OH. They collected data from each sensor, processed data separately, and fused the results using a fuzzy Bayesian approach to provide information about bird/bats. Recent studies rely more on data fusion of sensors than one type of sensor.

10.2.2 Offshore monitoring

Birds and bats monitoring under the wind turbines is performed by human observers walking around the area looking for carcasses. Electronic monitoring with radars, cameras, and acoustic recorders is also performed. This type of monitoring provides the absence or presence of any birds/bats activity. These monitoring of carcasses under the wind turbines work well only onshore wind farms. Finding carcasses is not possible when wind turbines are offshore. Human observers will not be able to retrieve carcasses, and determination of collision or effects of offshore wind turbines will need to be devised using electronic means. Niemi *et al.* [42] have developed an electronic monitoring system employing radar systems and video recordings. Their system uses Robin Radar Systems B. V., which consists of two radars, horizontal scanning S-band radar, and vertical scanning X-band radar. The system has a 3-D tracking capability. The S-band radar will provide longitude and altitude for the target bird. The X-band system provides altitude and can detect small birds up to a distance of 5 km. Once radars determine that birds on their way toward the wind turbine, radar signals will trigger a video camera to start recording. Video images and radar parameters are then used to identify the target bird. The software uses deep learning algorithms such as CNN to classify or identify target birds. When the birds' identity, their time, and 3-D location are provided to wildlife biologists, they can determine why this particular bird was at a particular location. The information collected for a long period of time will help identify collision risks and subsequently help develop mitigation techniques.

10.3 Impact of bird collision

Birds are at higher collision risk with wind turbines and with the airplanes or other flying objects, and power lines. Bird collision is dependent on their agility, aerodynamic capability, maneuverability, morphology, and other biomechanical abilities. It has been said that birds tried to avoid incoming airplane but were unable to do that due to their maneuverability and agility. Collision is also a function of how much time a particular species spend time flying. A higher percentage of flying time will result in higher collision probability. It is dependent and varies from species to species. Nocturnal migratory birds are at higher risk of colliding with wind turbines, especially during low visibility. Birds use their mainly four sensory capabilities to interact with the environment and specifically with wind turbines. Their sensory capabilities can be classified as [43]:

- vision
- hearing
- chemical sensing
- magneto-reception.

Vision and hearing are two main sensory aids for finding food, recognizing mates, and avoiding structures and predators. All four sensory capabilities differ from species to species. Birds' ear shapes and locations and their hearing canals have different anatomy and vary from species to species. Birds are sensitive to sound between 1 and 5 kHz and have directional sensitivity. Their songs are different, and calls may announce predators, food, mating information, and advertise their territory. Some commercial buildings use predatory calls to keep birds from nesting, and this approach may also be effective in the vicinity of wind turbines. Predatory calls may add to existing wind turbine noise and may result from complaints from neighbors. Birds use chemoreception or chemical sensing for direction finding, foraging, and avoiding predators. Birds also can use the earth's magnetic field for navigation during migration.

Bird collision fatalities force one to consider what needs to be done to reduce birds and bats mortality rates in a given wind farm. Researchers have discovered that bird mortalities are not the same and vary from one wind farm to another and species to species as they have spatial distribution. One needs to develop a mathematical model to predict future mortalities and help in devising some form of mitigation. It is well known that a mathematical model depends on assumptions, their relationship, and qualitative conclusions derived from a quantitative approach. The following are some of the collision models that have been developed.

Sebastian-Gonzalez *et al.* [44] have used network analysis to select specifies which effects monitoring of bird collisions. Their study was conducted in peninsular Spain. This area has wildlife and the presence of many endangered birds and bats species. Their model accepts fatality information, and monitoring programs involve identifying species, date, time, and wind turbine along with its geographical location. The model offers minimum mortality rate information that can be used in the planning of appropriate mitigation methods. The model follows three steps. The first step is making sure that the data has been organized in a nested pattern. The idea is to see if there are more fatalities of a particular species at a given turbine indicated by the nestedness matrix. The model allows predicting other species that will also be at higher risk at a given turbine. The process in the second step identifies the species that contributing most to the nestedness. The third step involves the appropriateness of its biological characteristics. Identified species may also signal other species that were hard to detect and other factors that should be considered in the model. Weighted Nestedness of Decreasing Fill (WNODF) [45, 46] metric provides a percentage of a nested matrix as variation in the number of fatalities relating to the degree of nestedness. The perfect nestedness matrix will correspond to a 100-percent value. They have compared their observed value of nestedness with 1 000 matrices that were generated using the null model. In this case, the number of fatalities per species will correspond to the nestedness. Finally, authors calculate each species' contribution to the nestedness and determine accuracy to predict the each species' mortality. Larger and positive values will indicate the species which has been mostly killed or are at higher risk.

Khosravifard *et al.* [47, 48] have described two common approaches to determining the home range and their utilization distribution. These approaches help in determining space use and birds movement. Birds use their home area for foraging, gathering, mating, and raising their families. Utilization distribution shows the probabilities of their using areas other than the home range. Home and space utilization have been quantified using various methods such as minimum convex polygon, bivariate normal method, grid square, population utilization distribution, kernel density, dynamic Brownian bridge movement model, etc. (see references in [47]). Most authors have considered the 2-D model, whereas Khosravifard *et al.* [47] and others have used the 3-D model, which provides additional information. They have performed volumetric analysis (3-D) using Kernel Utilization Distribution (KUD). The study area was Southern Spain. Their study was limited to one tagged Griffon vulture covering over 150 km^2 that included 269 wind turbines in twenty wind farms. They have used Bird Tracking System blogger [49], which was developed at the University of Amsterdam. Wind farm developers can use the 3-D model to calculate the risk before deciding to locate wind farms at a certain location using preconstruction data.

Band *et al.* [50, 51] proposed a method to estimate the number of bird collisions in wind farms. First, they proposed data collection by finding the time each selected species flies around the chosen area. Second, they determined the time spent by the selected species using a different area. They can then compute the proportion of time that is spent within the wind turbine area. The data are collected by observing the target species near wind turbines and other observable areas. They used a two-stage model. First, they estimated the number of birds' collisions. In the second stage, they computed the probability of birds flying through the wind turbines. These two parts are then multiplied together. They concluded that a higher degree of collision avoidance would lower mortality. Details of various steps are provided in reference [50, 51].

Eichorn *et al.* [52] provided a model-based approach to simulate the Red Kite's foraging behavior. They determined the collision risk factor as a function of distance. They determined that the risk factor exponentially decreases with the increase in distance. They derived an impact function for quantifying risks as a function of the distance between the turbine and the bird's nest. Everaert [53] calculated the collision risk and avoidance rate in five different wind farms in Belgium. Everaert modeled collision risks by estimating the number of collision fatalities with respect to the total number of birds crossing the wind farm. Collision risk rates are then used to compute the micro-avoidance rate. The author has estimated the proportion of birds flying through with wind turbine area but managed to avoid passing through the rotor swept area. Everaert used uniform distribution, and the model is based on the Band's model [50]. Mortality rates and actual collision risk in the study area were computed. The author claims that this model can be extended to other species and areas.

Heist [54] assessed bat and bird fatality risks using acoustic detectors. He recorded birds and bats calls for a three-year period and analyzed using automated bird and bat call identification software. His predictive model indicates pass rates. According to his conclusion, the approach was not a good predictor of fatalities, but his work is useful in finding birds and bats in this neighborhood. His study can be incorporated in siting wind farm facilities in low-impact areas.

Korner-Nievergelt *et al.* [55] proposed a Mixture model for estimating birds and bats mortality rates. They combined carcass searches with the survey of birds and bats at high risk of collisions. They estimated animals exposed to high risk by using acoustic data collected and bird migration traffic data. Under certain assumptions, the model will help predict collision rate. They have modeled the number of carcasses found using a binomial distribution. The number of the bats collision was modeled using Poisson distribution. They have estimated various parameters using Bayesian methods, and Monte Carlo Simulation was performed using Markov Chain (MCMC). Details of this method can be found in reference [56].

In general, collision models have been developed by first selecting study areas and species. A mathematical model is then created, and various studies are performed. Generally, the developed model should be applicable to other areas and species.

10.4 Mitigation

Construction of wind farms is a complex process and requires preconstruction evaluation and approval at various levels. This process is commonly known as siting and has four components [57, 58]. First of all, wind farm developers should consider many aspects, such as support from the community and local government. Second, it should study the environmental impact on birds and bats. Third, it should evaluate easy access to transmission of power. The fourth is to study the impact on national security. Since wind turbines are over 160 m then approval from the Federal Aviation Administration (FAA) will be required due to their height. FAA will automatically ask the Department of Defense (DoD) if wind turbines will cause any shadow on their radars. The DoD will in turn work with the developer to mitigate radar-related issues, and in most cases, it gets resolved with certain modifications to the radar systems. Above all these factors, wind farms are only developed where there is the availability of high winds.

Siting of wind farms is crucial from its impact on birds and bats and requires mitigation efforts. Mitigation is very complex and involves constant monitoring and time to time adaptation and adjustments. Mitigation has been looked at from different points of view, such as preconstruction vs. post-construction, species-specific vs. geographic-specific, and turbine-based vs. bird-based issues [43, 56]. Marques *et al.* [56] have looked into mitigation strategy from three points of view: avoidance, minimization, and compensation. Preconstruction mitigation for bird collision for proper siting includes avoiding migratory paths, endangered species habitat, and foraging areas or bird sanctuaries and habitats. Wind farms should also not be developed in hazardous or other high-risk areas from the birds' and bats' point of view. Preconstruction issues mostly relate to proper siting.

Mitigation is complex, and it is very hard to predict if the certain effort will result in any benefit in reducing collisions with wind turbines. There may also be an increase in the collision rate. Collision models described in the previous section will guide and help in predicting future collisions. These collision models will help in deciding the building of wind farms in certain geographic locations. One way to monitor benefits of any mitigation is to use either human observers or automatic monitoring with radars, video, and acoustic surveillance. Researchers have tried, experimented and recommended several mitigation approaches. The following are some of the approaches that have been used to mitigate collision risks. (1) Replace older lattice-based towers with tubular or cylindrical towers. This eliminates perching opportunities for birds and predators. Older turbines are being replaced with more powerful, larger, and fewer turbines. (2) Use of larger turbines which are higher and produce more energy and have lower fatalities rate. This may create a small cluster of wind turbines and eliminate certain areas for birds and bats. Large turbines rotate slowly, and birds have time to avoid the moving blade. The benefit will be that birds and bats may learn to avoid this area and reduce collision rates. Another benefit will be newer turbines will be more energy efficient.

Wind turbines need a certain wind speed to rotate their blades. Turbine speed is also related to the production of power. Temporarily shutting down turbines is another mitigation approach to reduce collision rates. Birds may avoid collision with stationary or slow-moving blades. There are some drawbacks of shutting down the turbines during low wind speed. One drawback will be loss of energy production resulting in loss of revenue. The second negative effect will be that birds may become habitual, treat as fixed structure if turbines are shut down for a longer period of time. One optimum way of operation is to shut turbines down only during high bird activity. This will require some automatic alert system via avian radar, indicating high bird activity. Shutting down turbines during low visibility will be another mitigation approach that will be beneficial.

To design post-construction mitigation strategies, one needs to consider birds' capabilities in terms of vision, hearing, chemical sensing abilities, and magnetoreception that helps in navigation capabilities [56]. Mitigating measures may be species and/or geographic area specific. Vision- and acoustic-based mitigation approaches can provide fruitful results to a certain degree of success. One should also consider that some of the wind turbines are on farmlands, and there are local residents. Any mitigation approach should not create a nuisance to the local neighborhood as both vision, and acoustic-based solutions will certainly also get the attention of local residents. The use of lighting should also be FAA compliant as wind turbines are high enough and may have an effect on air traffic. Blackwell *et al.* [59–62] placed lights on a toy aircraft and experimented with Canada geese. It was discovered that Canada geese took evasive action on approaching aircraft and considered dangerous. The study was done for bird strike applications and can be applied to wind turbines. It can be deduced that lighting may be one of the vehicles to get birds 'attention. If one can utilize a bird visionary sensor and use light as a collision reduction mechanism, then it can be a win-win situation. One should also be mindful of the interplay of species' sensory abilities, behavior, foraging, courtship, the territorial defense may limit detection and or perception and aerodynamic capability. Impulses may affect vigilance and can limit the detection and perception of potential danger. Impulsive behavior can lead to a potential collision with either the wind turbine structure or the moving blade.

In designing appropriate mitigation techniques, based on vision sensing, any visual effects should resemble predation situation and potential danger or fear. Two causes should be considered, namely species specific biomechanical and optical characteristics. Increasing the birds' awareness of the dangerous turbine or tower will allow them to maneuver and avoid any collision. Some of the vision-based approaches are making rotor blades more visible with passive and active visual cues such as painting or lighting. Hodos [63] marked patterns such as scarecrows, conspecific/raptor models, or conspecific corpses to see their effect on reducing collision. Blackwell *et al.* [60, 61] painted one of three blades as black for reducing motion smear. On the other hand, minimal use of lights for the benefit of bats and nocturnal migrating birds can reduce collision risk, especially during low visibility. One study has found that nocturnal migrating birds are less attracted to blue and green lights as they may not be visible during low visibility. Another study suggests the use of strobing, flashing, bright lights, and revolving lights, causing temporary blinding and thereby creating confusion. One drawback of using lights is that birds may learn to live with them and become habitual, therefore losing their effectiveness in deterrence. According to one researcher, randomized selection of at least two strobe frequencies may reduce habituation with light [63]. The use of marking, reducing motion smear, reflectors, or UV-coating can be another way in mitigating collision risk. Use of light may also be a nuisance for local residents

Acoustic-based deterrence or harassment approaches can also be used as one of the mitigation techniques among all others. Airports commonly fire gas cannons or other devices such as pyrotechnics and ultrasound to scare or deter birds hanging out at the airport. Sound-producing devices will need to make a sound periodically, and care should be taken that birds should not become habituated [43, 56]. They may benefit in the short term. The use of acoustic devices may increase noise levels and result in complaints from the neighborhood. Habituation is a form of learning that leads to decreased associability, which loses its effectiveness. Therefore, it is essential to minimize the risk of habituation. Some commercial buildings use predatory, or distress calls for stopping birds from nesting.

Another form of mitigation is called habitat alterations, which can be on or off-site. Areas near the wind turbine or entire wind farms can mitigate collision risk by making them unattractive to birds. For example, one vendor from New Zealand (may be Grasslanz Technology Limited) was promoting a type of grass that can be used in airport areas and a wind farm in a bird strike conference. This particular artificial grass will upset the birds' stomachs. Birds may get sick and will have the motivation to move away from airport or wind farm areas. Another approach is to have open vegetation near the turbine less attractive for either birds or their prey. Clear cutting forests may reduce prey, food, and perching opportunities. Knowledge of neighborhood birds and predators may help in incorporating vegetation which may not be attractive to native birds. The development of off-site habitat is crucial and can be an effective mitigation approach. Development of perching, roosting, foraging, and nesting opportunities away from the wind farm can be very helpful. The increasing availability of food supply and breeding sites will make the area attractive to local birds, and they can move away from wind farm areas. This approach may also be attractive to other birds and predators. Developing habitats in other areas may attract birds to eventually move away from wind farm areas.

Mitigation approaches can be chosen from vision- or acoustic-based and alteration of habitat. The selected approach will vary from geographical areas to another geographical area. This is due to different birds and bats habitats in a particular area. A study should be conducted to use appropriate mitigation approach which will help in reducing collision rates. Employed mitigation approach should be monitored on a regular basis to determine their effectiveness at the desired level. The use of strong intermittent stimuli of vision or acoustic type should be used to reduce habituation and keep its effectiveness. May *et al.* [43] have provided six qualitative criteria to evaluate mitigation. This evaluation technique can be an effective tool to arrive at a series of mitigation approaches.

Development of habitat to enhance bird and bat populations is another form of mitigation as it can compensate for loss of population due to collision with wind turbines. Researchers have proposed the creation of new habitat, foraging, resting, increasing availability of prey, and reducing predator population can be beneficial [43, 56].

10.5 Conclusions

Birds/bats monitoring and mitigation techniques in wind farms or area potential for future construction of wind farms are essential for preserving the birds and bats. There are different traditional and advanced approaches for the monitoring of birds and bats. Traditional techniques are time-consuming, labor-intensive, and weather-dependent. Therefore, these techniques may not be effective for long-term monitoring period. Advanced monitoring techniques use one or multiple sensors to collect data from the biological targets and process data to provide useful information about the distribution and movements of the birds and bats. The most common advanced sensors for bird and bat monitoring are acoustic detector, thermal or video cameras, and radar. Each of these sensors collects data separately, and the data will be used for further processing to provide useful information about the birds/bats. Acoustics provide identification of the birds/bats at the species level. IR camera is a useful tool for providing information about the number of passes, and radar is a useful tool for estimating direction, number of targets, and flight height. Data fusion of the sensors (acoustics, IR camera, and radar) can provide complementary and more accurate results about birds and bats' movements. Monitoring techniques will provide collision information near wind turbines. The results of monitoring techniques will be used by decision-makers to create and apply appropriate mitigation techniques to preserve birds/bats population, especially the bird/bat species on the endangered list.

Acknowledgments

We would like to thank all researchers working in the area of collision modeling and developing monitoring and mitigation approaches to live in harmony with nature. We cannot list all the works that are available due to lack of space. We encourage readers to look at references within references to get a bigger picture of birds' and bats' interaction with wind turbines.

References

[1] *Wind Turbine Interactions with Wildlife and Their Habitats*. American Wind Wildlife Publication. 2019. Available from www.awwi.org.

[2] Kunz T.H., Braun de Torrez E., Bauer D., Lobova T., Fleming T.H. 'Ecosystem services provided by bats'. *Annals of the New York Academy of Sciences*. 2011;**1223**:1–38.

[3] Loeb S.C., Rodhouse T.J., Ellison L.E. *A plan for the North American bat monitoring program (NABat)*. Gen. Tech. Rep. SRS-208. Asheville, NC: U.S. Department of Agriculture Forest Service, Southern Research Station; 2015. p. 100.

[4] Peeples G. Bat monitoring protocol. USFW. 2015. Available from https://www.fws.gov/southeast/podcasts/2015/11/bat-monitoring-protocol/ [Accessed 5/19/2021].

[5] Parisé J., Walker T.R. 'Industrial wind turbine post-construction bird and bat monitoring: A policy framework for Canada'. *Journal of Environmental Management*. 2017;**201**:252–9.

[6] Russo D., Voigt C.C. 'The use of automated identification of bat echolocation calls in acoustic monitoring: a cautionary note for a sound analysis'. *Ecological Indicators*. 2016;**66**:598–602.

[7] Mirzaei G. 'The BIO-acoustic feature extraction and classification of bat echolocation calls'. *2012 IEEE International Conference on Electro/Information Technology*. Indianapolis, IN; 2012. pp. 1–4.

[8] Bastas S., Majid M.W., Mirzaei G., *et al.* 'A novel feature extraction algorithm for classification of bird flight calls'. *2012 IEEE International Symposium on Circuits and Systems (ISCAS)*. Seoul; 2012. pp. 1676–9.

[9] Neece B.D., Loeb S.C., Jachowski D.S. 'Implementing and assessing the efficacy of the North American bat monitoring program'. *Journal of Fish and Wildlife Management*. 2019;**10**(2):391–409.

[10] Qian K., Zhang Z., Ringeval F., Schuller B. 'Bird sounds classification by large scale acoustic features and extreme learning machine'. *2015 IEEE Global Conference on Signal and Information Processing (GlobalSIP)*. Orlando, FL; 2015. pp. 1317–21.

[11] Hsu S.-B., Lee C.-H., Chang P.-C., Han C.-C., Fan K.-C. 'Local wavelet acoustic pattern: a novel time–frequency descriptor for birdsong recognition'. *IEEE Transactions on Multimedia*. 2018;**20**(12):3187–99.

[12] Mac Aodha O., Gibb R., Barlow K.E., *et al.* 'Bat detective-deep learning tools for bat acoustic signal detection'. *PLOS Computational Biology*. 2018;**14**(3):e1005995.

[13] Mirzaei G. 'Acoustic monitoring techniques for avian detection and classification'. *2012 Conference Record of the Forty Sixth Asilomar Conference on Signals, Systems and Computers (ASILOMAR)*. Pacific Grove, CA; 2012. pp. 1835–8.

[14] *Wildlife Acoustics [online]*. Available from https://www.wildlifeacoustics.com/.

[15] *Binary Acoustics Technology [online]*. Available from http://binaryacoustictech.com/.

[16] Mirzaei G. 'The application of evolutionary neural network for bat echolocation calls recognition'. *The 2011 International Joint Conference on Neural Networks*. San Jose, CA; 2011. pp. 1106–11.

[17] Proakis J.G., Manolakis D.G. *Digital Signal Processing*. Prentice Hall, New Jersey: Pearson Prentice Hall; 2007.

[18] Polikar R. *The Wavelet Tutorial, Rowan University, College of Engineering [online]*. Available from http://users.rowan.edu/~polikar/WAVELETS/WTtutorial.html.

[19] Zbancioc M., Costin M. *Using Neural Networks and LPCC to Improve Speech Recognition. Signals, Circuits and Systems*. Iasi, Romania: International Symposium; 2003. pp. 445–8.

[20] Priyadarshani N., Marsland S., Castro I. 'Automated birdsong recognition in complex acoustic environments: a review'. *Journal of Avian Biology*. 2018;**49**(5):jav-01447.

[21] Clement M.J., Murray K.L., Solick D.I., Gruver J.C. 'The effect of call libraries and acoustic filters on the identification of bat echolocation'. *Ecology and Evolution*. 2014;**4**(17):3482–93.

[22] Huang G.-B., Zhu Q.-Y., Siew C.-K. 'Extreme learning machine: theory and applications'. *Neurocomputing*. 2006;**70**(1-3):489–501.

[23] *A Comprehensive Guide to Convolutional Neural Networks—the ELI5 way [online]*. Available from https://towardsdatascience.com/

a-comprehensive-guide-to-convolutional-neural-networks-the-eli5-way-3bd-2b1164a53.

[24] Toth B.P., Czeba B. 'Convolutional neural networks for large-scale bird song classification in noisy environment'. *Working Notes of CLEF 2016 – Conference and Labs of the Evaluation Forum*. Evora, Portugal: CEUR-WS Proceedings Notes; 2016. pp. 560–8.

[25] Alerstam T., Rosén M., Bäckman J., Ericson P.G.P., Hellgren O. 'Flight speeds among bird species: allometric and phylogenetic effects'. *PLoS Biology*. 2007;**5**(8):e197.

[26] Matzner S., Cullinan V.I., Duberstein C.A. 'Two-dimensional thermal video analysis of offshore bird and bat flight'. *Ecological Informatics*. 2015;**30**(8):20–8.

[27] Mirzaei G. 'Implementation of ant clustering algorithm for IR imagery in wind turbine applications'. *2012 IEEE 55th International Midwest Symposium on Circuits and Systems (MWSCAS)*. Boise, ID; 2012. pp. 868–71.

[28] Mirzaei G., Jamali M.M., Gorsevski P.V., Frizado J., Bingman V.P. 'Regularized logistic regression based classification for infrared images'. *2014 48th Asilomar Conference on Signals, Systems and Computers*. Pacific Grove, CA; 2014. pp. 487–90.

[29] Shan C., Porikli F., Xiang T., Gong S. *Video Analytics for Business Intelligence*. Berlin Heidelberg: Springer-Verlag; 2012.

[30] Porikli F., Yilmaz A.. ' 'Object detection and tracking' in Shan C., Porikli F., Xiang T., Gong S. (eds.). *Video Analytics for Business Intelligence. Studies in Computational Intelligence*. Berlin, Heidelber: Springer; 2012.

[31] Mirzaei G. 'Avian detection & tracking algorithm using infrared imaging'. *2012 IEEE International Conference on Electro/Information Technology*. Indianapolis, IN; 2012. pp. 1–4.

[32] Wei L. 'Birds/bats movement tracking with IR camera for wind farm applications'. *2014 IEEE International Symposium on Circuits and Systems (ISCAS)*. Melbourne VIC; 2014. pp. 341–4.

[33] Mitchell W.F., Clarke R.H. 'Using infrared thermography to detect night-roosting birds'. *Journal of Ornithology*. 2019;**90**(1):39–51.

[34] *DeTect MERLIN Radar [online]*. Available from https://detect-inc.com/avian-bat-radar-systems/ [Accessed 5/19/2021].

[35] *US Fish and Wildlife Services [online]*. Available from https://www.fws.gov/radar/radarunits/index.html [Accessed 5/19/2021].

[36] Komatineni P. 'Quantification of bird migration using Doppler Weather Surveillance Radars (NEXRAD)'. *2013 IEEE Radar Conference (RadarCon13)*. Ottawa, ON; 2013. pp. 1–4.

[37] *Furuno Product Solutions [online]*. Available from https://www.furuno.com/en/products/radar?market=recreational.

[38] Mirzaei G., Jamali M., Ross J.D., Gorsevski P.V., Frizado J.P., Bingman V.P. 'Radar-based monitoring system for nocturnal assessment'. *IEEE International Conference on Electro/Information Technology*. Milwaukee, WI; 2014. pp. 252–5.

[39] Mirzaei G., Jamali M.M., Gorsevski P.V., Frizado J., Bingman V.P. 'Data fusion of IR and marine radar data'. *2013 Asilomar Conference on Signals, Systems and Computers*. Pacific Grove, CA; 2013. pp. 565–8.

[40] New Jersey Audubon Old Bird, Incorporated North East Ecological Services. 'Post-construction wildlife monitoring at Maple Ridge wind farm'. New York State Energy Research and Development Authority; January 2013.

[41] Mirzaei G., Jamali M.M., Ross J., Gorsevski P.V., Bingman V.P. 'Data fusion of acoustics, infrared, and marine radar for avian study'. *IEEE Sensors Journal*. 2015;**15**(11):6625–32.

[42] Niemi J., Tantttu J.T. (edseds.) 'Automatic bird identification for offshore wind farms' in Bispo R., Bernardino J., Coelho H., Costa J.L. (eds.). *In Wind Energy and Wildlife Impacts*. Cham: Springer; 2019. pp. 131–51.

[43] May R., Reitan O., Bevanger K., Lorenstein S.H., Nygard T. 'Mitigating wind-turbine induced avian mortality: sensory, aerodynamic and cognitive constraints and options'. *Renewal and Sustainable Energy Reviews* 42: elsevier. 2015;42:170 -81. Available from http://lithornis.nmsu.edu/~phoude/May_et_al_2015_Mitigating_wind-turbine_induced_avian_mortality_sensory_aerodynamic_and_cognitive_constraints_and_options.pdf [Accessed 5 May 2020].

[44] Sebastián-González E., Pérez-García J.M., Carrete M., Donázar J.A., Sánchez-Zapata J.A. 'Using network analysis to identify indicator species and reduce collision fatalities at wind farms'. *Biological Conservation*. 2018;**224**(4):209–12.

[45] Nicholas B., Mário A.-N., Gilberto C. 'Which matrices show perfect nestedness or the absence of nestedness? an analytical study on the performance of NODF and WNODF'. 2016;**4**.

[46] Vázquez D.P., Melián C.J., Williams N.M., Blüthgen N., Krasnov B.R., Poulin R. 'Species abundance and asymmetric interaction strength in ecological networks'. *Oikos*. 2007;**116**(7):1120–7.

[47] Khosravifard S., Skidmore A.K., Naimi B., Venus V., Muñoz A.R., Toxopeus A.G. 'Identifying birds' collision risk with wind turbines using a multidimensional utilization distribution method'. *Wildlife Society Bulletin*. 2020;**44**(1):191–9.

[48] Wilcove D.S. 'Endangered species management: The US experience' in Sodhi N.S., Ehrlich P.R. (eds.). *Conservation Biology for All*. Oxford, England: Oxford University Press; 2010. pp. 220–35.

[49] Bouten W., Baaij E.W., Shamoun-Baranes J., Camphuysen K.C.J. 'A flexible GPs tracking system for studying bird behaviour at multiple scales'. *Journal of Ornithology*. 2013;**154**(2):571–80.

[50] Band W., Madders M., Whitfield D.P. *Developing Field and Analytical Methods to Assess Avian Collision Risk at Wind Farms [online]*. Available from https://www.natural-research.org/application/files/4114/9182/2839/Band_et_al_2007.pdf [Accessed 15 May 2020].

[51] McGrady M.J., Whitfield D.R., Petty S.J., Grand J.R., Bainbridge I.P. 'Golden eagles and forestry'. *Forestry Commission Research Information Not 292*. Roslin; 1997.

[52] Eichhorn M., Johst K., Seppelt R., Drechsler M. 'Model-based estimation of collision risks of predatory birds with wind turbines'. *Ecology and Society*. 2012;**17**(2).

[53] Everaert J. 'Collision risk and micro-avoidance rates of birds with wind turbines in Flanders'. *Bird Study*. 2014;**61**(2):220–30.

[54] Heist K.W. *Assessing Bat and Bird Fatality Risk at Wind Farm Sites using Acoustic Detectors*. Ph.D. thesis, University of Minnesota; 2014.

[55] Korner-Nievergelt F., Brinkmann R., Niermann I., Behr O. 'Estimating bat and bird mortality occurring at wind energy turbines from covariates and carcass searches using mixture models'. *PLoS ONE*. 2013;**8**(7):e67997.

[56] Marques A.T., Batalha H., Rodrigues S., *et al*. 'Understanding bird collisions at wind farms: an updated review on the causes and possible mitigation strategies'. *Biological Conservation*. 2014;**179**(2):40–52.

[57] *Siting Fact Sheet [online]*. Available from https://poweringtexas.com/wp-content/uploads/2018/12/Siting-Fact-Sheets.pdf [Accessed 27 May 2020].

[58] *Building It The Right Way: Siting A Wind Farm [online]*. Available from https://poweringtexas.com/wp-content/uploads/2018/12/Siting-Talking-Points.pdf [Accessed 27 May 2020].

[59] Rechtenwald J. *NWCC Mitigation Subgroup, Mitigation Tool Box*. National Wind Cooperating Collaborative (NWCC); 2007. p. 114.

[60] Blackwell B.F., DeVault T.L., Seamans T.W., Lima S.L., Baumhardt P., Fernández-Juricic E. 'Exploiting avian vision with aircraft lighting to reduce bird strikes'. *Journal of Applied Ecology*. 2012;**49**(4):758–66.

[61] Blackwell B.F., Bernhardt G.E., Dolbeer R.A. 'Lasers as nonlethal avian repellents'. *The Journal of Wildlife Management*. 2002;**66**(1):250–8.

[62] Bishop J., McKay H., Parrott D., Allen J. 'Review of international research literature regarding the effectiveness of auditory bird Scaring techniques and potential alternatives'. *Dep. Food Rural Aff.* 2003:1–52.

[63] Hodos W. *Minimization of Motion Smear: Reducing Avian Collision with Wind Turbines. Period of Performance*. Golden (CO): National Renewable Energy Laboratory; 2003. p. 35.

Part IV

The Intermittency Issue

Chapter 11

Overview of state-of-the-art of wind power forecasting

Rongfu Sun[1], Mingjian Cui[2], Haixiang Xu[1], Qing He[3], and Xinnan Zhou[1]

11.1 Introduction

Due to the time-varying characteristics of wind resources, wind power has strong randomness, intermittency, and uncertainty. As more and more wind power is integrated into power systems, the uncertainty of power system operation increases. The contradiction between the safe operation of power systems and the efficient consumption of renewable energy becomes increasingly prominent [1].

To cope with the challenge of wind power forecasting, many countries worldwide have successively researched wind power forecasting technologies. In recent years, the wind power prediction technology and its application have made significant progress in the areas of high penetrations of wind power [2, 3], such as economic dispatch, unit commitment, and market trading. In terms of wind power forecasting accuracy, by improving core technologies of numerical weather prediction (NWP) and optimizing the wind power conversion model, the prediction error has been significantly reduced to ensure the balanced supply and demand of electricity. The prediction error of the wind power of the most advanced international level is as low as 2.5 percent [4]. The forecast errors of different signs for different regions are mainly reduced by combining many wind turbines. In terms of prediction products, models of event forecasting, probabilistic forecasting, and clustering forecasting are derived by combining with extreme meteorological conditions such as high wind, low temperature, freezing rain, and haze. Consequently, forecasting products are becoming more diversified. In the aspect of prediction application and evaluation, the prediction evaluation index system has been significantly improved by combining with the requirement of economic dispatch and electricity market, which promotes the overall improvement of forecasting performance.

[1]State Grid Jibei Electric Power Company, Hebei, China
[2]Southern Methodist University, Dallas, Texas, USA
[3]State Grid Corporation of China, Beijing, China

Based on the technical development and practical application of wind power forecasting in recent years, the state-of-the-art progress in wind power forecasting is investigated and summarized. This chapter summarizes the whole process of NWP, wind power conversion model, wind power forecasting applications, and the prediction performance evaluation. The existing problems and future research directions are discussed at the end of this chapter to provide a reference for the sustainable development of wind power forecasting.

11.2 Technical routine of wind power forecasting

Based on the research and practice worldwide, the technical routine of wind power forecasting has been solidified. According to this technical routine, the four main components of wind power forecasting include NWP, wind power conversion model, application of prediction results, and evaluation of prediction. Figure 11.1 illustrates how these key steps relate to each other.

First, the global model background field is generated into NWP products using the meteorological mesoscale model. Subsequently, meteorological elements related to NWP stroke resources are input into the wind power conversion model to obtain the forecasted wind power time series. The wind power prediction results are then applied to the combination arrangement of day-ahead unit commitment and day-ahead generation planning adjustment. Finally, the prediction results of wind power are evaluated, and the defects of the prediction model are optimized.

11.3 Numerical weather prediction

NWP is a method to predict the atmospheric motion state and weather phenomena in the future period by solving the operating equations. It describes the law of atmospheric motion change through large high-performance computing with given atmospheric initial and boundary conditions. There are two main types of NWP. One is the global model, which seeks to solve the global weather conditions and provide the necessary background field data for the regional model prediction. The other one is the regional model, which can more accurately simulate the microtopography

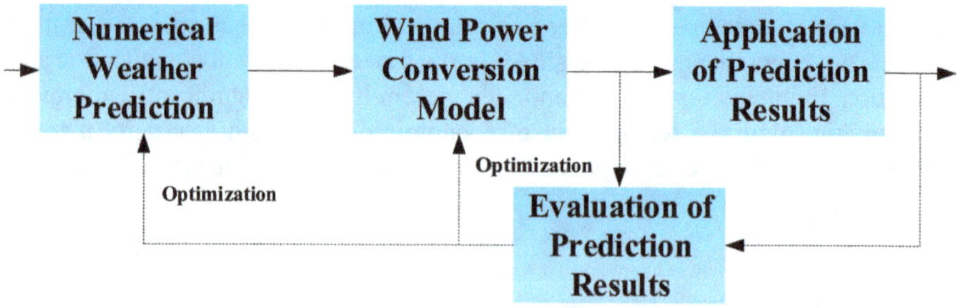

Figure 11.1 The main steps of wind power forecasting

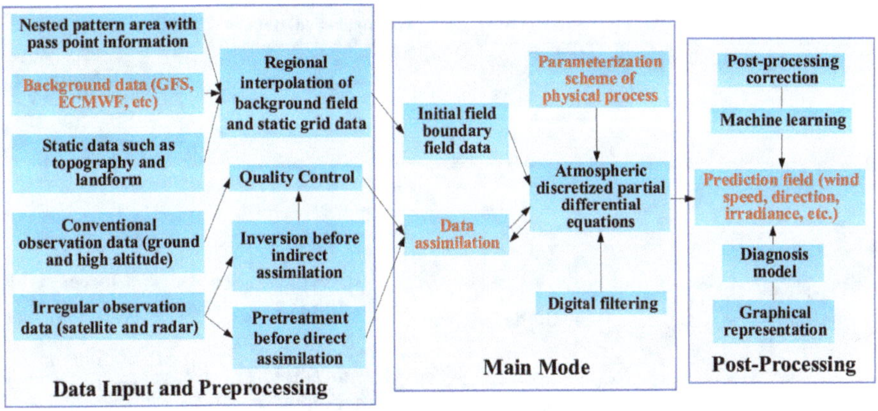

Figure 11.2 The flowchart of the mesoscale NWP model

and microclimate of wind farms. Moreover, the local satellite and radar data can be assimilated with high frequency, and the prediction results are more accurate than the global model. Figure 11.2 shows the general operation process of the regional mesoscale numerical model. It includes processing of the background field data, data assimilation, physical process parameterization, and other key links.

11.4 Wind power conversion model

The existing commercial wind power conversion model has formed a mature technical system. According to forecasting and application scenarios, the wind power conversion model can be divided into the deterministic forecast, event forecast, probabilistic forecast, and other models. Figure 11.3 shows the data flow of various wind power conversion models. First, the input data mainly include physical information of the wind farm (terrain and fan distribution), historical power data, historical meteorological data (wind speed and wind direction), and NWP data. Second, the forecast methods consist of deterministic forecast methods, event forecast methods, probabilistic forecast methods, and clustering forecast methods. Different information combinations are put into the corresponding forecast methods, as shown in Figure 11.3.

11.4.1 Deterministic forecast

Deterministic forecast refers to the prediction of the specific point value of wind speed or wind power in the future [5]. According to different time scales, the ultra-short-term prediction is used to optimize the spinning reserve capacity and real-time economic dispatch of power systems. Short-term forecasting is used to arrange the conventional unit commitment and schedule the daily generation plan. Medium- and long-term forecast is used to make annual maintenance and generation plans.

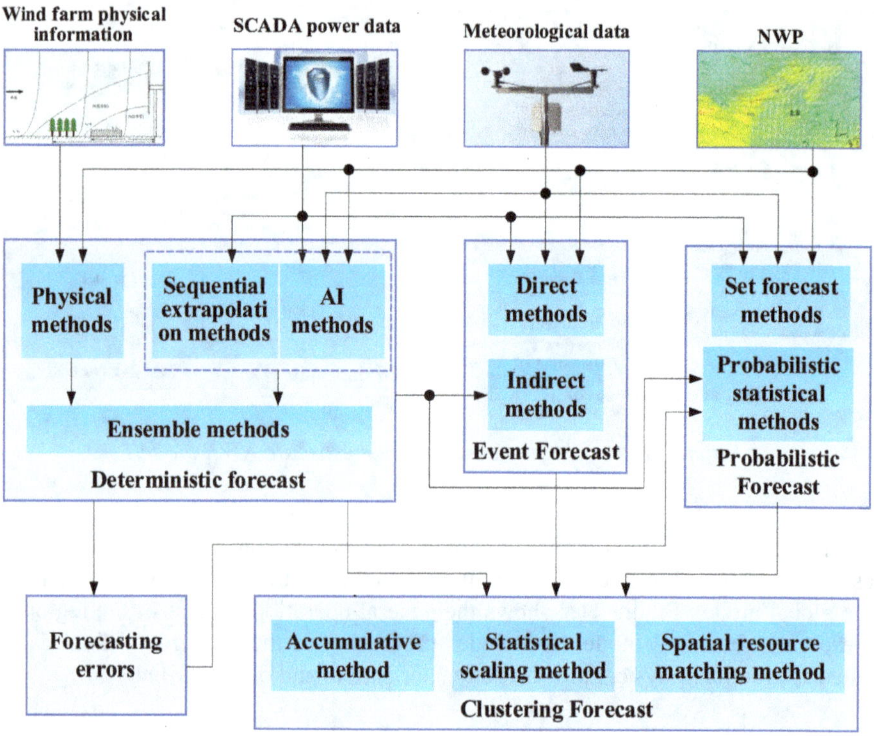

Figure 11.3 Information flow of wind power forecasting

According to the different prediction, it can be divided into physical methods and statistical methods, and the ensemble of physical and statistical methods.

11.4.1.1 Physical forecasting method

The physical forecasting method is the first method used in wind power forecasts. Denmark's Risø national renewable energy laboratory developed the first set of relatively complete wind power prediction system named Prediktor [6]. The physical forecasting method mainly includes three technical components: the introduction of NWP data, the acquisition of wind speed and direction at the hub height of the wind turbine, and the wind speed and power conversion. Considering wind farm landform and other information, the NWP data represents the actual wind resource information that can be used for wind forecasts considering the specific environment. The predicted power of the wind turbine is calculated according to the performance parameters. This method employs complex calculations. The prediction is made by solving the multidimensional partial differential equations. However, the technical requirements for wind farms are relatively low. Only the wind farm location, terrain, landform, and wind turbine information are needed for physical forecasting methods, which can be applied to new wind farms.

However, NWP data is updated at a low frequency (generally 1–3 h), making it challenging to meet the requirements of 15 min ultra-short-term prediction. Also, the accuracy of the physical forecasting model directly using NWP usually fails to meet the application requirements. Therefore, combining machine learning methods and data analysis to improve the prediction accuracy is the main research direction of the physical forecasting method. Cassola and Burlando [7] used the Kalman filtering algorithm to filter the NWP output, which effectively reduced the systematic error of weather prediction and significantly improved the accuracy of the NWP model. Men *et al.* [8] used the Gaussian hybrid model to construct the mapping relationship between measured wind speed and NWP data to correct the NWP wind speed. This correction significantly improved the accuracy of both NWP data and power prediction.

11.4.1.2 Statistical forecasting method

The statistical forecasting method does not consider the process of the physical change of wind speed. It learns the characteristics of the wind farm's historical meteorological and power data using a statistical algorithm. By combining the measured power data or NWP data, statistical forecasting methods can predict the future power generation of wind farms [9].

This method avoids the complex physical modeling process. However, it requires a large amount of historical data for learning modeling, which is suitable for running wind farms.

(1) Time-series extrapolation-based forecasting method

Forecasting methods using the time-series extrapolation do not need to use the meteorological information, such as wind speed and direction. Wind power can be predicted by extrapolating directly from the historical sequence characteristics of wind farm power generation (such as autocorrelation, standard deviation, and other statistical characteristics) using the continuous method, auto-regressive and moving average (ARMA) method, or Kalman filtering algorithm.

The persistence method is the simplest time-series extrapolation prediction method. It takes the measured wind power at the current moment as the predicted power at the next moment. The prediction accuracy is fair on a short time scale. However, with the increase of time scale, the prediction accuracy drops sharply, which is only suitable for the ultra-short-term wind power prediction.

The ARMA model is separately constructed to improve the prediction results. Though this method is simple to use, it has poor adaptability, making it challenging to deal with nonstationary abrupt changes such as wind gusts. The prediction accuracy decreases with the increase of time scale, which is suitable for the ultra-short-term and short-term power forecasting.

To cope with the above-described shortcomings, some scholars combine the time-series with other mathematical methods, such as Kalman filter and chaos theory, to gradually optimize and improve the accuracy and adaptability of the wind power forecasting model. Thus, these methods can make up for the shortcomings of the time series-based forecasting methods. Song *et al.* [10] utilized a hybrid model that integrated the ARIMA model into a Kalman filter. The hybrid model performed well in terms of the three performance indexes and different types of wind direction time series. Rodriguez *et al.* [11] proposed a technique that used an artificial neural network (ANN) and *k*-Nearest Neighbors (KNNs) to reconstruct and forecast an incomplete and chaotic time series.

(2) Artificial intelligence (AI)-based forecasting method

Using the time-series extrapolation-based forecasting method of historical power data, the prediction accuracy decreases significantly with the increase of time scale. One or more AI algorithms are used to train them considering the related factors between weather and electrical power. The connection is extracted between NWP historical data and historical wind power generation. Based on this relationship, using NWP data to predict wind power is an important research direction for wind power forecasting. It mainly consists of the ANN, support vector machine (SVM), and deep learning techniques.

The ANN method has the characteristics of nonlinear mapping, distributed parallel processing, robust fault-tolerant adaptive learning, and generalization ability. It has become the most widely used statistical method in wind power forecasting [12]. Common neural network models include backpropagation (BP) network, radial basis function (RBF) network, and general regression neural network (GRNN) network. The nonlinear mapping ability, generalization ability, and fault tolerance of the BP network make it widely popular. However, the learning convergence is slow and easy to fall into the local minimum. RBF network has good applicability without oscillation and falling into the local minimum during the training process. However, the center of its basis function is not easy to choose. GRNN model is simple, fast, and suitable for small sample data.

Based on the theory of Vapnik–Chervonenkis dimension and the principle of structural risk minimization, SVM can effectively overcome the problem of a local minimum of training results and effectively solve the learning problem for wind power forecast. Hu *et al.* [13] derived a general loss function and developed a technique of the uniform model of v-support vector regression for the general noise model using the augmented Lagrange multiplier method. However, the selection of SVM parameters is relatively arbitrary as it depends on the empirical judgment. One can use a variety of methods, including the genetic algorithm, particle swarm optimization, ant colony optimization, and fruit fly optimization algorithm to optimize the selection of SVM parameters.

Data mining algorithms such as KNN [14] and fuzzy logic algorithm [15] have also been used for wind power forecasts. With the development of big data technology, AI-based prediction methods have been developed from the machine learning algorithm to the deep learning algorithm. AI-based prediction method does not need to establish a specific analytical relationship between the wind power forecasts and the meteorological data (or NWP data). Good prediction results can be obtained by training mass data, which is suitable for short- and medium-term wind power forecasts.

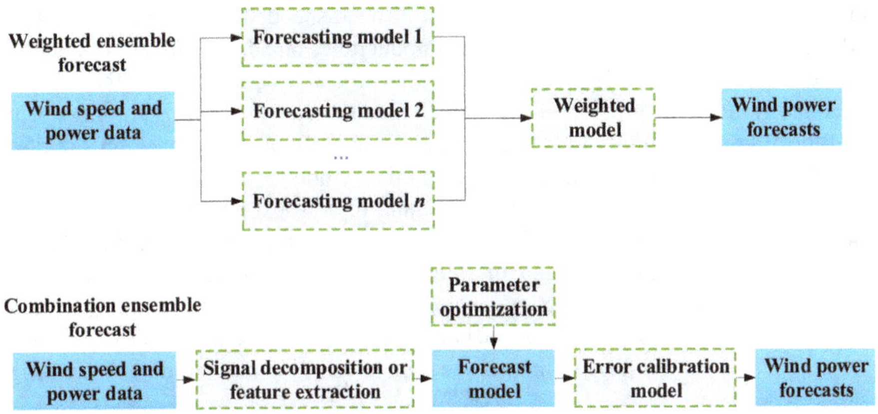

Figure 11.4 Wind power ensemble forecasting method

11.4.1.3 Ensemble forecasting method

Wind power forecasts can be affected by many factors. The traditional single prediction model cannot fully contain all kinds of factors. Especially, the single prediction model does not fully learn the historical information under the extreme weather conditions, which may lead to a large deviation in the prediction. The ensemble forecasting method considers the advantages of different models. The optimal combination of multiple single models can significantly improve the accuracy of wind power prediction. It mainly includes a weighted ensemble forecast and a combination ensemble forecast. The flowchart of ensemble forecasting methods is shown in Figure 11.4.

11.4.2 Ramping event forecast

Large fluctuations in wind power in a short period of time are called wind power ramping events, which will destroy the balance of power supply and power consumption of the power system and threaten the operation of the power grid [16, 17]. Ramp events can generally be divided into two basic types based on the direction: up-ramps and down-ramps. Up-ramps commonly have the characteristic that the wind power increases sharply. Strong low-pressure air systems (or cyclones), low-level jets, thunderstorms, wind gusts, or similar atmospheric phenomena can induce up-ramps. Down-ramps are the reverse physical processes of up-ramps and are generally caused by the same aforementioned meteorological phenomena [18].

Depending on the used prediction model, one can divide ramping event forecasts into the indirect prediction and direct prediction. The direct prediction method mines the mapping relation between the meteorological information and the features of ramping events in the historical data. Furthermore, the weather prediction information can directly predict the features of the ramping event without predicting the wind power in advance. Zareipour [19] mined the features of the ramping events in the historical data based on SVM and verified the effectiveness of the method by combining the simulation of the actual data of wind speed. The method is accurate and straightforward. However, its training model relies on a large number of historical observation samples. Besides, the completeness and correctness of the sample data directly affect the prediction accuracy. In addition, a wind power ramping event is a typical low-probability event. The inherent scarcity of its observation samples restricts the application and development of this method.

The indirect prediction method is the main topic for the prediction of current ramping events. First, the deterministic prediction method is used to predict future wind power. Then, the definition of ramping events is combined to extract the ramping events from the prediction sequence. The corresponding ramping feature information is extracted. Cui *et al.* [20] analyzed the characteristics of wind power ramping events from different time scales such as ramping magnitude and ramping rate.

However, the indirect prediction method relies too much on the wind power prediction results, which may ignore the extreme weather samples and lead to the missed report of ramping events. Generally, one can use the ensemble prediction theory to improve event prediction accuracy. One way is to decompose the original wind power sequence into multiple components in the model's input part. Then, the principal components of the ramping event are extracted. Finally, the differences in component characteristics are used for prediction. The other way is to take into account the probability characteristics of ramping events under the same meteorological conditions in the model output. The correlation is used to amend the prediction of wind power ramps.

11.4.3 Probabilistic forecast

Due to the strong random fluctuation of wind resources, the prediction results discussed in Sections 11.4.1 and 11.4.2, such as deterministic forecast and event forecast, inevitably contain uncertainties caused by the incomplete data and inaccurate models. It is necessary to develop a probabilistic forecast based on the above forecasting methods. It can reveal the probability distribution of forecasting errors, quantitatively describe the uncertainty in the prediction, and provide more comprehensive information for power system regulation and decision-making. Depending on the prediction model, the probability prediction can be divided into probability and statistical methods based on historical data and the NWP-based ensemble forecasting method.

The probability and statistical methods are generally called statistical methods, which combine a large amount of historical data and form the probability density function of wind power by modeling the wind power or forecasting errors. The probabilistic prediction model can be divided into a parametric model and a nonparametric model according to the assumption that the modeling object obeys the known distribution. The parametric model describes the probability density function with known probability distributions (such as Gaussian distribution and exponential distribution). However, the inherent modeling error will affect the accuracy of the predicted distribution. The nonparametric model does not assume the distribution characteristics of the modeling object in advance. It directly calculates the probability density function through data analysis methods such as quantile regression and kernel density estimation. Wan *et al.* [21] proposed a direct quantile regression approach to efficiently generate nonparametric probabilistic forecasting of wind power generation combining extreme learning machine and quantile regression. He *et al.* [22] presented a hybrid wind power probability density prediction method based on quantile regression neural network and Epanechnikov kernel function using unbiased cross-validation.

Another widely used probabilistic forecasting method is the NWP-based ensemble forecasting method. The probability density function of wind power is constructed by combining multiple NWP prediction results, which is generally called the physical method [23]. Compared with the single NWP method, it can provide more abundant weather forecast approaches, including filtering method, dimension reduction method, and direct method. The filter method first converts the set of NWP data into a set of single-point power prediction results and then obtains the distributed prediction results through post-processing. The dimension reduction method first extracts the information of trend and dispersion from the set NWP data and then inputs the probability prediction model. The direct method seeks to input all the collected NWP data into the probability prediction model for direct prediction. With the increase of sample complexity, this method will reduce forecasting efficiency and consume many computational resources.

11.5 Predictive performance evaluation

Wind power prediction and evaluation can comprehensively grasp the prediction system's operation and profoundly mitigate the defects, which is the key to the optimization of wind power forecasting models, performance improvement, and selection of the prediction system.

11.5.1 Deterministic forecast evaluation

The deterministic prediction error can be divided into horizontal error and vertical error. The vertical error describes the deviation between the predicted result and the actual measurement on the vertical power axis in a given period of time. The horizontal error describes the deviation between the predicted results and the actual measurement on the horizontal time axis in a given period of time.

11.5.1.1 Vertical error evaluation

The commonly used vertical error evaluation metrics in existing literature include mean error (ME), mean absolute error (MAE), root mean squared error (RMSE), mean absolute percentage error (MAPE), and error frequency distribution indexes. The mathematical expressions of MAE and RMSE are given by

$$ME = \frac{y_i - y_i^p}{N} \tag{11.1}$$

$$MAE = \frac{\sum_{i=1}^{N} |y_i - y_i^p|}{N} \tag{11.2}$$

$$RMSE = \sqrt{\frac{\sum_{i=1}^{N} (y_i - y_i^p)^2}{N}} \tag{11.3}$$

$$MAPE = \frac{\sum_{i=1}^{N} \left| \frac{y_i - y_i^p}{y_i} \right|}{N} \tag{11.4}$$

where y_i is the actual power at time i. y_i^p is the predicted power at time i. N is the total number of testing samples.

11.5.1.2 Horizontal error evaluation

The correlation coefficient (CC) is commonly used to evaluate the similarity between the measured wind power data and predicted values. The mathematical expression is given by

$$CC = \frac{\sum_{i=1}^{N} \left[(y_i - \bar{y}_i) \cdot \left(y_i^p - \bar{y}_i^p \right) \right]}{\sqrt{\sum_{i=1}^{N} (y_i - \bar{y}_i)^2} \cdot \sqrt{\sum_{i=1}^{N} \left(y_i^p - \bar{y}_i^p \right)^2}} \tag{11.5}$$

where $\bar{y}_i$ is the average value of all the measured power data samples. $\bar{y}_i^p$ is the average of all the predicted power data samples. The horizontal deviation degree is proposed to quantify the degree of horizontal deviation of the measured and predicted wind power data. The mathematical expression is given by

$$CC(l) = \frac{\sum_{i=l+1}^{N} \left[(y_i - \bar{y}_i) \cdot \left(y_{i-l}^p - \bar{y}_i^p \right) \right]}{\sqrt{\sum_{i=l+1}^{N} (y_i - \bar{y}_i)^2} \cdot \sqrt{\sum_{i=1}^{N-l} \left(y_i^p - \bar{y}_i^p \right)^2}} \tag{11.6}$$

where CC(l) is the CC between the predicted sequence of delay l and the measured wind power sequence.

11.5.2 Event forecast evaluation

The event forecast method can be divided into two categories: predicting event state and prognosis of event features. There are significant differences between these forecasting categories; thus, the evaluation method of their corresponding forecast results is also different.

(1) Evaluation of state prediction

There are four types of predicted results of ramping events, which are shown in Table 11.1. According to the different physical meanings expressed by the four results, two metrics describing the comprehensiveness or accuracy of the forecast can be established [24]. In this table, True Positive (TP) represents the number of forecasted ramping events (predicted YES) that are observed in the actual ramping events (observed YES); False Positive (FP) is the number of forecasted ramping events (predicted YES) that are not observed in the actual wind power (observed

Table 11.1 The result of event state prediction

Observation prediction	**YES**	**NO**
YES	TP	FP
NO	FN	TN

NO); False Negative (FN) represents the number of observed ramping events (observed YES) that are not predicted by the wind forecasting system (predicted NO); and True Negative (TN) is the number of non-observed and non-predicted ramping events.

a. Hit rate (HR)
This metric can describe the comprehensiveness of event prediction and represent the probability of an accurate forecast of actual events.

$$HR = \frac{N_{TP}}{N_{TP} + N_{FN}} \tag{11.7}$$

b. Accuracy rate (AR)
This metric describes the accuracy of the forecast. It represents the probability that the predicted result actually occurs.

$$AR = \frac{N_{TP}}{N_{TP} + N_{FP}} \tag{11.8}$$

In addition to the above metrics, frequency bias score [25], critical success index [25], Peirce skill score [26], and other indicators can evaluate event state prediction results' accuracy.

(2) Evaluation of feature prediction

The prediction of feature quantities of ramping events is similar to that of the deterministic forecast. Most of the evaluation metrics of deterministic prediction (such as MAE and RMSE) can be used to evaluate the prediction results of event features. However, it cannot accurately describe part of the features of the ramping event, such as the degree of time lag and the degree of ramping capture.

a. Ramping occurrence

The detected ramping event occurs with the time lag. Some scholars propose to use the temporal distortion index (TDI) to divide the prediction error of ramping occurrence into two parts [27]. One part is the distortion of time series, that is, the error caused by the deviation of time calibration. The other part is the magnitude error.

$$T_{\mathrm{DI}} = \frac{1}{N^2} \sum_{l=1}^{N-1} \left| \left(i_{l+1} - i_l\right) i_{l+1} + i_l - j_{l+1} - j_l \right| \tag{11.9}$$

where (i, j) represent the local distance between the ith element in the test sequence T_i and the jth element in the reference sequence R_j. N is the sample size.

b. Ramping score

Based on the swinging door algorithm, the prediction accuracy of ramping magnitude is evaluated by the ramping score through setting the door width ϵ to adjust the sensitivity of ramping events. It can reflect the capture degree of forecasting results.

$$S_{\text{ramp score}} = \frac{1}{t_{\max} - t_{\min}} \int_{t_{\min}}^{t_{\max}} \left| S_D(T(t)) - S_D(R(t)) \right| dt \tag{11.10}$$

where $t_{\min}$ and $t_{\max}$ denote the occurrence and termination of one ramping event. S_D represents the ramping rate after the process of the swinging door algorithm. $T(t)$ and $R(t)$ represent the values of the test sequence and the reference sequence at time t, respectively.

11.5.3 Probabilistic forecast evaluation

The results of the probabilistic forecast are generally evaluated by reliability and sensitivity metrics. The reliability metric refers to the statistical consistency between the distribution of predicted variables and the distribution of observed variables, reflecting the predicted results' ability to match the observed values. The sensitivity metric demonstrates the concentration of the forecasting distribution. A more compact distribution contains more available information.

11.5.3.1 Prediction intervals coverage probability (PICP)

This metric describes the reliability of the probabilistic forecasting model. The probability prediction interval under the confidence $1-\alpha$ obtained by the prediction model is expressed as $[L^{\alpha}, U^{\alpha}]$. Given N test points y_i^* $(i = 1, 2, \cdots, N)$, the corresponding N prediction intervals are $[L_i^{\alpha}, U_i^{\alpha}]$. Then, the coverage probability of the predicted results can be expressed as:

$$P_{\text{ICP}} = \frac{1}{N} \sum_{i=1}^{N} 1 \left(y_i^* \in \left[L_i^{\alpha}, U_i^{\alpha} \right] \right) \tag{11.11}$$

where the indicator is 1 if $y_i^* \in [L_i^{\alpha}, U_i^{\alpha}]$; otherwise, the indicator is 0.

11.5.3.2 Prediction intervals relative width (PIRW)

If the relative width of the prediction interval is larger, less information will be provided, and the probability prediction results will also be affected. The model sensitivity can be evaluated by the relative width index of the prediction interval, which is defined as:

$$\text{PIRW} = \frac{1}{N \cdot R} \sum_{i=1}^{N} \left(U_i^{\alpha} - L_i^{\alpha} \right) \tag{11.12}$$

where R represents the difference between the maximum and minimum of the target value. The aim of using R is to normalize the average interval width. A smaller PIRW denotes that the prediction model is more sensitive.

11.5.3.3 Daily accumulated deviation index (DADI)

In practice, the coverage probability and the relative width of the prediction interval are still unable to assess the impact of probabilistic forecast accurately. Therefore, the DADI metric, defined by (11.13) and (11.14), is introduced to accurately and intuitively reflect the deviation degree between the actual value and the predicted interval.

$$D = \sum_{i=1}^{N} d_i \tag{11.13}$$

$$d_i = \begin{cases} \frac{y_i - U_i^{\alpha}}{y_i} & y_i > U_i^{\alpha} \\ 0 & L_i^{\alpha} \le y_i \le U_i^{\alpha} \\ \frac{L_i^{\alpha} - y_i}{y_i} & y_i < L_i^{\alpha} \end{cases} \tag{11.14}$$

where d_i is the deviation value between the actual power and the predicted interval at time i. A smaller d_i denotes a better forecasting result. The probabilistic prediction evaluation index also includes coverage bandwidth criterion, marginal calibration, and continuous ranking probability score.

11.6 Conclusion and future work

Wind power forecasting is interdisciplinary knowledge combining meteorology, electrical engineering, and statistics. Wind power forecasting is vital to the safe operation of the power grid and increased penetration of wind power. This chapter summarized the technical components of the wind power forecast. Although wind power prediction technology has made great progress, a considerable improvement is still needed in resolving the uncertainty of wind power.

(*i*) *AI modeling.* With the development of emerging technologies such as AI and big data, it is expected to break through the technical bottleneck of wind power forecast and significantly improve prediction accuracy. Due to the self-learning ability of AI models, including deep belief network, deep neural network, and recurrent neural network, it has higher accuracy of prediction results compared with the physical and statistical models. For instance, machine learning and deep learning approaches will significantly improve the accuracy of short-term wind power forecasts.

(*ii*) *Online interactive prediction.* Currently, the operation mode of "off-line modeling, model encapsulation, and model embedding" is widely used in wind power forecast. The forecast model is not updated in time and cannot respond to the operation status of the wind farm in real-time. Thus, the establishment of an online interactive prediction method of "online modeling, real-time adjustment, and centralized distribution" is of great significance for improving the responsiveness and accuracy of the wind power forecast model.

References

[1] Chen H., Zhang J., Tao Y., Tan F. 'Asymmetric GARCH type models for asymmetric volatility characteristics analysis and wind power forecasting'. *Protection and Control of Modern Power Systems*. 2019;**4**(4):356–66.

[2] Kou P., Gao F., Guan X. 'Stochastic predictive control of battery energy storage for wind farm dispatching: using probabilistic wind power forecasts'. *Renewable Energy*. 2015;**80**(8):286–300.

[3] Lujano-Rojas J.M., Osório G.J., Matias J.C.O., Catalão J.P.S. 'A heuristic methodology to economic dispatch problem incorporating renewable power forecasting error and system reliability'. *Renewable Energy*. 2016;**87**(3):731–43.

[4] Ernst B., Oakleaf B., Ahlstrom M., *et al*. 'Predicting the wind'. *IEEE Power and Energy Magazine*. 2007;**5**(6):78–89.

[5] Cui M., Ke D., Gan D., Sun Y. 'Statistical scenarios forecasting method for wind power ramp events using modified neural networks'. *Journal of Modern Power Systems and Clean Energy*. 2015;**3**(3):371–80.

[6] Prediktor A.S. Unlock the full potential of Industrial IoT. 2021. Available from www.prediktor.no/ [Accessed 5/19/2021].

[7] Cassola F., Burlando M. 'Wind speed and wind energy forecast through Kalman filtering of numerical weather prediction model output'. *Applied Energy*. 2012;**99**(6):154–66.

[8] Men Z., Yee E., Lien F.-S., Wen D., Chen Y. 'Short-term wind speed and power forecasting using an ensemble of mixture density neural networks'. *Renewable Energy*. 2016;**87**(3):203–11.

[9] Cui M., Feng C., Wang Z., Zhang J. 'Statistical representation of wind power RAMPs using a generalized Gaussian mixture model'. *IEEE Transactions on Sustainable Energy*. 2018;**9**(no. 1):261–72.

[10] Song D., Yang J., Liu Y., Su M., Liu A., Joo Y.H. 'Wind direction prediction for yaw control of wind turbines'. *International Journal of Control, Automation and Systems*. 2017;**15**(4):1720–8.

[11] Rodriguez H., Flores J.J., Morales L.A., Lara C., Guerra A., Manjarrez G. 'Forecasting from incomplete and chaotic wind speed data'. *Soft Computing*. 2019;**23**(20):10119–27.

[12] Cui M., Feng C., Wang Z., Zhang J. 'Statistical representation of wind power RAMPs using a generalized Gaussian mixture model'. *IEEE transactions on sustainable energy*. 2018;**9**(1):261–72.

[13] Hu Q., Zhang S., Xie Z., Mi J., Wan J. 'Noise model based ν-support vector regression with its application to short-term wind speed forecasting'. *Neural Networks: The Official Journal of the International Neural Network Society*. 2014;**57**(9):1–11.

[14] Wu W., Peng M. 'A data mining approach combining k-means clustering with bagging neural network for short-term wind power forecasting'. *IEEE Internet of Things Journal*. 2017;**4**(4):979–86.

[15] Zheng D., Eseye A.T., Zhang J., Li H. 'Short-term wind power forecasting using a double-stage hierarchical ANFIS approach for energy management in microgrids'. *Protection and Control of Modern Power Systems*. 2017;**2**(2):136–45.

[16] Cui M., Ke D., Sun Y., Di Gan J.Z., Hodge B.-M. 'Wind power ramp events forecasting using a stochastic scenario generation method'. *IEEE Transactions on Sustainable Energy*. 2015;**6**(2):422–33.

[17] Cui M., Krishnan V., Hodge B.-M., Zhang J. 'A Copula-based conditional probabilistic forecast model for wind power RAMPs'. *IEEE Transactions on Smart Grid*. 2019;**10**(4):3870–82.

[18] Cui M., Zhang J., Wang Q., Krishnan V., Hodge B.-M. 'A data-driven methodology for probabilistic wind power ramp forecasting'. *IEEE Transactions on Smart Grid*. 2019;**10**(2):1326–38.

[19] Zareipour H., Huang D., Rosehart W. 'Wind power ramp events classification and forecasting: a data mining approach'. *IEEE Power & Energy Society General Meeting*. 2011:1–3.

[20] Cui M., Zhang J., Florita A.R., Hodge B.-M., Ke D., Sun Y. 'An optimized swinging door algorithm for identifying wind Ramping events'. *IEEE Transactions on Sustainable Energy*. 2016;**7**(1):150–62.

[21] Wan C., Lin J., Wang J., Song Y., Dong Z.Y. 'Direct quantile regression for non-parametric probabilistic forecasting of wind power generation'. *IEEE Transactions on Power Systems*. 2016;**32.4**:2767–78.

[22] He Y., Li H. 'Probability density forecasting of wind power using quantile regression neural network and kernel density estimation'. *Energy Conversion and Management*. 2018;**164**(9):374–84.

[23] Zhao J., Guo Z.H., Z Y S., *et al.* 'An improved multi-step forecasting model based on WRF ensembles and creative fuzzy systems for wind speed'. *Applied Energy*. 2015;**155**(19):808–26.

[24] Gallego-Castillo C., Cuerva-Tejero A., Lopez-Garcia O. 'A review on the recent history of wind power ramp forecasting'. *Renewable and Sustainable Energy Reviews*. 2015;**52**(12):1148–57.

[25] Cui M., Zhang J., Feng C., Florita A.R., Sun Y., Hodge B.-M. 'Characterizing and analyzing ramping events in wind power, solar power, load, and netload'. *Renewable Energy*. 2017;**111**(4):227–44.

[26] Ferreira C., Gama J., Matias L., *et al.* 'A survey on wind power ramp forecasting'. *Argonne National Lab.(ANL), Argonne, IL*. 2011.

[27] Florita A., Hodge B., Orwig K. 'Identifying wind and solar ramping events'. *2013 IEEE Green Technologies Conference (GreenTech)*. Denver, CO; 2013. pp. 147–52.

Chapter 12

Storage-integrated wind farms

Javad Khazaei[1], Dinh Hoa Nguyen[2], and Arash Asrari[3]

Abstract

This chapter develops a foundation for stability analysis of storage-integrated wind farms and delves into the concept of distributed control design to provide ancillary services to the grid using storage-integrated wind turbines. The stability analysis is performed for a double-fed induction generator (DFIG)-based wind turbine integrated with a stand-alone battery energy storage system (BESS). Dynamics of the induction generator, AC-side filters, phase-locked loop (PLL), rotor-side converter (RSC), grid side converter (GSC), and BESS controllers are considered. Eigenvalue analysis is performed to evaluate the stability of the integrated systems. A distributed controller is designed to provide active and reactive power-sharing and energy synchronization capabilities to the BESS units in an IEEE 14-bus system integrated with a wind farm composed of 10 storage-integrated wind turbines.

12.1 Introduction

Future integration of wind generation sources to smart grids requires advanced control technologies that enable wide ranges of ancillary services to the grid. These requirements include voltage and frequency regulation, output power smoothing, low voltage ride through (LVRT), or inertia emulation [1–3]. Output power smoothing and the capability of wind turbines in providing active/reactive power support have been considered the most crucial ancillary services [4, 5]. Double-fed induction generator (DFIG) is a more popular generator for wind turbine applications due to their low implementation cost [1, 2].

By integrating an energy storage system to wind turbines, the steady-state and dynamic response of wind turbines can significantly be improved [6, 7]. Therefore, storage-integrated wind turbines have gained popularity over the past few years. For example, a hybrid energy storage (HES) system was supplemented to a remote area

[1]Pennsylvania State University, State College, Pennsylvania, USA
[2]Kyushu University, Fukuoka, Japan
[3]Southern Illinois University, Carbondale, IL, USA

that was dominated by a wind generation system to manage the fluctuations between the generated wind power and fixed demand [6]. In another study, a supercapacitor energy storage was used to enhance the power-sharing capability of DFIG-based wind turbines [7]. Commercial application of storage-integrated wind turbines has recently been demonstrated through the Notrees project funded by the department of defense (DoD) [8], and it is expected that more storage-integrated wind turbines will be distributed in the smart grid of the future.

So far, centralized controllers have been designed and implemented for the storage-integrated wind turbines to provide ancillary grid services. This means a centralized control unit is devoted to controlling all wind turbines and storage units in a wind farm by receiving information from the individual units and sending the appropriate commands to individual elements. Consequently, a two-way communication link is needed (for) each storage-integrated wind turbine and the corresponding central control unit. This process significantly increases the amount of data that the centralized unit needs to process and may result in a single point of failure in case the centralized unit fails. Therefore, centralized control is not a suitable strategy for controlling a smart grid with a massive amount of storage-integrated wind turbines [9, 10]. Distributed controllers, on the other hand, rely on peer-to-peer communications between the neighbors and do not suffer from a single point of failure. Therefore, they are a better solution to control storage-integrated wind turbines of smart grid systems [11–14].

Distributed control of storage-integrated wind generation systems with distributed BESSs has been reported in [11–19]. In [11], a distributed control method was implemented using a model predictive control (MPC) methodology for battery energy storage systems (BESSs) to enhance the output power regulation of wind turbines. A planning technique was introduced in [12] to identify the optimal placement, capacity, and power rating of storage units in smart grids using a short-term optimal power flow algorithm. A dispatching framework using a distributed control design was developed in [13] to provide charging and discharging for the flywheel storage-integrated wind farms using directed and undirected communication topologies. In two recent studies, battery energy storage integration into wind turbines was considered [14, 15]. It was shown that by the integration of storage units to the DC-link of the wind turbines, output power, and torque control can be achieved. This method did not consider standalone storage units and directly integrated the storage units to the DC-link, which limits the controllability and flexibility of storage units in supporting wind turbines. In the authors' recent research, the application of standalone battery energy storage units in providing ancillary services to DFIG- and permanent magnet synchronous generator (PMSG)-based wind turbines were studied [16–19]. It was demonstrated that a distributed controller could provide voltage and frequency regulation as well as active/reactive power support to the wind tribunes in smart grids. While the research and development in distributed control of storage-integrated wind turbines are ongoing, limited research has been devoted to analyzing the stability issues that might be introduced by the integration of storage units to the wind turbines. None of the existing studies in the distributed control design of storage-integrated considered the importance of stability margins that might be compromised by storage units. The existing studies assumed a wide operating condition for the distributed control design. Therefore, to address this gap, the primary motivation behind this work is to

- develop a detailed small-signal model for a BESS-integrated wind turbine based on DFIG,
- include the dynamics of the AC-side, filters, control loops, phase-locked loop (PLL), storage unit and controller, and induction generator in the small-signal modeling,
- conduct a stability analysis to analyze the eigenvalues of a storage-integrated wind turbine,
- derive a simplified model for storage-integrated wind turbines for secondary control design purposes,
- design a distributed controller for storage-integrated wind turbines that responds to the automatic generation control (AGC) signals and shares the load demand between storage units and wind turbines,
- include inherent communication delays in the control design and detailed proof for the designed controllers.

The rest of the chapter is organized as follows: Section 12.2 includes a detailed modeling procedure for small-signal analysis of storage-integrated wind turbines. Section 12.3 elaborates on the distributed control design for storage-integrated wind turbines. Section 12.4 details the case studies, and Section 12.5 concludes the chapter.

12.2 Modeling of storage-integrated wind farms

A typical structure of a storage-integrated wind farm is illustrated in Figure 12.1. In this study, wind turbines with DFIGs are considered, and storage units are assumed to be BESSs. In the following subsections, modeling of components of storage-integrated wind farms will be included.

12.2.1 Modeling of wind generator and drive train

The dynamics of the voltages and flux linkages are considered, and the equations are transformed into a synchronously rotating frame, also known as *dqo* frame, to model the DFIG in state-space form. Eventually, a sixth-order dynamic model is derived for the DFIG, which is adopted from [20] and is represented in (12.1).

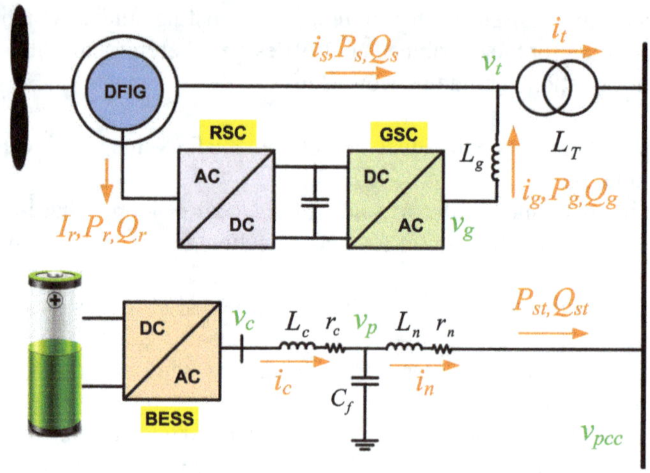

Figure 12.1 Structure of a storage-integrated wind turbine

$$\frac{d}{dt}\begin{bmatrix} i_{qs} \\ i_{ds} \\ i_{os} \\ i'_{qr} \\ i'_{dr} \\ i'_{or} \end{bmatrix} = A\begin{bmatrix} i_{qs} \\ i_{ds} \\ i_{os} \\ i'_{qr} \\ i'_{dr} \\ i'_{or} \end{bmatrix} + B\begin{bmatrix} v_{qs} \\ v_{ds} \\ v_{os} \\ v'_{qr} \\ v'_{dr} \\ v'_{or} \end{bmatrix} \tag{12.1}$$

where i_{qs}, i_{ds}, i_{os} are stator currents converted to the dq frame and $i'_{qr}, i'_{dr}, i'_{or}$ are rotor currents converted to dq frame. In addition, v_{qs}, v_{ds}, v_{os} and $v'_{qr}, v'_{dr}, v'_{or}$ are rotor and stator voltages in dqo frame, respectively, and A, B are state-space matrices of the DFIG model defined in (12.2). In (12.2), ω_e is the angular velocity of the arbitrary reference frame, which is normally equal to the base frequency, for example, $\omega_e = \omega_b$, also known as a synchronous reference frame. To account for the transformation of variables to an arbitrary reference frame, a prime symbol is used for rotor variables. In addition, ω_r is the angular velocity of the rotor, and $X_{ss}, X_{ls}, X_{rr}, X_M, X_{lr}$ are the stator and rotor's self and mutual reactances [21]. A two-mass representation is usually considered to model the drive train dynamics of the DFIG [22]. Dynamics of the turbine's drive train can be expressed as

$$B=\begin{bmatrix} \frac{X_{ss}}{\omega_b} & 0 & 0 & \frac{X_M}{\omega_b} & 0 & 0 \\ 0 & \frac{X_{ss}}{\omega_b} & 0 & 0 & \frac{X_{ss}}{\omega_b} & 0 \\ 0 & 0 & \frac{X_{ls}}{\omega_b} & 0 & 0 & 0 \\ \frac{X_{ss}}{\omega_b} & 0 & 0 & \frac{X'_{rr}}{\omega_b} & 0 & 0 \\ 0 & \frac{X_{ss}}{\omega_b} & 0 & 0 & \frac{X'_{rr}}{\omega_b} & 0 \\ 0 & 0 & 0 & 0 & 0 & \frac{X'_{lr}}{\omega_b} \end{bmatrix},$$

$$A=\text{-}B\times\begin{bmatrix} r_s & \frac{\omega_e}{\omega_b}X_{ss} & 0 & 0 & \frac{\omega_e}{\omega_b}X_M & 0 \\ -\frac{\omega_e}{\omega_b}X_{ss} & r_s & 0 & -\frac{\omega_e}{\omega_b}X_M & 0 & 0 \\ 0 & 0 & r_s & 0 & 0 & 0 \\ 0 & \frac{\omega_e-\omega_r}{\omega_b}X_M & 0 & r'_r & \frac{\omega_e-\omega_r}{\omega_b}X'_{rr} & 0 \\ -\frac{\omega_e-\omega_r}{\omega_b}X_M & 0 & 0 & \frac{\omega_e-\omega_r}{\omega_b}X'_{rr} & r'_r & 0 \\ 0 & 0 & 0 & 0 & 0 & r'_r \end{bmatrix} \tag{12.2}$$

$$\begin{bmatrix} \Delta\dot{\omega}_t \\ \Delta\dot{\omega}_r \\ \Delta\dot{T}_g \end{bmatrix}=\begin{bmatrix} -\frac{D_t+D_{tg}}{2H_t} & \frac{D_{tg}}{2H_t} & \frac{-1}{2H_t} \\ \frac{D_{tg}}{2H_g} & -\frac{D_g+D_{tg}}{2H_g} & \frac{1}{2H_g} \\ K_{tg}\omega_e & -K_{tg}\omega_e & 0 \end{bmatrix}\begin{bmatrix} \Delta\omega_t \\ \Delta\omega_r \\ \Delta T_g \end{bmatrix}+\begin{bmatrix} \frac{T_m}{2H_t} \\ \frac{-T_e}{2H_g} \\ 0 \end{bmatrix} \tag{12.3}$$

where ω_t and ω_r are angular frequencies of the turbine and generator, respectively, H_g and H_t are the inertia coefficients of the DFIG and the turbine. Also, T_g is an integral torque of the DFIG model, D_t and D_g are the mechanical damping coefficients of two masses, and K_{tg} is the shaft stiffness. The electrical torque of the DFIG wind turbine, T_e, is expressed as

$$T_e=\lambda_{qm}i'_{dr}-\lambda_{dm}i'_{qr} \tag{12.4}$$

where $\lambda_{qm}=L_M(i_{qs}+i'_{qr})$ and $\lambda_{dm}=L_M(i_{ds}+i'_{dr})$ are the air gap flux linkages. The power transfer between the GSC and RSC is enabled through a common DC-link capacitor, whose dynamics is presented by (12.5).

$$CV_{dc}\frac{dv_{dc}}{dt}=P_r-P_g \tag{12.5}$$

where $P_r=\frac{1}{2}(v'_{qr}i'_{qr}+v'_{dr}i'_{dr})$ and $P_g=\frac{1}{2}(v_{qg}i_{qg}+v_{dg}i_{dg})$ are the active powers of the RSC and GSC as shown in Figure 12.1.

12.2.2 Modeling of DFIG's transmission system

Assuming that the DFIG is connected to the point of common coupling (PCC) through a transformer with an impedance of $Z_t = r_t + j\omega_e L_t$, the AC-side dynamics of the DFIG wind turbine can be written in space-phasor form as

$$\vec{v}_t - \vec{v}_{pcc} = r_t \vec{i}_t + L_t \frac{d}{dt}\vec{i}_t \tag{12.6}$$

$$\vec{v}_g - \vec{v}_t = j\omega_e L_g \vec{i}_g \tag{12.7}$$

where variables with an arrow on top denote the space-phasor representation [23]. By replacing (12.7) in (12.6) and converting the result to dq-frame, dynamics of the DFIG on the AC-side in dq-frame can be presented in (12.8).

$$\frac{d}{dt}\begin{bmatrix} i_{td} \\ i_{tq} \end{bmatrix} = \begin{bmatrix} -\frac{r_t}{L_t} & \omega_e \\ -\omega_e & -\frac{r_t}{L_t} \end{bmatrix} \begin{bmatrix} i_{td} \\ i_{tq} \end{bmatrix} + \begin{bmatrix} \frac{\omega_e L_g}{L_t} & 0 \\ 0 & -\frac{\omega_e L_g}{L_t} \end{bmatrix} \begin{bmatrix} i_{gd} \\ i_{gq} \end{bmatrix} + \begin{bmatrix} \frac{v_{gd} - v_{pcc,d}}{L_t} \\ \frac{v_{gq} - v_{pcc,q}}{L_t} \end{bmatrix} \tag{12.8}$$

12.2.3 Modeling of controllers for the wind turbine

By looking at the DFIG diagram in Figure 12.1, it is observed that the DFIG is linked to the grid through both stator and rotor terminals. The stator is connected to the grid directly at the terminal, point v_t, and the rotor is connected to the same point through two back-to-back converters known as grid-side converter (GSC) and rotor-side converter (RSC). The DFIG delivers active and reactive powers, $P_{s,i}$ and $Q_{s,i}$, to the grid, and the GSC delivers $P_{g,i}$, $Q_{g,i}$ to the interconnection point (terminal) through independent active and reactive power control. It is worth mentioning that the nominal rating of the GSC is usually less than 30% of DFIG's nominal power. The DC link between the RSC and GSC should be regulated to enable active power transfer between the DFIG and grid. The DC link is regulated through GSC control loops, and the active power is controlled by the RSC through maximum point tracking (MPPT) controller.

12.2.3.1 RSC control and dynamics

The control block of the RSC is depicted in Figure 12.2. The control is achieved through a reference frame that is synchronously rotating, also known as stator flux-oriented reference frame. The d-axis of the stator flux-oriented frame is normally aligned with the stator flux vector to achieve a decoupled torque (power) and reactive power regulation. An inner current control loop is also added to regulate the rotor current and avoid overcurrents in the rotor winding due to faults or contingencies. The dynamics of the inner current control loop are presented by (12.9) and (12.10).

$$V_{dr}^* = \left(k_{p1} + \frac{k_{i1}}{s}\right)(i_{dr}^* - i_{dr}) - \omega_{slip} L_r i_{qr} \tag{12.9}$$

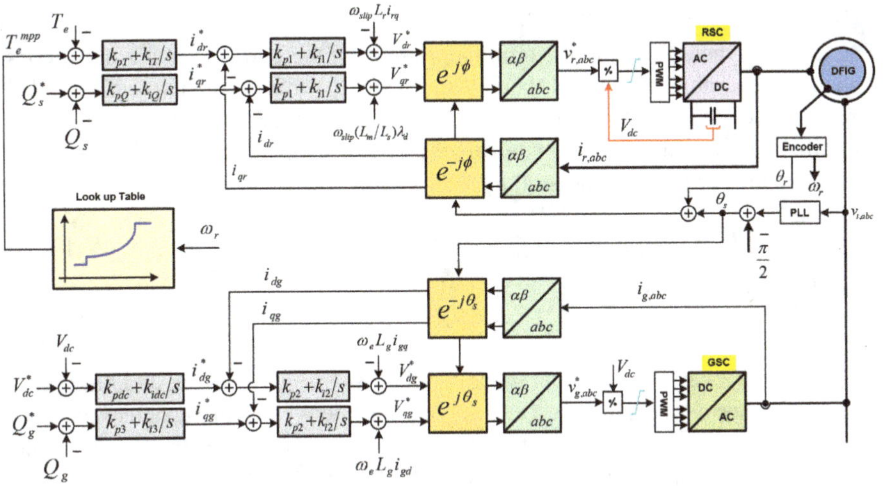

Figure 12.2 Schematic of DFIG controllers

$$V_{qr}^{*} = \left(k_{p1} + \frac{k_{i1}}{s}\right)(i_{qr}^{*} - i_{qr}) - \omega_{slip}\frac{L_m}{L_s}\lambda_d \tag{12.10}$$

where $\omega_{slip} = \omega_s - \omega_r$, $\lambda_d = L_s i_{sd} + L_m i_{rd}$, and $L_s = L_{ls} + L_m$ [24]. It is noted that the reference-frame angle (ϕ) is selected in such a way to force $\lambda_q = 0$ and simplify the control design. More information about the RSC control loops can be found in [24]. Also, k_{p1} and k_{i1} are the gains of the proportional integral (PI) controllers for the current control loop. In this study, since the current control loop bandwidth is much higher than the outer loops, the small-signal model can be simplified by only considering the outer loop dynamics of torque and reactive power [25]. Referring to Figure 12.2, the dynamics of the outer loops can be represented by (12.11)-(12.14) [26].

$$\frac{d\psi_{Te}}{dt} = k_{iT}(T_e^{mpp} - T_e) \tag{12.11}$$

$$\frac{d\psi_{Qs}}{dt} = k_{pQ}(Q_s^{*} - Q_s) \tag{12.12}$$

$$i_{dr} = i_{dr}^{*} = k_{pT}(T_e^{mpp} - T_e) + \psi_{Te} \tag{12.13}$$

$$i_{qr} = i_{qr}^{*} = k_{iQ}(Q_s^{*} - Q_s) + \psi_{Qs} \tag{12.14}$$

where T_e^{mpp} is the reference electromagnetic torque that results in the maximum power point of the wind turbine, k_{pT} and k_{iT} are PI controller gains for the electromagnetic torque regulator, k_{pQ} and k_{iQ} are the PI regulator gains for the stator's reactive power controller, and ψ_{T_e} and ψ_{Q_s} are two intermediate variables to linearize the dynamics of the outer loops.

12.2.3.2 Grid-side converter dynamics

The GSC is synchronized to the terminal voltage on the grid side, $v_{t,abc}$, and is controlled in dq-frame [26]. A PLL is used to extract the transformation angle needed for dq conversion (θ_s). Similar to the RSC's inner control loop, two PI regulators are utilized for the inner current control loop of the GSC. The dynamics of the inner current controller can be expressed by (12.15) and (12.16).

$$V_{dg}^{*} = \left(k_{p2} + \frac{k_{i2}}{s}\right)(i_{dg}^{*} - i_{dg}) - \omega_e L_g i_{qg} \tag{12.15}$$

$$V_{qg}^{*} = \left(k_{p2} + \frac{k_{i2}}{s}\right)(i_{qg}^{*} - i_{qg}) + \omega_e L_g i_{dg} \tag{12.16}$$

where k_{p2} and k_{i2} are PI controller gains for GSC's current regulator, and the i_{dg}^{*} and i_{qg}^{*} are reference d-axis and q-axis current setpoints derived from the outer loop controllers. Referring to Figure 12.2, the outer loop in GSC regulates the DC-link voltage and the reactive power of the GSC converter. Similar to the RSC, fast dynamics of the inner current loop in the GSC can be ignored and therefore, the GSC dynamics in the outer loop can be presented by (12.17)–(12.20).

$$\frac{d\eta_V}{dt} = k_{idc}(V_{dc}^{*} - V_{dc}) \tag{12.17}$$

$$\frac{d\eta_{Qg}}{dt} = k_{i3}(Q_g^{*} - Q_g) \tag{12.18}$$

$$i_{dg}^{*} = i_{dg} = k_{pdc}(V_{dc}^{*} - V_{dc}) + \eta_V \tag{12.19}$$

$$i_{qg}^{*} = i_{qg} = k_{p3}(Q_g^{*} - Q_g) + \eta_{Qg} \tag{12.20}$$

where k_{pdc} and k_{idc} are PI regulator gains of DC voltage controller, k_{p3} and k_{i3} are PI regulator gains of GSC's reactive power controller, and η_V and η_{Qg} are two intermediate variables defined to linearize the outer loop dynamics of the GSC.

12.2.4 Modeling of battery energy storage unit

Referring to Figure 12.1, the BESS is connected to the wind turbine at the PCC through an LCL filter. The detailed schematic of the BESS with its output filter and control blocks is depicted in Figure 12.3. It is shown that the BESS includes a battery pack, an LCL filter, and two cascaded control loops. Control loops regulate the converter currents, reactive power, and active power in a synchronously rotating

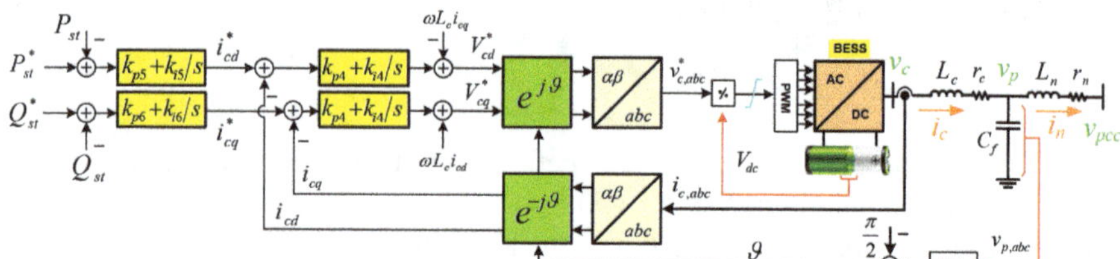

Figure 12.3 Control diagram of a BESS connected to the grid

reference frame. A PLL is used to synchronize the converter to the point p as shown in Figure 12.3.

12.2.4.1 AC-side dynamics

Dynamics of the LCL filter can be written in space-phasor form as

$$\begin{cases} \vec{v}_c - \vec{v}_p & = L_c \frac{\vec{i}_c}{dt} + r_c \vec{i}_c \\ \vec{v}_p - \vec{v}_{pcc} & = L_n \frac{\vec{i}_n}{dt} + r_n \vec{i}_n \\ \vec{i}_c - \vec{i}_n & = C_f \frac{\vec{v}_p}{dt} \end{cases} \tag{12.21}$$

where r_c, L_c, C_f are the parameters of the converter output filter, r_n, L_n are the resistance and inductance of transmission line connecting the BESS to the PCC, respectively, $\vec{v}_c$ is the space vector of the converter output voltage, $\vec{v}_p$ is the space vector of the voltage at point p, and $\vec{v}_{pcc}$ is the space vector of the PCC voltage, where the wind turbine and BESS are linked to. Knowing the relationship between the space-phasors and synchronous rotating frame (dq) frame, that is, $\vec{x} = (x_d + jx_q)e^{j\epsilon(t)}$ and assuming that $\epsilon(t) = \omega t + \theta_0$ [23], the LCL dynamics in (12.21) can be converted to dq frame as expressed in the following:

$$v_{cd} - v_{pd} + \omega L_c i_{cq} = (L_c s + r_c) i_{cd} \tag{12.22}$$

$$v_{cq} - v_{pq} - \omega L_c i_{cd} = (L_c s + r_c) i_{cq} \tag{12.23}$$

$$v_{pd} - v_{pcc,d} + \omega L_n i_{nq} = (L_n s + r_n) i_{nd} \tag{12.24}$$

$$v_{pq} - v_{pcc,q} - \omega L_n i_{nd} = (L_n s + r_n) i_{nq} \tag{12.25}$$

$$i_{cd} - i_{nd} + \omega C_f v_{pq} = C_f s v_{pd} \tag{12.26}$$

$$i_{cq} - i_{nq} - \omega C_f v_{pd} = C_f s v_{pq} \tag{12.27}$$

where $s = \frac{d}{dt}$ is the Laplace operator, v_{cd}, v_{cq}, i_{cd}, and i_{cq} are the dq-frame components of the converter output voltage and current, respectively. Also, v_{pd}, v_{pq}, $v_{pcc,d}$, and $v_{pcc,q}$ are the dq-frame components of the voltage at the point p and PCC, respectively, and i_{nd} and i_{nq} are the dq-frame components of the current at the output of the filter.

12.2.4.2 Modeling storage controllers

Referring to Figure 12.3, the BESS controller includes inner current control and the outer power loops. The inner current controller generates the dq-frame converter's setpoint voltages, which are then converted to the abc frame using transformation angle, ϑ, from the PLL. The reference abc voltages are then fed to the pulse width modulation (PWM) generation unit that generates the required gate signals for the BESS converter. Referring to Figure 12.3, the inner current control loop's dynamics can be presented by

$$V_{cd}^{*} = \left(k_{p4} + \frac{k_{i4}}{s}\right)(i_{cd}^{*} - i_{cd}) - \omega L_c i_{cq} + v_{pd} V_{cq}^{*} = \left(k_{p4} + \frac{k_{i4}}{s}\right)(i_{cq}^{*} - i_{cq}) + \omega L_c i_{cd} + v_{l} \tag{12.28}$$

where k_{p4} and k_{i4} are the gains of PI regulator for BESS's inner loop and i_{cd}^{*} and i_{cq}^{*} are reference converter currents that are derived from outer loop control. Assuming an ideal operation of the converter and ignoring the PWM dynamics, the converter reference voltages can be tracked accurately, and therefore, the converter's measured voltages at the terminal, v_{cd} and v_{cq} are the same as reference voltages, that is, $v_{cd}^{*} \approx v_{cd}$ and $v_{cq}^{*} \approx v_{cq}$.

The amount of active and reactive powers transferred to the PCC by the BESS (e.g., P_{st} and Q_{st}) is regulated by the outer loops through two PI controllers. It is noted that in the grid-connected mode of operation, independent active and reactive power regulation is achieved by the BESS. The outer loop dynamics are presented by

$$i_{cd}^{*} = \left(k_{p5} + \frac{k_{i5}}{s}\right)(P_{st}^{*} - P_{st}),\ i_{cq}^{*} = \left(k_{p6} + \frac{k_{i6}}{s}\right)(Q_{st}^{*} - Q_{st}) \tag{12.29}$$

where P_{st}^{*} and Q_{st}^{*} are the active and reactive power setpoints of the BESS decided by the secondary control algorithms (to be discussed in the next section), P_{st} and Q_{st} are the measured active and reactive powers of the BESS, which are calculated by (12.30) and (12.31) [27].

$$P_{st} = \frac{3}{2}(v_{cd} i_{cd} + v_{cq} i_{cq}) \tag{12.30}$$

$$Q_{st} = \frac{3}{2}(-v_{cd} i_{cq} + v_{cq} i_{cd}) \tag{12.31}$$

12.2.4.3 Phase-locked loop

The PLL block measures the real-time frequency of the system and decouples the *dq*-frame components so that they can be independently controlled [28]. A compensator to the form of a PI controller is used to force the *q* component of the voltage at point *p* to zero and align the converter's *d*-axis with the *d*-axis at point *p*. The PLL dynamics is expressed by (12.32):

$$\omega_{PLL} = \left(k_p^{PLL} + \frac{k_i^{PLL}}{s}\right) v_{pq} + \omega_0 \tag{12.32}$$

since the above equation is not in a standard linear form, it can be linearized using the following equations:

$$\frac{d}{dt}\Pi_{PLL} = k_i^{PLL} v_{pq}, \tag{12.33}$$

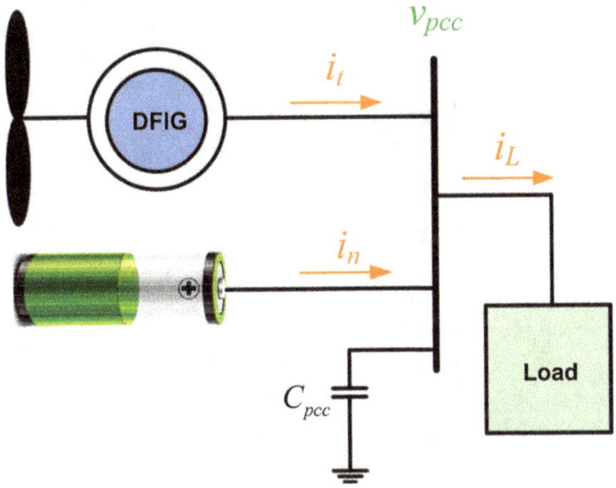

Figure 12.4 *Integration of BESS and wind turbine to the load through PCC*

12.2.4.4 Integration of BESS and wind turbine at PCC

Since the storage unit and the DFIG are integrated at the PCC, the interconnection between the small-signal models can be feasible by considering a capacitor at the point of interconnection, C_{pcc}. Therefore, referring to Figure 12.4, a KCL at PCC provides the dynamics of the integrated model such that

$$\frac{d}{dt}v_{pcc,d} = \frac{1}{C_{pcc}}\left(i_{nd} + i_{td} - i_{Ld}\right) + \omega v_{pcc,q}\ \frac{d}{dt}v_{pcc,q} = \frac{1}{C_{pcc}}(i_{tq} + i_{nq} - i_{Lq}) - \omega v_{pcc,d} \tag{12.34}$$

The eigenvalue analysis will be performed in the case studies to validate the developed small-signal model. In the next step, a secondary controller is designed for a storage-integrated wind farm to provide ancillary services to the farm using minimum communications between the neighboring storage-integrated wind turbines.

12.3 Stability analysis

In this section, the small-signal model of the storage-integrated wind turbine is linearized and integrated using (12.34) to form the overall small-signal model. The state-space model was developed in MATLAB® Simulink® and using MATLAB simulink and was then linearized using the "linmod" function. Using the state-space linearized matrices of the overall system, the eigenvalues of the closed-loop model were found. Parameters of the DFIG can be found in [29] and parameters of the BESS are given in [30]. Eigenvalues of the integrated model are included in Table 12.1. As it can be seen, the integrated system has 29 modes. To investigate the damping of each mode (e.g., $\lambda_i = a_i \pm jb_i$), the damping ratio, ζ (%), of an underdamped oscillation mode is calculated by $\zeta = -a/\sqrt{a^2 + b^2} \times 100$ [31]. As can be seen from Table 12.1, the system is stable, but λ_1, λ_2, λ_{19}, λ_{20}, λ_{23}, λ_{24}, λ_{27}, and

Table 12.1 Eigenvalue analysis of a storage-integrated DFIG wind turbine.

Eigenvalue	f (Hz)	ζ (%)	Eigenvalue	f (Hz)	ζ (%)	Eigenvalue	f (Hz)	ζ (%)
$\lambda_1=-10.2+j377$	60	2.7	$\lambda_{11}=-20$	0	100	$\lambda_{21}=-39.6$	0	100
$\lambda_2=-10.2-j377$	60	2.7	$\lambda_{12}=-3.89$	0	100	$\lambda_{22}=-20$	0	100
$\lambda_3=-33.2+j125$	20	25	$\lambda_{13}=-1$	0	100	$\lambda_{23}=-9.83+j942$	150	1
$\lambda_4=-33.2-j125$	20	25	$\lambda_{14}=-0.9$	0	100	$\lambda_{24}=-9.83-j942$	150	1
$\lambda_5=-78.7$	0	100	$\lambda_{15}=-0.001$	0	100	$\lambda_{25}=-3.6+j37$	6	9.6
$\lambda_6=-108.9+j31$	5	96	$\lambda_{16}=-1$	0	100	$\lambda_{26}=-3.6-j37$	6	9.6
$\lambda_7=-108.9-j31$	5	96	$\lambda_{17}=-20$	0	100	$\lambda_{27}=-0.1+j37$	6	0.27
$\lambda_8=-20.79$	0	100	$\lambda_{18}=-48.9$	0	100	$\lambda_{28}=-0.1-j37$	6	0.27
$\lambda_9=-15.1+j62$	10	23.6	$\lambda_{19}=-35.1+j12\,440$	1980	0.28	$\lambda=-19.2$	0	100
$\lambda_{10}=-15.1-j62$	10	23.6	$\lambda_{20}=-35.1-j12\,440$	1980	0.28			

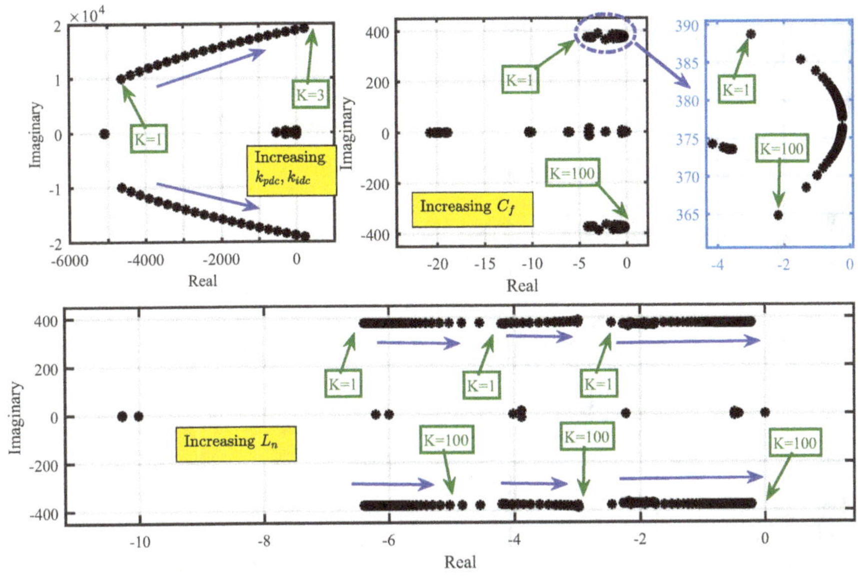

Figure 12.5 *Sensitivity of parameters on stability of storage-integrated wind turbine*

λ_{28} provide low damping that might destabilize the system with operating point changes or changes in the parameters of the controllers.

12.3.1 Sensitivity analysis

A case study was performed to examine the sensitivity of parameters of the design on the stability of the storage-integrated wind turbine. Eigenvalue analysis results are shown in Figure 12.5. The first subplot on the left shows the sensitivity of eigenvalues toward increasing the DFIG's DC voltage regulator gains. It is observed that the overall system is very sensitive toward changing the DC-link regulator gains, and increasing the gains by three times results in instability of the integrated system. The two subplots on the right side of first row illustrate the effect of changing the storage's capacitive component of the LCL filter on the overall stability of the system. As can be seen, the system is not sensitive toward selecting a large capacitive component for the LCL filter of the BESS and remains stable. The last scenario studies the impact of inductive component of the BESS's LCL filter on the PCC side, L_n, on the stability of the overall system. As can be seen, by increasing the inductor value, the eigenvalues with 60 Hz natural frequency (λ_1 and λ_2) move toward the origin, but the system remains stable. This case study demonstrates the effectiveness of the small-signal stability analysis in designing storage-integrated wind tribunes and the selection of parameters of the design that validates the stable operation for ancillary services. A similar procedure can be done to analyze the stability of storage-integrated PMSG-based wind turbines.

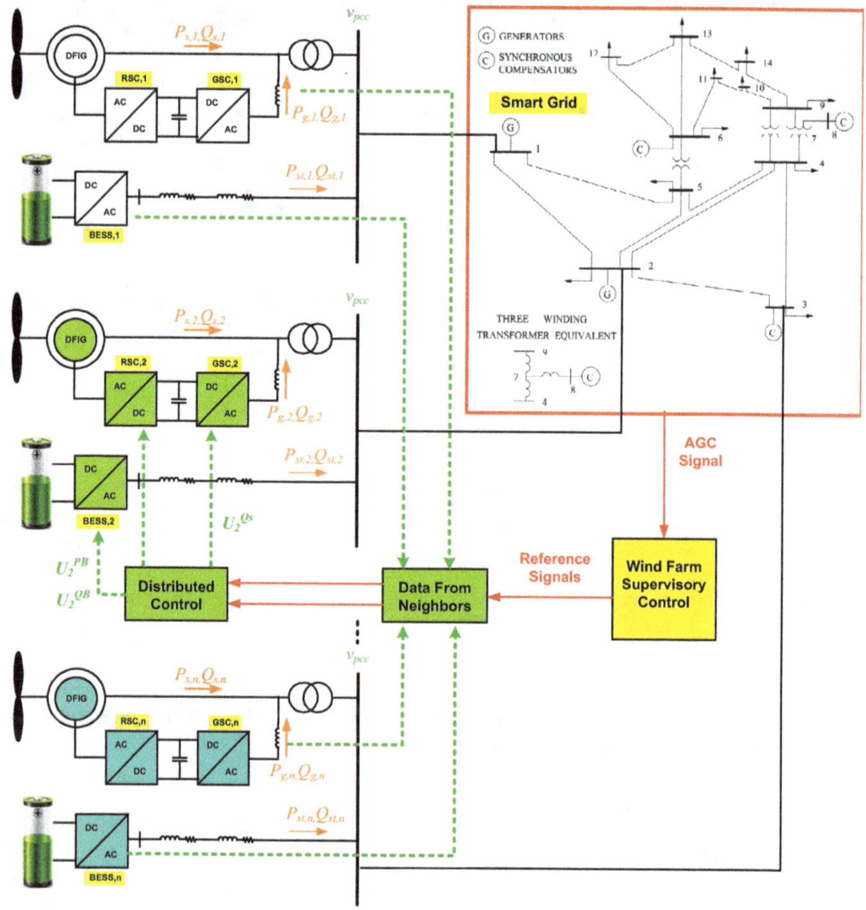

Figure 12.6 Concept of distributed control for storage-integrated wind farms

12.4 Distributed control design for storage-integrated wind farms

In this section, the application of distributed control for storage-integrated wind farms will be elaborated to highlight the importance of storage units in providing ancillary services to the wind farms using minimum communications between the neighboring wind turbines integrated with storage. Figure 12.6 illustrates the concept, where *n* storage-integrated wind turbines are connected to a smart-grid system. The AGC control signal will be shared the with wind farm's supervisory controller and that would be the reference signals (e.g., the total amount of active/reactive power needed from the wind farm or voltage/frequency setpoints) to be sent to the distributed controllers. Each storage-integrated wind turbine will be supplemented with a distributed controller that receives data from neighboring storage-integrated wind turbines and reference signals from AGC (if that storage-integrated wind turbine is identified as the leader) and will generate input signals to be sent to the GSC, RSC, or BESS controllers. In this section, a simple example will be demonstrated to provide load sharing and energy synchronization for the storage-integrated wind turbines using distributed controllers, which can easily be expanded to other control objectives.

12.4.1 Control objectives

The main task of distributed control is to coordinate the operation of BESS converters, GSCs, and RSCs to achieve the following objectives:

Equal power-sharing: Power-sharing is the main objective of a storage-integrated wind farm. The total active and reactive power demand, for example, P_d and Q_d, will be shared to the leader through AGC signal. The first objective is to ensure the summation of the storage-integrated wind turbine's active and reactive powers is equal to the demand. Since not all the wind turbines have the same size, it is expected that a wind turbine with a higher capacity contributes more to the load support compared to a smaller-scale wind turbine. Therefore, this problem is defined as power-sharing and a gain, $K_{st,i}$, is added to denote the heterogeneous nature of wind turbines in the smart grid. Since the storage unit that is integrated into each wind turbine is normally designed based on the capacity of the wind turbines, the same heterogeneous gain can be defined for the wind turbines [16, 17]. The power-sharing objective is formulated as

$$\begin{aligned} &= P_d - \textstyle\sum_{i=1}^{n} K_{st,i}(P_{g,i} + P_{s,i}) \\ &\Leftrightarrow \sum_{i=1}^{n} K_{st,i} P_{st,i} = P_d - \sum_{i=1}^{n} K_{st,i}(P_{g,i} + P_{s,i}) \end{aligned} \tag{12.35}$$

It is assumed that the wind turbines operate at MPP and do not contribute in active power-sharing (curtailment) and only storage units will participate in compensating the mismatch between the demand, and overall power generated by the wind turbines. Moreover,

$$Q_d = \sum_{i=1}^{n} K_{st,i}(Q_{g,i} + Q_{s,i} + Q_{st,i}). \tag{12.36}$$

The induction generator normally operates at unity power factor and therefore, $Q^*_{s,i} \approx 0$. Assuming the outer loop controller of the RSC regulates the reference reactive power setpoint, the measured reactive power of the stator terminal is also zero, that is, $Q_{s,i} \approx 0$. Thus,

$$Q_d = \sum_{i=1}^{n} K_{st,i} Q_{st,i} + \sum_{i=1}^{n} K_{st,i} Q_{g,i}. \tag{12.37}$$

Equal storage power-sharing: BESSs will share the active power mismatch between the demand and summation of wind active powers based on their capacities.

In addition, BESSs and GSCs will contribute in sharing the reactive power demand based on (12.37), these objectives are formulated by

$$\begin{aligned} K_{st,i}P_{st,i} &= K_{st,j}P_{st,j} \\ K_{st,i}(Q_{g,i} + Q_{st,i}) &= K_{st,j}(Q_{g,j} + Q_{st,j}). \end{aligned} \tag{12.38}$$

As the GSCs are usually rated at 30% of the DFIG power while storage units can be the same as the capacity of the DFIG, the control objective assigns 30% of the load's reactive power to be compensated by the GSCs and the remaining portion will be shared by the BESSs such that

$$\sum_{i=1}^{n} K_{st,i}Q_{st,i} = 70\ Q_d, \sum_{i=1}^{n} K_{st,i}Q_{g,i} = 30\ Q_d. \tag{12.39}$$

BESSs energy synchronization: One of the main aspects of integrating storage units to wind turbines is to ensure the BESSs are always available for demand response. Therefore, a state-of-charge synchronization objective is added to the distributed control design to synchronize the energy levels of storage units along with the demand response objectives defined in the previous section. The energy synchronization of the storage units is formulated by

$$E_{st,i} = E_{st,j}. \tag{12.40}$$

12.4.2 Simplified model of storage-integrated wind turbines for secondary control design

Since the bandwidth of secondary controllers is much slower than bandwidth of outer loops and inner loops in the DFIG or BESS, simplified dynamic models can be used to facilitate the control design [16–18].

12.4.2.1 Wind turbine's simplified model

Assuming ΔQ_{G_i} as the output of distributed controller that is supplemented to the GSC control loop together with the initial reactive power reference of the GSC ($Q^0_{g,i}$), the reference setpoint reactive power of the GSC in the ith storage-integrated wind turbine can be formulated as

$$Q^*_{g,i} = Q^0_{g,i} + \Delta Q_{G_i} \tag{12.41}$$

The modeling procedure can be simplified by ignoring the fast inner current loop dynamics and assuming that the GSC's outer control loops can track the references fast enough, that is $V^*_{dc,i} \approx V_{dc,i}$ and $Q^*_{g,i} \approx Q_{g,i}$. Using a consensus integral controller theory, an input can be designed to represent the dynamics of the ΔQ_{G_i} such that

$$\Delta \dot{Q}_{G_i} = \frac{1}{K_{st,i}} u_i^{Qg} \tag{12.42}$$

where u_i^{Qg} is an input designed for the distributed reactive power-sharing of DFIG WT+BESS using GSC. By replacing $Q_{g,i}$ instead of $Q^*_{g,i}B$ in (12.41) and differentiating the result,

$$\frac{d}{dt}Q_{g,i} = \frac{d}{dt}Q_{g,i}^{0} + \frac{d}{dt}\Delta Q_{G_i} \tag{12.43}$$

Knowing the fact that $\frac{d}{dt}Q_{g,i}^{0} = 0$ and replacing the dynamics of ΔQ_{G_i} from (12.42) in (12.43), simplified dynamics of the GSC is derived.

$$K_{st,i}\dot{Q}_{g,i} = u_i^{Qg} \tag{12.44}$$

Since the distributed controller will only add an input to the DFIG's GSC, dynamics of the induction generator and RSC can be ignored [16–18].

12.4.2.2 Simplified model of BESSs

It was shown in [16–18, 30] that the simplified model of BESS provides an accurate approximation for the distributed control design purposes in smart grids and, therefore, is adopted in this section. The model takes into the account the heterogeneous nature of storage-integrated wind turbines in smart grids and provides an ability for the users to develop BESS models with different capacities and parameters. The simplified model is represented as

$$\dot{E}_{st,i} = \frac{-K_{st,i}}{3600}P_{st,i} \tag{12.45a}$$

$$K_{st,i}\dot{P}_{st,i} = u_i^{PB} \tag{12.45b}$$

$$K_{st,i}\dot{Q}_{st,i} = u_i^{QB} \tag{12.45c}$$

where $K_{st,i} > 0$ is a gain to represent the heterogeneous nature of different BESSs, $E_{st,i}$, $P_{st,i}$, and $Q_{st,i}$ present the energy level, active power, and reactive power of ith BESS unit. In addition, u_i^{PB} is the input control signal for the active power-sharing, u_i^{QB} is the input control signal for the reactive power-sharing of ith storage unit. Consequently, the simplified dynamics of the storage-integrated wind turbine can be presented by

$$\frac{d}{dt}\begin{bmatrix} E_{st,i} \\ K_{st,i}P_{st,i} \\ K_{st,i}Q_{st,i} \\ K_{st,i}Q_{g,i} \end{bmatrix} = \begin{bmatrix} 0 & \frac{-1}{3600} & 0 & 0 \\ 0 & 0 & 0 & 0 \\ 0 & 0 & 0 & 0 \\ 0 & 0 & 0 & 0 \end{bmatrix} \begin{bmatrix} E_{st,i} \\ K_{st,i}P_{st,i} \\ K_{st,i}Q_{st,i} \\ K_{st,i}Q_{g,i} \end{bmatrix} + \begin{bmatrix} 0 & 0 & 0 \\ 1 & 0 & 0 \\ 0 & 1 & 0 \\ 0 & 0 & 1 \end{bmatrix} \begin{bmatrix} u_i^{PB} \\ u_i^{QB} \\ u_i^{Qg} \end{bmatrix} \tag{12.46}$$

12.4.3 Graph theory

A brief background in the graph theory is included in this section to facilitate the distributed control design theory. Using multi-agent theory, each storage-integrated wind turbine is treated as an agent and an undirected graph noted as $\mathcal{G}$ is defined that has a vertex set of $\mathcal{V}$ and edge set of $\mathcal{E}$. Each storage-integrated wind turbine is a vertex and the interconnection between storage-integrated wind turbines i and j is denoted as edge $(i,j) \in \mathcal{E}$. The neighboring set for each storage-integrated wind turbine i is identified by $\mathcal{N}_i \triangleq \{j \in \mathcal{V} : (i,j) \in \mathcal{E}\}$. In addition, an adjacency matrix $\mathcal{A}$ of $\mathcal{G}$ is defined whose a_{ij} elements are calculated by:

$$\mathcal{A} \longrightarrow \begin{cases} a_{ij} = 1 & \text{if } (i,j) \in \mathcal{E} \\ a_{ij} = 0 & \text{if } (i,j) \notin \mathcal{E} \end{cases} \tag{12.47}$$

Also, the degree matrix of $\mathcal{G}$ is denoted by

$$D = \begin{bmatrix} d_1 & 0 & \dots & 0 \\ 0 & d_2 & \dots & 0 \\ \vdots & \vdots & \ddots & \vdots \\ 0 & 0 & \dots & d_n \end{bmatrix} \tag{12.48}$$

where $d_i \triangleq \sum_{i \in \mathcal{N}_k} a_{ij}$. Finally, the Laplacian matrix L associated with $\mathcal{G}$ is expressed by $L = \mathcal{D} - \mathcal{A}$. To share the AGC signals with the storage-integrated wind turbines, one of the wind turbines is assigned as a virtual leader that can receive the AGC signals through communication links. This minimizes the number of communication links from AGC to wind turbines. The AGC commands are then shared with neighboring storage-integrated wind turbines through communication links through communication graph of the system. Assume the sub-index 0 defines the leader and $\mathcal{N}_0$ is the set of leader's neighboring storage-integrated wind turbines. Then, $a_{0k} = 1$ if $k \in \mathcal{N}_0$, that is, the kth storage-integrated wind turbine can communicate to the leader, and $a_{0k} = 0$ if $k \notin \mathcal{N}_0$, or the kth wind turbine is not connected to the leader, $k = 1, \dots, N$. Next, define

$$\Gamma = \begin{bmatrix} a_{01} & 0 & \dots & 0 \\ 0 & a_{02} & \dots & 0 \\ \vdots & \vdots & \ddots & \vdots \\ 0 & 0 & \dots & a_{0n} \end{bmatrix} \tag{12.49}$$

12.4.4 Control design without communication delays

In order for the storage-integrated turbines to track the AGC commands, dynamics of the virtual leader is considered as

$$\dot{x}_0 = Ax_0 \tag{12.50}$$

where $x_0 \in \mathbb{R}^4$ is the virtual leader's state vector that receives information from the AGC. The following theorem initiates the control design for the storage-integrated wind turbines.

Theorem 1 Let the communication graph $\mathcal{G}$ among storage-integrated wind turbines be connected, which means at least one storage-integrated wind turbine should be connected to the virtual leader. Assume the initial state vector of the virtual leader is defined as

$$x_0(0) = \left[0, \frac{P_d - \sum K_{st,i}(P_{g,i} + P_{s,i})}{n}, \frac{0.7Q_d}{n}, \frac{0.3Q_d}{n}\right]^T. \tag{12.51}$$

Then to ensure energy synchronization and active/reactive power-sharing among storage-integrated wind turbines, the control input is designed by

$$u_i = -cK\sum_{j\in\mathcal{N}_i} a_{ij}(x_i - x_j) - c_0 a_{i0} K(x_i - x_0) \tag{12.52}$$

for any $c > 0, c_0 > 0$; $K = \begin{bmatrix} k_1 & k_2 & 0 & 0 \\ 0 & 0 & k_3 & 0 \\ 0 & 0 & 0 & k_4 \end{bmatrix}$, $k_1 < 0, k_2 > 0, k_3 > 0, k_4 > 0$, and $i = 1, 2, \ldots, n$.

Proof. Define

$$\zeta = x - \mathbf{1}_n \otimes x_0. \tag{12.53}$$

Then, (12.52) can be written in a compact form as

$$u = -c(L \otimes K)x - (c_0\Gamma \otimes K)\zeta. \tag{12.54}$$

Subsequently, knowing that $L\mathbf{1}_n = 0$,

$$\begin{aligned}\dot{\zeta} &= (I_n \otimes A)\zeta + (I_n \otimes B)u \\ &= (I_n \otimes A)\zeta - c(I_n \otimes B)[L \otimes K]x \\ &\quad -(I_N \otimes B)(c_0\Gamma \otimes K)\zeta \\ &= [I_n \otimes A - c_0\Gamma \otimes (BK) - cL \otimes (BK)]\zeta \\ &\quad -c[L \otimes (BK)](\mathbf{1}_n \otimes x_0), \\ &= [I_n \otimes A - (c_0\Gamma + cL) \otimes (BK)]\zeta,\end{aligned} \tag{12.55}$$

Let us denote

$$\mathcal{A} \triangleq I_n \otimes A - (c_0\Gamma + cL) \otimes (BK), \tag{12.56}$$

Equation (12.55) can then be rewritten as,

$$\dot{\zeta} = \mathcal{A}\zeta. \tag{12.57}$$

Since the communication graph of the system is connected, $c_0\Gamma + cL$ is positive definite [32], which means all the eigenvalues of $c_0\Gamma + cL$ are positive. Assume $\lambda_i, i = 1, \ldots, n$ as eigenvalues of $c_0\Gamma + cL$. Define $U \in \mathbb{R}^{n\times n}$ as an orthogonal matrix that diagonalizes $c_0\Gamma + cL$, then it can be concluded that $\mathcal{A} = (U \otimes I_4)\text{diag}\{A - \lambda_i BK\}_{i=1,\ldots,n}(U^T \otimes I_4)$, that is, eigenvalues of $\mathcal{A}$ are equal to the eigenvalues of $A - \lambda_i BK$ for $i = 1, \ldots, n$. In order for the power-sharing and energy levels of storage-integrated wind turbines to track the states of the leader, all eigenvalues of $\mathcal{A}$ and $A - \lambda_i BK$ must have negative real parts. Knowing that the characteristic equation of the closed-loop matrix $A - \lambda_i BK$ is

$$(s + \lambda_i k_4)(s + \lambda_i k_3)(s^2 + \lambda_i k_2 s - \frac{1}{3600}\lambda_i k_1) = 0. \tag{12.58}$$

The system is stable if all eigenvalues of $A - \lambda_i BK$ have negative real parts. This can only happen if $\lambda_i k_4 > 0, \lambda_i k_3 > 0, \lambda_i k_2 > 0$ and $\frac{1}{3600}\lambda_i k_1 < 0$, which are equivalent to

$k_4 > 0, k_3 > 0, k_2 > 0$ and $k_1 < 0$. Therefore, the energy synchronization and power-sharing is achieved using the above conditions and proper initial condition selection for the leader.

12.4.5 Control design with communication delays

Since no communication system is delay-free, inherent communication delays should be incorporated in the distributed control design. If a unified delay of $\tau > 0$ is considered for all the channels in the communication system, the following consensus design is proposed to achieve the control objectives defined earlier.

Theorem 2 If the communication graph $\mathcal{G}$ among storage-integrated wind turbines is connected and the initial state vector of the virtual leader is the same as (12.51), the energy synchronization and equal power-sharing are achieved if the controller is designed as

$$u_i(t) = cK\sum_{j\in\mathcal{N}_i} a_{ij}[x_i(t-\tau) - x_j(t-\tau)] - c_0 a_{i0} K(x_i(t-\tau) - x_0(t-\tau)) \tag{12.59}$$

for any $c > 0, c_0 > 0$, K as defined in Theorem 1, and

$$\tau < \frac{\arccos \dfrac{1}{\frac{k_2^2\lambda_n}{2\alpha k_1} + \sqrt{(\frac{k_2^2\lambda_n}{2\alpha k_1})^2 + 1}}}{\sqrt{\frac{k_2^2\lambda_n^2}{2} + \sqrt{(\frac{k_2^2\lambda_n^2}{2})^2 + \alpha^2 k_1^2\lambda_n^2}}} \tag{12.60}$$

where $\alpha = -\frac{1}{3\,600}$ and λ_n are the maximum eigenvalues of $c_0\Gamma + cL$ [16, 17].

Proof. Using the controller defined in (12.59), the closed-loop system will have the form of

$$\dot{\zeta}(t) = (I_n \otimes A)\zeta(t) - (c_0\Gamma + cL) \otimes (BK)\zeta(t-\tau). \tag{12.61}$$

By transferring (12.61) into the Laplace domain with the frequency variable, s, one can obtain

$$[sI_{4n} - (I_n \otimes A - (c_0\Gamma + cL) \otimes (BK)e^{-\tau s})]\zeta(s) = 0 \tag{12.62}$$

which results in the characteristic equation $sI_{4n} - (I_n \otimes A - (c_0\Gamma + cL) \otimes (BK)e^{-\tau s}) = 0$. By multiplying $(U \otimes I_4)$ and $(U^T \otimes I_4)$ to the characteristic equation,

$$(U \otimes I_4)\text{diag}\{sI_4 - (A - \lambda_i BKe^{-\tau s})\}_{i=1,\dots,n}(U^T \otimes I_4) = 0 \tag{12.63}$$

which is equivalent to $sI_4 - (A - \lambda_i BKe^{-\tau s}) = 0$ for all $i = 1, \dots, n$. Therefore, (12.61) is stable if and only if all eigenvalues of the matrices $sI_4 - (A - \lambda_i BKe^{-\tau s})$ belong to the open left half complex plane. Equivalently, all roots of the characteristic equations

$$(s + \lambda_i k_4 e^{-\tau s})(s + \lambda_i k_3 e^{-\tau s}) \times (s^2 + \lambda_i k_2 e^{-\tau s} s - \frac{1}{3600}\lambda_i k_1 e^{-\tau s}) = 0 \tag{12.64}$$

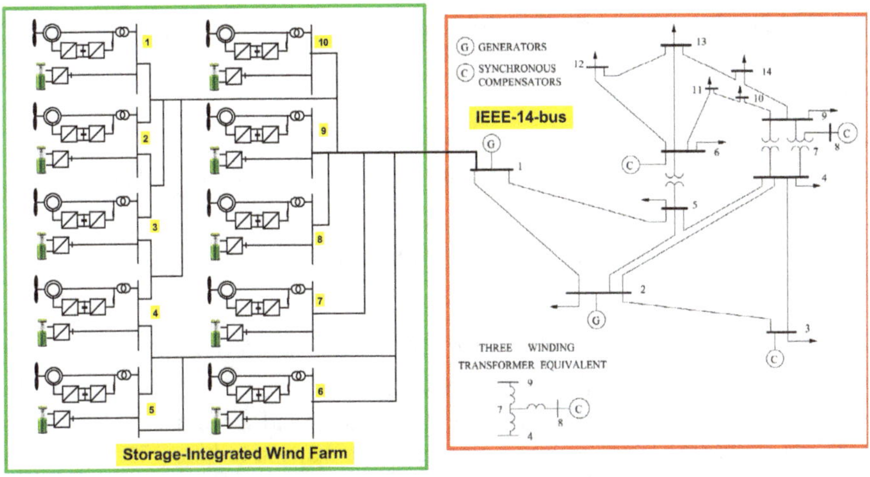

Figure 12.7 *Modified IEEE 14-bus power system aggregated with a wind farm consisting of 10 storage-integrated wind turbines*

have negative real parts. This means, the terms $s + \lambda_i k_4 e^{-\tau s} = 0$ and $s + \lambda_i k_4 e^{-\tau s} = 0$ will have negative roots if and only if $k_4 > 0$ and $k_3 > 0$. In addition, the term $s^2 + \lambda_i k_2 e^{-\tau s} s - \frac{1}{3600}\lambda_i k_1 e^{-\tau s} = 0$ has roots with negative real parts if the condition (12.60) is satisfied, more information can be found in [33].

Remark 1 Note that the bound of delay in (12.60) does not account for the sampling time in the simulation, so in order to compute the delay bound in the simulation, it should be multiplied with the sampling time.

12.5 Case studies

To validate the effectiveness of the designed distributed controller in providing ancillary services to the smart grids, an IEEE 14-bus power system is considered [34]. A MATLAB open-source toolbox named MatDyn is used for dynamic simulations of the IEEE 14-bus system [35]. The model is then modified to incorporate 10 storage-integrated wind turbines at bus 1 as shown in Figure 12.7. The secondary controllers are then designed and supplemented to each storage-integrated wind turbines referring to (12.52) for the delay-free case study and (12.59) for the case study with communication delays. Parameters of the design are expressed in Table 12.2.

The communication graph for the storage-integrated wind turbines is shown in Figure 12.8. It is noted that the green circles denote the followers and the yellow circle denotes the leader that receives information from AGC with regard to the demands and setpoints needed for the entire wind far. Two case studies are conducted to test the performance of the proposed controllers in providing ancillary services and energy synchronization with and without communication delays.

Table 12.2 Parameters of the control design

Parameter	BESSi, i=1, 2, ..., 10
Nominal power	400 kW
Nominal voltage	13.8 kV
$K_{st,i}$(p.u.)	[0.1,0.11,0.08,0.1, 0.07, 0.09, 0.1, 0.12, 0.11, 0.09]
c_0,c_1	0.5, 1 p.u.
n	10
k_1, k_2, k_3, k_4	−0.1, 0.4, 0.2, 0.2 p.u.

12.5.1 Ancillary services without communication delays

In this case study, no communication delays were considered and the distributed controller is activated at the beginning of the simulation. Wind speed randomly changes to create a variation in the output generated power of wind turbines. The mismatch between a constant demand (active and reactive powers) should be compensated by battery energy storage units. Simulation results for this case study are illustrated in Figure 12.9. It is observed from the first subplot that equal power-sharing ($K_{st}P_{st}$) is achieved in less than 2 000 s. This means the BESSs could cooperatively share the active power mismatch between generated wind power and demand based on their capacities. Furthermore, energy synchronization is achieved as confirmed by the second subplot. By taking a look at the third subplot, it is observed that the reactive power-sharing of storage units is also achieved although two step changes were applied after 1 500 and 4 000 s (0.25 p.u. and −0.3 p.u.). The last subplot shows the active power mismatch and summation of active powers generated by the BESSs. As it can be observed, the overall BESSs contributions could successfully compensate the demand mismatch and provide ancillary services to the grid. It is noted that a step change of 1 p.u. in active power demand occurred after 4 000 s, and the BESSs could successfully respond to the load change event within a few seconds.

12.5.2 Ancillary services with communication delays

This case investigates the performance of the designed distributed controller in providing power-sharing and energy synchronization capabilities to storage-integrated wind turbines when communication delays exist. Simulation results for this case study are depicted in Figure 12.10 for 10, 20, and 30 milliseconds (ms) communication delays. The first three subplots on the left demonstrate the reactive power-sharing of BESS units for three different communication delays. It is observed that

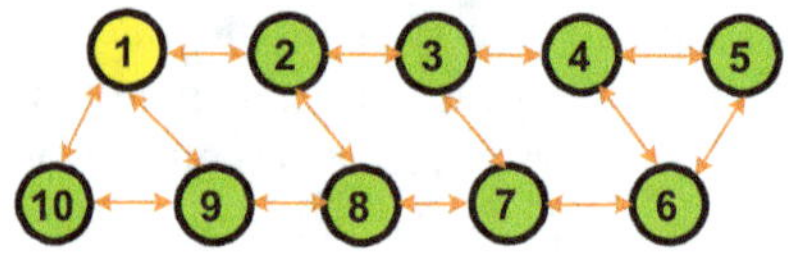

Figure 12.8 Communication graph of the storage-integrated wind turbines

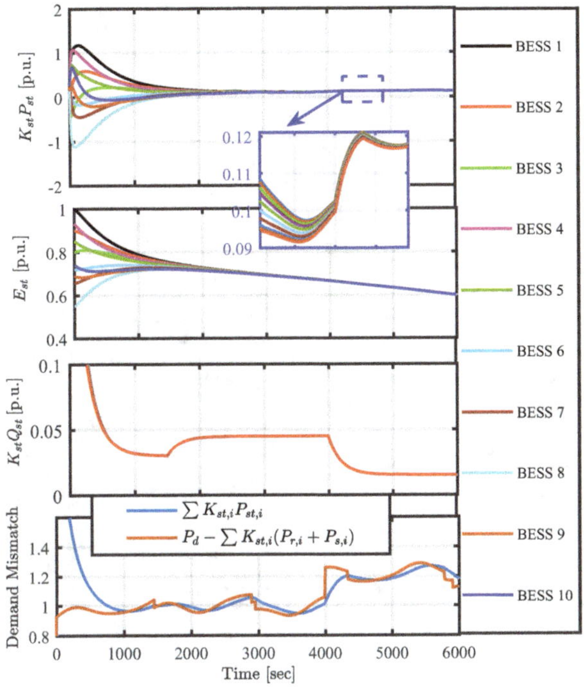

Figure 12.9 Simulation results of the distributed controller performance without communication delays

the designed controller can successfully respond to up to 20 ms delays, but will be unstable for 30 ms communication delay. The first three subplots on the right show the reactive power-sharing of GSCs in the wind turbines. It can be observed that GSCs compensate only 30% of the total load, and therefore, their setpoint reactive power is lower compared to the BESSs. Similar to the BESSs, the distributed controller on GSCs fail to operate with 30 ms communication delay. The last three subplots illustrate the active power, active power-sharing, and energy levels of BESSs for three different communication delays. The communication delay settings were updated during the simulation. It is observed that the BESSs could successfully share their energy levels and the active power of the load for up to 20 ms communication delays but fail to operate for 30 ms communication delay. Referring to the IEEE and the international electrotechnical commission (IEC) standard with regard to the maximum allowable communication delay in smart grids, 16 ms is the standard communication delay for command and control in smart grids. This means, any communication delays for command and control signals that experience delays more than 16 ms are not within the standard and should not be considered [36]. Therefore, the proposed controller in this chapter that tolerates up to 20 ms can successfully operate within standard communication systems.

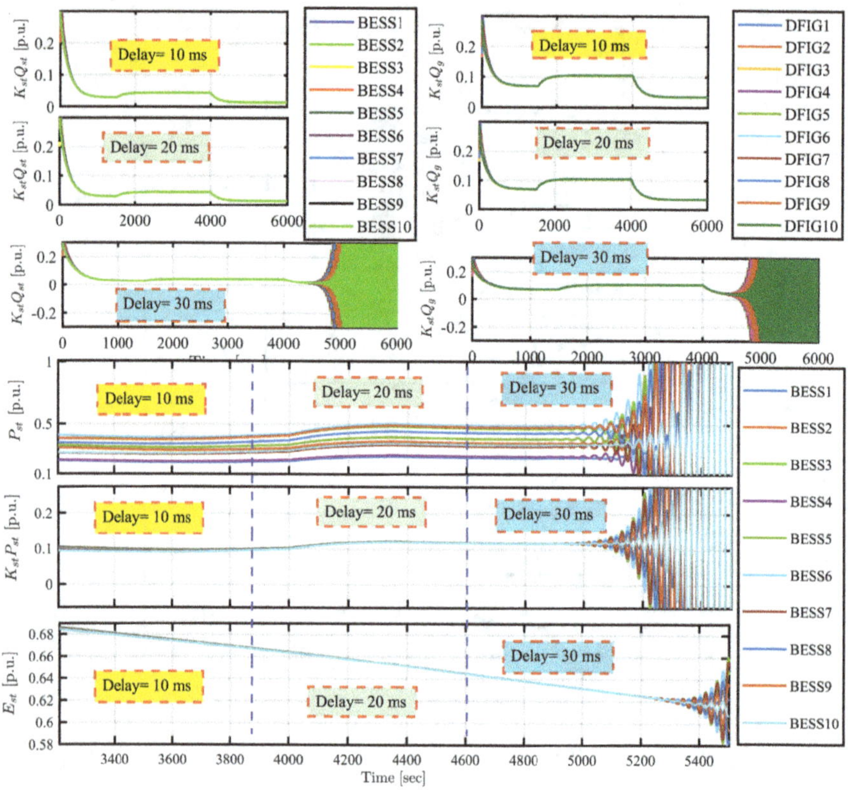

Figure 12.10 Simulation results of the distributed controller performance with communication delays

12.6 Conclusion

In this chapter, small-signal modeling and distributed control design of storage-integrated wind turbines based on DFIG units were elaborated. Dynamics of induction generator with drive train, DFIG controllers, PLL, AC filters, and integration of storage and DFIG were considered in the modeling procedure. The eigenvalue analysis resulted in identification of modes with low damping. Sensitivity analysis of the small-signal model revealed stability issues related to the DC-link regulator gains of the DFIG. It was concluded that the stability of storage-integrated wind turbine is not adversely impacted by the filter parameters of the BESS. A distributed controller was designed to provide energy synchronization to the storage units, equally share the active power mismatch between wind turbines and demand, and reactive power support through storage units and GSC of DFIGs. The designed controller successfully shares the active and reactive power mismatch of the system between storage units and synchronizes the energy levels of the storage units. The designed controller is robust against communication delays up to 20 ms, which is above the

16 ms standard communication systems. Future work will include a detailed stability analysis of storage-integrated wind turbines with various types of controllers, including droop control.

References

[1] Mauricio J.M., Marano A., Gómez-Expósito A., Martinez Ramos J.L., Ramos J.L.M. 'Frequency regulation contribution through variable-speed wind energy conversion systems'. *IEEE Transactions on Power Systems*. 2009;**24**(1):173–80.

[2] Howlader A.M., Urasaki N., Yona A., Senjyu T., Saber A.Y. 'A review of output power smoothing methods for wind energy conversion systems'. *Renewable and Sustainable Energy Reviews*. 2013;**26**(113):135–46.

[3] Molinas M., Jon Are Suul., Undeland T. 'Low voltage ride through of wind farms with cage generators: STATCOM versus SVC'. *IEEE Transactions on Power Electronics*. 2008;**23**(3):1104–17.

[4] Muyeen S., Ali M.H., Takahashi R., Murata T., Tamura J. 'Wind generator output power smoothing and terminal voltage regulation by using statcom/ess'. *Power Tech, 2007 IEEE Lausanne*. IEEE; 2007. pp. 1232–7.

[5] Østergaard P.A. 'Ancillary services and the integration of substantial quantities of wind power'. *Applied Energy*. 2006;**83**(5):451–63.

[6] Mendis N., Muttaqi K.M., Perera S. 'Management of low- and high-frequency power components in demand-generation fluctuations of a DFIG-based wind-dominated RAPS system using hybrid energy storage'. *IEEE Transactions on Industry Applications*. 2014;**50**(3):2258–68.

[7] Qu L., Qiao W. 'Constant power control of DFIG wind turbines with supercapacitor energy storage'. *IEEE Transactions on Industry Applications*. 2011;**47**(1):359–67.

[8] Wehner J., Mohler D., Gibson S. 'Technology performance report: Duke energy Notrees wind storage demonstration project'. *Duke Energy Renewables*. Charlotte, NC (United States): Technical Report; 2015.

[9] Dragičević T., Guerrero J.M., Vasquez J.C., Skrlec D. 'Supervisory control of an adaptive-droop regulated DC microgrid with battery management capability'. *IEEE Transactions on Power Electronics*. 2014;**29**(2):695–706.

[10] Xu Y., Zhang W., Hug G., Kar S., Li Z. 'Cooperative control of distributed energy storage systems in a microgrid'. *IEEE Transactions on Smart Grid*. 2015;**6**(1):238–48.

[11] Khalid M., Savkin A.V. 'Minimization and control of battery energy storage for wind power smoothing: aggregated, distributed and semi-distributed storage'. *Renewable Energy*. 2014;**64**(3):105–12.

[12] Sedghi M., Ahmadian A., Aliakbar-Golkar M. 'Optimal storage planning in active distribution network considering uncertainty of wind power distributed generation'. *IEEE Transactions on Power Systems*. 2016;**31**(1):304–16.

[13] Cao Q., Song Y.-D., Guerrero J.M., Tian S. 'Coordinated control for flywheel energy storage matrix systems for wind farm based on charging/discharging ratio consensus algorithms'. *IEEE Transactions on Smart Grid.* ;**7**(3):1259–67.

[14] Baros S., Ilic M.D. 'A consensus approach to real-time distributed control of energy storage systems in wind farms'. *IEEE Transactions on Smart Grid.* 2019;**10**(1):613–25.

[15] Baros S., Ilić M.D. 'Distributed torque control of deloaded wind DFIGs for wind farm power output regulation'. *IEEE Transactions on Power Systems.* 2017;**32**(6):4590–9.

[16] Khazaei J., Nguyen D.H., Asrari A. 'Consensus-based demand response of PMSG wind turbines with distributed energy storage considering capability curves'. *IEEE Transactions on Sustainable Energy.* 2019;**11**(4):2315–25.

[17] Hguyen D.H., Khazaei J., Stewart S.W., King J.R. 'Distributed cooperative control of wind farms with on-site battery energy storage systems'. *National Renewable Energy Laboratory (NREL).* Golden, CO (United States): Technical Report; 2019.

[18] Khazaei J., Nguyen D.H. 'Distributed consensus for output power regulation of DFIGs with on-site energy storage'. *IEEE Transactions on Energy Conversion.* 2018;**34**(2):1043–51.

[19] Khazaei J., Nguyen D.H., Asrari A., Mohammadpour H.A. 'Small-signal stability evaluation of DFIG wind farms with on-site battery energy storage'. *2019 IEEE PES GTD Grand International Conference and Exposition Asia (GTD Asia).* IEEE; 2019. pp. 274–9.

[20] Miao Z., Fan L. 'The art of modeling and simulation of induction generator in wind generation applications using high-order model'. *Simulation Modelling Practice and Theory.* 2008;**16**(9):1239–53.

[21] Krause P., Wasynczuk O., Sudhoff S.D., Pekarek S. *Analysis of Electric Machinery and Drive Systems.* **75**. John Wiley & Sons; 2013.

[22] Akhmatov V. *Induction generators for wind power.* Multi-Science Publishing; 2005.

[23] Yazdani A., Iravani R. *Voltage-sourced Converters in Power Systems: Modeling, Control, and Applications.* John Wiley & Sons; 2010.

[24] Ostadi A., Yazdani A., Varma R.K. 'Modeling and stability analysis of a DFIG-based wind-power generator interfaced with a series-compensated line'. *IEEE Transactions on Power Delivery.* 2009;**24**(3):1504–14.

[25] Pena R., Clare J.C., Asher G.M. 'Doubly fed induction generator using back-to-back PWM converters and its application to variable-speed wind-energy generation'. *IEE Proceedings – Electric Power Applications.* 1996;**143**(3):231–41.

[26] Yang L., Xu Z., Østergaard J., Dong Z.Y., Wong K.P., Ma X. 'Oscillatory stability and eigenvalue sensitivity analysis of a DFIG wind turbine system'. *IEEE Transactions on Energy Conversion.* 2011;**26**(1):328–39.

[27] Lorduy A., LÃ¡zaro, A A., Barrado C., Quesada FernÃ¡ndez I.., Lucena C. 'Simplified synchronous reference frame control of the three phase

grid connected inverter'. *2010 Twenty-Fifth Annual IEEE Applied Power Electronics Conference and Exposition (APEC)*. 2010:1026–33.

[28] Du W., Chen X., Wang H. 'Parameter tuning of the PLL to consider the effect on power system small-signal angular stability'. *IET Renewable Power Generation*. 2018;**12**(1):1–8.

[29] Mohammadpour H.A., Santi E. 'Analysis of sub-synchronous resonance (SSR) in doubly-fed induction generator (DFIG)-based wind farms'. *Synthesis Lectures on Power Electronics*. 2015;**5**(3):1–64.

[30] Khazaei J., Miao Z. 'Consensus control for energy storage systems'. *IEEE Transactions on Smart Grid*. 2018;**9**(4):3009–17.

[31] Kundur P., Balu N.J., Lauby M.G. *Power System Stability and Control*. **7**. New York: McGraw-H; 1994.

[32] Nguyen D.H. 'A sub-optimal consensus design for multi-agent systems based on hierarchical LQR'. *Automatica*. 2015;**55**(8):88–94.

[33] Peng L., Yingmin J., Junping D., Shiying Y. 'Distributed consensus control for second-order agents with fixed topology and time-delay'. *Proceedings of 26th Chinese Control Conference*. 2007:577–81.

[34] Milano F. 'An open source power system analysis toolbox'. *IEEE Transactions on Power Systems*. 2005;**20**(3):1199–206.

[35] Cole S., Belmans R. 'MatDyn, a new Matlab-based toolbox for power system dynamic simulation'. *IEEE Transactions on Power Systems*. 2011;**26**(3):1129–36.

[36] Gungor V.C., Sahin D., Kocak T., *et al.* 'A survey on smart grid potential applications and communication requirements'. *IEEE Transactions on Industrial Informatics*. 2013;**9**(1):28–42.

Chapter 13

Current status of research on the design of hybrid wind and solar plants

Piyali Ganguly[1], Akhtar Kalam[1], and Aladin Zayegh[1]

Abstract

Due to the rapid depletion and significant environmental impact of conventional energy sources, research is being conducted worldwide to find out alternative energy sources. The use of renewable energy technologies is being explored around the world as an alternative to conventional sources. These sources are non-depletable and environment-friendly. They use natural resources like solar and wind to generate the required energy. They can be used as single sources or as a combination of more than one type of renewable energy source along with storage devices like battery or fuel cells. Renewable energy systems (RES) can be used as grid-connected or standalone systems. However, the unpredictable nature of these natural resources reduces their reliability to a great extent. Hence, proper selection and sizing of the sources are necessary to avoid excessive establishment and maintenance costs or compromised power reliability. This chapter discusses different methodologies available in the literature on optimum sizing of the renewable energy systems, including wind power plants.

13.1 Introduction

The global energy demand has considerably increased in the last century. Currently, the major part of the global energy demand is provided by coal and natural gas. The increased energy demands result in rapid growth in the requirement of conventional fossil fuels [1].

The drawback of the conventional energy sources is that they are finite and fast depleting, which might impact the balance of upcoming energy requirement/generation [2]. Furthermore, energy production using conventional energy sources raises greenhouse gas releases contributing to global warming [3]. On the other hand, being environment-friendly and non-depletable, the RES can provide an alternative

[1]College of Engineering and Science, Victoria University, Melbourne, Australia

solution to fossil fuel. Renewable energy can be attained from natural resources like sun, wind, biomass, water, tides, ocean waves, and geothermal heat. The power produced by the RES is the cleanest form of energy [4].

Numerous types of renewable technologies are available in present days, that is, solar, hydroelectric, biomass, wind, ocean, and geothermal energy. Each type of RES is exclusively suitable for a particular application and has distinct advantages. The improved renewable energy penetration reduces the dependence on external energy sources. The substantial characteristic of the RES is that almost all the renewable energy sources do not release gaseous or liquid pollutants during operation. This is significantly advantageous over conventional systems.

Presently, renewables are now recognized as mainstream sources of energy [5]. The fast growth of RES in the power sector is motivated by numerous factors that include improving the cost-efficiency of renewable technologies and dedicated policy initiatives. Due to their modular nature, solar and wind energies are the most commonly used renewable energy sources [6, 7]. Hence, the research and technologies in the field of solar and wind have experienced commendable growth for the past two decades, especially the prospect of using them in the standalone application has been explored [6–8].

RES can be used in two different types of applications, grid-connected, and standalone.

13.1.1 Grid-connected systems

Grid-connected RES are the ones that are connected to the utility grid. When the power generated by the renewable sources is less than the load demand, the additional power can be drawn from the grid. Any excess generation for the renewable sources can be fed back to the grid. Grid-connected systems can be of two types, the first one mostly supplies the local electricity demand, and they feed any surplus generation into the grid. Any shortage in electricity generator from the RES due to unfavorable climatic conditions can be drawn from the grid. The other type of grid-connected systems is the utility-scale. In these systems, the decentralized stations are maintained by the utilities in a similar manner as the large-scale electrical power plants. The power produced by these types of grid-connected system is fed into the central grid without catering to the local load demand.

Some important characteristics of grid-connected systems can be listed as follows [9]:

- A grid-connected RES is an independent decentralized power generation system connected to the grid. This kind of system is ideal for applications that are close to the grid.
- The capacity of operation of these systems is determined by the supply source (Grid). The system would not operate unless the supply sources are available.
- Being connected to the grid, this kind of systems permits setting up relatively large-scale systems and enables them to operate at high plant load factors, which results in an enhancement in the economic viability of the operation.

- In these kinds of RES, the grid acts like a battery with an infinite storage capacity and can recompense the load variations due to climatic changes. This improves the overall efficiency of a grid-connected system over a standalone system.

13.1.2 Standalone RES systems

Standalone systems are the special application of RES, which works indecently of the utility grid. This kind of system is more suitable for remote locations where extension of the grid is expensive and meaningfully hard with no other power sources available. Often the standalone systems consist of PV panels are the most economical choice for powering remote areas like military bases, hill stations, and lighthouses [9]. However, the standalone systems have some disadvantages, including low capacity factor, the excess cost of the storage batteries, and limited storage capacity for the excess electricity, which causes the waste of the excess generation.

Some significant characteristics of standalone systems can be listed as follows.

- In this kind of RES, the operational capacity is considered to meet the load demand.
- Standalone systems prioritize the local demand.
- These systems are ideal for remote applications, where the system is designated to operate at low plant load factors.
- A typical standalone RES is classically consisting of renewable sources like solar PV, which is not consistently available throughout the year, limiting the operation of this kind of system to seasonal in most cases.
- These systems need storage to store any excess electricity produced during off-peak demand periods, which leads to an extra storage cost. In the absence of proper storage, excess generation needs to be thrown away.

Both grid-connected and standalone RES have their own strength and weaknesses. Hence, the selection of grid-connected or standalone RES is subject to several factors. However, both types of systems primarily function on renewable energy sources resulting in a substantial decrease in greenhouse emissions and climatic changes. The standalone RES is favored in applications where grid accessibility is a huge issue. They are also increasingly popular in supplying electricity to hill stations and remote villages.

13.2 Hybrid renewable energy systems

Both solar and wind applications have a significant drawback. That would be the unpredictability of their nature and their dependence on climatic changes. Which results in the mismatch of time distribution and load demand [10].

This limitation impacts the energy performance of the system and results in reduced battery life. The autonomous use of both types of renewable sources may result in significant oversizing, which ends up the system design being less

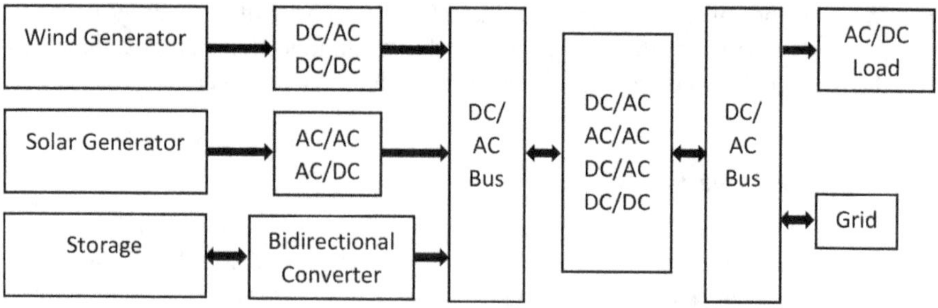

Figure 13.1 The basic structure of an HRES comprising of solar–wind generator

cost-effective. Hence, neither of the standalone PV nor the wind energy system can deliver a continuous power supply due to their dependency on climatic and periodical variations for standalone applications [11].

Hybrid renewable energy system (HRES) can be a combination of renewable and conventional energy sources. To overcome the issues caused by the variable nature of these resources, that can be overcome by incorporating more than one type of renewable energy source in proper combination along with storage, where the weakness of one source can be the strength of the other one. This incorporation of multiple sources results in improved system efficiency and reliability and reduces storage requirements compared to a system comprising of a single type of renewable sources [10]. The complementary characteristics of solar energy and wind energy for certain locations, the RES that comprises both solar and wind inputs equipped with storage can offer a reliable power generation.

HRES can operate as standalone or grid-connected applications. The basic structure of an HRES comprising of solar–wind sources plus storage can be represented as in Figure 13.17 [7]. The wind-solar generators plus the storage module are connected to an AC/DC bus via converters. As per network requirements, the AC/DC bus may be connected to another AC/DC bus via converters. In the case of the grid-connected systems, the AC/DC bus may be connected to the grid or else to load.

Due to the presence of more than one type of renewable source, the design of HRES becomes more complicated than single-source systems. The optimum design of the components of an HRES becomes complex due to numerous factors, including the inconsistent supply of the renewable energy sources, complex load demand, and non-linear characteristics of the system gears. Hence, multiple variables and parameters are needed to be considered for the optimum system design. One needs to remember the fact that the optimum configuration and optimum control strategy of the system are interdependent. This complication contributes to difficulties in designing and analyzing HRES [10].

Proper sizing of HRES is very important as an oversized or undersized system might lead to incapacity to supply the load demand and increased establishment cost. Hence, extensive research is being conducted to design the system components optimally and improve the performance, reliability, and accuracy of predicting the generated power of HRES to make them more acceptable [12].

The optimum sizing can ensure minimum investment with adequate and full use of the solar/wind sources and storage devices.

13.3 HRES comprising of solar/wind and storage system components

A typical standalone HRES comprising of solar–wind system consists of PV arrays, wind turbines, storage devices like battery and fuel cells, inverter, controller, and other accessory devices and cables. The solar and wind generators work as sources to cater to the load demand.

The design of HRES comprising of solar–wind system is primarily dependent on the performance of its individual components. Hence, it is important to model the individual components correctly to predict the system performance. Their combination can be assessed further to meet the required reliability. If the prediction of the generated power from these individual components is accurate enough, the successive combination will supply the load demand at the least cost [10].

13.3.1 Solar energy

Sun is the primary source of renewable energy. As a result of the thermonuclear reaction of Hydrogen and Helium atoms, enormous energy is produced by the Sun. This incident energy of the Sun can be harvested efficiently from the solar photovoltaics (SPV) or concentrated solar power (CSP) systems [13].

Harvesting solar energy using PV has increased considerably over the years. Globally, a 33% (402 GW) increase in the PV installations for generating solar energy has been observed. This indicates that equivalent to 40 000 PV panels are being installed every hour, which is supposedly greater than the combined capacity of fossil fuels and nuclear power installation every hour. In other words, it can be stated that more solar PV was installed than the net capacity additions of fossil fuels and nuclear power combined every hour [14].

Energy harnessed from CSP is increasing significantly and has reached 4.9 GW in 2017. The reduction in the cost and improvement of technology driven by the competition has contributed to this significant growth. Spain and the USA have substantial contributions to this technology. Presently, roughly 2 GW of new plants are planned to be established, including the 300 MW and 350 MW of CSP plants by China and Morocco, respectively.

To select the accurate components for an HRES and precise forecast of its energy production, it is important to analyze their performance under different operating conditions. The performance of a PV module depends on the material of the PV module, the temperature of the module, and the solar radiance on the surface of the PV module [10]. Numerous researches have been reported in the literature where researchers have analyzed the environmental factors that influence the PV module/array's performance [15–22].

13.3.2 Wind energy

A massive potential source of renewable energy is wind. The wind is caused by complex mechanisms that involve the Earth's rotation, Sun's heat capacity, polar ice caps, the cooling effect of the oceans, the temperature gradient between land and sea, and the impact of the physical presence of mountains and other obstacles present on Earth.

Wind energy is harvested by converting the wind currents into electrical energy using the wind turbines. Turbines have the mechanism to extract energy from the wind currents by converting the kinetic energy into electrical energy via rotational movement of a rotor attached to a generator. The efficiency of conversion can be measured by the energy density, as a capacity factor. Wind energy can be used in both grid-connected and standalone applications.

Wind energy is a rapidly expanding renewable energy source in many countries and is expected to grow rapidly over the period. Presently, the production of wind energy is concerted mainly in Europe, the USA, China, and India.

Wind energy generators can be installed and transmitted at a reasonably quick pace. They can be installed in distant, mountainous, and inaccessible areas. Electricity produced from wind generators is non-depletable, and their cost does not increase. Wind energy generators can help to save much fossil fuel used for the same purpose, and a huge amount of greenhouse emissions can be avoided.

In recent years, wind energy contribution to global RE generation has been impressive both offshore and onshore. Wind generation capacity is estimated to reach 839 GW by 2023, where an increase of 10% is expected in the offshore wind [23].

Compared to 2016, the addition of wind energy is reduced by 4%; however, 52 GW of wind power was still added. Total wind energy generation by the end of 2017 was around 539 GW.

Even though numerous countries like UK, Uruguay, Costa Rica, and Nicaragua are able to meet at least 10% of their electricity demand by generating wind energy in 2017. A decline has been observed in China, which is considered as the global leader in wind energy generation. A decline has also been observed in 2018 in the global onshore and offshore wind additions primarily due to lower contribution by China and the USA [23]. This global decline is caused by imminent regulatory changes despite the existing wind energy utilization benefits.

13.3.3 Energy storage systems used in standalone HRES

The incorporation of different renewable energy raises some technical and economic problems. These types of energies are not easily storable or transportable like the energies generated by conventional sources. Additionally, their supply can be enormously uneven and variable. Hence, the incorporation of storage devices is required to recompense the irregularities in the solar and wind power distributions.

Energy storage systems improve the power quality and energy management functions of the HRES, especially in standalone mode. Their presence also improves the efficiency of the HRES. Different energy storage systems used in HRES are discussed in the following sections.

13.3.3.1 Batteries

Batteries are the most commonly used storage systems in standalone HRES. Lithium-ion (Li-ion) batteries are presently considered one of the foremost storage solutions for HRES because they have higher energy density compared to other batteries, that is, Ni-Cd, Ni-MH, and lead-acid. Apart from high energy density, they also have a long cycle life, and their cost is reasonable compared to the other battery systems [24, 25].

13.3.3.2 Fuel cells

Fuel cells (FC) are the indirect energy storage systems used in HRES. FC are electrochemical energy devices with similar constructional features to the battery storage systems; however, their operation principle is different. The energy produced by the FC is a result of the fuel consumptions (hydrogen, ethanol, methanol, and hydrocarbons). However, the principle of working of FC can be compared to the working principle of a battery. In FC, hydrogen or fuel stored in an external system is unconfined in accordance with the requirement for electricity production. The electrochemical reaction that takes place within the reactants also produces water and heat. During this reaction process, the reactants flow in, and the products of this reaction flow out, leaving the electrolytes in the FC.

Different varieties of FC are available depending upon the types of electrolyte and the fuel of the cell; some of them are proton exchange membrane FC known as PEMFC, direct methanol known as DMFC, direct ethanol known as DEFC, phosphoric acid known as PAFC, molten carbonate known as MCFC, alkaline fuel cell known as AFC, and solid oxide known as SOFC.

13.4 Optimization techniques for the components of HRES

To utilize the HRES most efficiently and economically, the optimum sizing of the components is crucial. This is important to ensure that the maximum utilization of the components with minimum investment. This is also important to ensure that the HRES can work optimally in terms of investment and power reliability constraints. An optimum design of the components of HRES is the best configuration compromising between power reliability and system cost. Hence, while selecting an optimum configuration of the components for an HRES, one must evaluate the power reliability and system life-cycle cost. The implicit criteria or methodologies are generally used to optimally size the components of HRES are presented in Figure 13.2 [26].

13.4.1 Criteria for hybrid solar–wind system optimization

A. Power reliability analysis

The power generation from the HRES is highly affected by solar radiation and wind speed. Power reliability analysis is a crucial step for optimum system design due

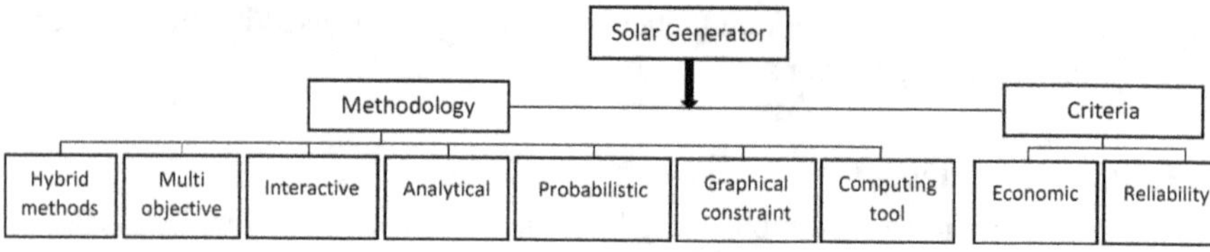

Figure 13.2 Optimum sizing criteria/methodology of solar wind HRES

to the irregular features of these parameters. Numerous methodologies have been reported in the literature to compute the reliability of the HRES. Loss of power supply probability (LPSP) is the most common one among them.

The LPSP measures the probability of deficiency in power supply as a result of insufficient generation from the hybrid system (PV, wind power, and energy storage), which is not enough to cater to the load demand [27, 28]. LPSP can be presented as in (13.1).

$$\mathrm{LPSP} = \frac{\sum_{t=1}^{T} DE(t)}{\sum_{t=1}^{T} P_{load}(t)\,\Delta t} \tag{13.1}$$

where $DE(t)$ is the deficit energy at hour t and $P_{load}(t)$ is the load demand.

Two methods have been reported in the literature to apply LPSP to design a standalone HRES comprising solar/wind generators. The first method is found on chronological simulations. This is complicated to calculate and needs the obtainability of data spanning for a definite time period. The other method is based on probabilistic methods that include the changeable nature of the power generation and the load demand, avoiding the requirement for time-series data [29].

In literature, some other power reliability criteria have also been reported, including loss of load probability (LOLP), deficiency of power supply probability (DPSP), system performance level (SPL), and loss of load hours (LOLH). The probability of a system demand exceeding system's power supply capacity in each period is measured by the LOLP. This is expressed in terms of the estimated number of days over an extensive period [10]. The probability that the load cannot be fulfilled is defined by SPL. Research also investigated the methods founded on the LOLR (loss of load risk) to determine the optimum combination of solar and wind energy in a hybrid system [30]. LCE can be expressed as in (13.2) [31]

$$\mathrm{LCE} = \frac{TAC}{E_{tot}} \tag{13.2}$$

The total annualized cost is represented by TAC, and total yearly energy production is E_{tot}. One can calculate TAC considering the present costs. The present value of costs can include the initial setup charge, the present value of maintenance expenses, the present value of replacement charges, and the capital recovery factor (CRF) of the HRES where CRF can be presented as [32]

$$\mathrm{CRF} = \frac{d(1+d)^{t}}{(1+d)^{t}-1} \tag{13.3}$$

where the discount rate is presented by d and the useful lifetime of the RES is presented by t. Research also has been conducted on other economic approaches, such as the Levelized Cost of System [11] and life-cycle cost [33].

B. System cost analysis

Levelized cost of energy (LCE), net present cost (NPC), and life-cycle cost are a few of the several economic criteria available in the literature. The total present value of a time series of cash flows is represented by NPC. The setup cost of the system components, the replacement cost of any component, and the maintenance cost are included in NPC [34–36]. The salvage costs (if there are any) are added to the NPC. Hence, salvage costs remain as a component of the system [10]. A well-known optimization software called HOMER (Hybrid Optimization Model for Electric Renewable) utilizes the total NPC to optimally size the renewable energy sources.

The ratio of the annual cost of the system to the annual electricity supplied by the system is represented by the LCE defined as in [29, 31]. This has been widely explored by academics to assess the hybrid PV/wind system configurations [37]. The following equation represents the LCE

$$\text{LCE} = \frac{TAC}{E_{tot}} \tag{13.4}$$

where the total annualized cost is presented by TAC and the total annual energy produced is E_{tot}. To calculate TAC, the present value of costs and the capital recovery factor (CRF) of HRES are considered. The present value of the costs is the value on a given date of a forthcoming expense or series of forthcoming payments, reduced to represent the time value of money for an HRES. CRF can be presented as [32]

$$\text{CRF} = \frac{d(1+d)^t}{(1+d)^t-1} \tag{13.5}$$

13.4.2 Methodologies for optimum sizing for the components of HRES

Some methods for optimum sizing for the components of HRES comprising of a solar–wind system are discussed below.

A. Probabilistic methods

One simple methodology for sizing is probabilistic methods. However, the solutions obtained using this method might not be the most appropriate ones. Usually, one or two system performance indicators are considered in this technique to optimally size the components of the considered HRES. The effects of the solar radiation and wind speed variabilities are considered for optimum sizing of the system components of the HRES in the probabilistic approach.

Bucciarelli [38] presented a procedure to size the components of HRES considering storage energy variation as a random walk. Some studies estimated the probability density for daily increment or decrement of storage level by a two-event probability distribution [39]. This method was further investigated and extended by Bucciarelli [40] considering the consequence of the correlation between day-to-day radiation values.

The method proposed by Bucciarelli [41] is further modified by Gordon. In their work, they approximated the storage energy transitions by the three-event probabilistic method, which overcomes the restrictions of the traditional two-event method in matching the actual distribution of the energy produced by HRES.

B. Analytical methods

The HRES is presented as a mathematical model, which is a function of its feasibility in analytical methods [31]. Hence, the performance of the system is measurable for a set of probable system configurations and/or the size of system components. Numerous performance indexes are analyzed to determine the optimum configuration of an HRES. This method allows the analysis of the performance of numerous HRES configurations. However, this technique needs data for long time series, typically one year of weather conditions data (solar and wind) for the calculations. The performance assessment of HRES can be performed using mathematical models. Numerous computer tools are available for evaluating the performance of HRES that helps the designers to investigate the incorporation of renewable sources.

Connolly *et al.* [42] analyzed various computational tools for simulation of HRES and compared their performances. In their review, they reported a widely accepted analytical tool for assessing the performance of the HRES named HOMER, which was developed by the National Renewable Energy Laboratory (NREL), USA. Researchers have widely used HOMER for optimal sizing of the components of the HRES [43–46].

C. Iterative methods

The iterative method uses a recursive procedure for performance assessment of HRES that ends when the optimum configuration is achieved as per design conditions [31].

Yang *et al.* [29] proposed an optimization model for an HRES comprising of wind and PV generators utilizing the iterative optimization techniques LPSP model and LCE model to analyze the power reliability and system cost. The analysis studies three sizing specifications, the size of the PV system, rated power of wind generator, and the battery bank size. The optimum sizing of the components can be obtained by iteratively searching all the possible system combinations to attain the lowest LCE for a desired LPSP value.

Ashok [47] designed an optimal HRES among different renewable energy configurations for a rural community. In their study, they minimized the total life cycle cost, confirming the reliability of the system. A numerical algorithm based on the Quasi-Newton method has been used in this study to calculate optimum system configuration.

Interactive artificial intelligence (AI) techniques are extensively used to optimally size the components of HRES to maximizing their commercial benefits [10].

Some of the AI techniques used for these purposes are Genetic Algorithms (GA), fuzzy logic, and artificial neural networks (ANN). AI techniques enable a machine or object to function like human thought [48].

GA is a stochastic global search and optimization procedure based on the natural evolution of species. GA can find global optimum solutions in the multi-modal and multi-optimization procedure [31]. GA is extremely appropriate to cases of non-linear systems, where it is difficult to find the global optimum. GA is not constrained by local optimum due to the probabilistic development of solutions. GA can find the global optimum system combination with simple calculations compared to other traditional optimization techniques such as dynamic programming and gradient methods.

Koutroulis *et al.* [49] suggested an optimum design methodology for HRES using GA. In their work, the optimum number and variety of units were selected among the list of system devices available commercially. This ensures that the total system cost for 20-year round is minimized bound by the constraint that the load supplies are completely met. Yang *et al.* designed a pilot HRES comprising of solar–wind power generation project using GA. The project aimed to supply power for a telecommunication relay station using RES on a remote island of China [11].

ANN is a well-known computational model that is based on biological neural networks. ANN consists of an interconnected group of artificial neurons that enables the model to process information using a connectionist method of computation [10]. Kalogirou [49] suggested an optimization model of PV systems using ANN and GA that uses a computer program called TRNSYS and considers the climatic conditions of Cyprus. A Typical Meteorological Year (TMY) file is used in this study. A limited number of TRNSYS simulations were used to train the ANN. The next step uses GA to estimate the optimum system configurations to maximize the life-cycle of the system, ensuring a meaningfully reduced design time.

Particle swarm optimization (PSO) is another interactive optimization method that has been reported in the literature [49]. PSO, which is inspired by the social behavior of bird flocking or fish schooling, is a stochastic optimization practice based on population. In this technique, bird flocking or fish schooling is the collective motion of multiple self-propelled objects [49]. In their work, Dehgan *et al.* [50] used PSO to size a hydrogen-based wind/PV plant for reliability indices. A modification for the PSO algorithm was suggested by Wang *et al.* [51] for developing the multicultural design of an integrated power generation system. In their work, a sensitivity study was conducted to examine the effects of various system parameters on the design performance.

D. Graphic construction technique

The graphical construction method is designed for PV generators and wind turbines based on the satisfaction of the average value of load demand by average values of production from PV and wind power generators. This method considers only two decision variables in the optimization method. Usually, this method considers PV-wind or PV-battery combination [52, 53].

The optimum sizing of the sources and battery for a standalone HRES is designed using long-term climatological data such as hourly solar radiation and wind speed documented for very long years in this method [54]. Some investigators have used the monthly average of the meteorological data [55]. For a specified load requirement and a specific LPSP, the optimum sizing of the sources of HRES comprising of PV and wind generators can be achieved if the total cost of the system is linearly connected to a number of PV modules and batteries [56]. At the point of tangency, the curve representing the relationship between the number of PV modules and the number of batteries will be the minimum cost of the system.

A seasonal study is done for the demand changes and resource availability for the generators during the winter and summer months. A sizing curve is established based on this between the various available sizes of wind generators and PV modules. A more refined curve can be obtained when a larger amount of data are accessible. Long-term data of solar radiation and wind speed that were recorded for every hour of the day for 30 years were used by some researchers [43]. In that work, the load consumption data of a typical household located in Massachusetts was fed as the load demand for the design of the HRES. In this work, for a given load demand and a chosen LPSP, based on the minimum cost of the system, the optimum configuration or number of batteries and PV modules was calculated.

In their research [19] for a given DPSP, the researchers presented the number of PV panels as a non-linear function of the number of wind turbines. The cost function given below was used to obtain a PV/wind turbine configuration to minimize the system cost:

$$C = \alpha \cdot N_{PV} + \beta \cdot N_{wind} + C_0 \tag{13.6}$$

C – Represents the capital cost of the HRES
α – Represents the cost of a PV module ($/kW)
β – Represents the cost of a wind turbine ($/kW)
N_{PV} – Capacity of PV generator unit (kW),
N_{wind} – Capacity of a wind generator (kW)
C_0 – Represents the total constant costs, including the cost of design, installation, and battery bank cost.

An optimum system combination is achieved when

$$\frac{\delta N_{PV}}{\delta N_{wind}} = -\frac{\beta}{\alpha} \tag{13.7}$$

The solution was illustrated using the graph presented in Figure 13.3. In this graph, point *S* denotes the optimum system configuration. The inclination of the line is equal to $(-\frac{\beta}{\alpha})$ [19].

E. Multi-objective design

Before carrying out any design in the engineering field, the designer considers multiple objectives simultaneously. In some cases, there might be a conflict between the objective [57]. To optimally design the components of HRES, a designer needs to consider at least two objectives minimization of the system costs and emission

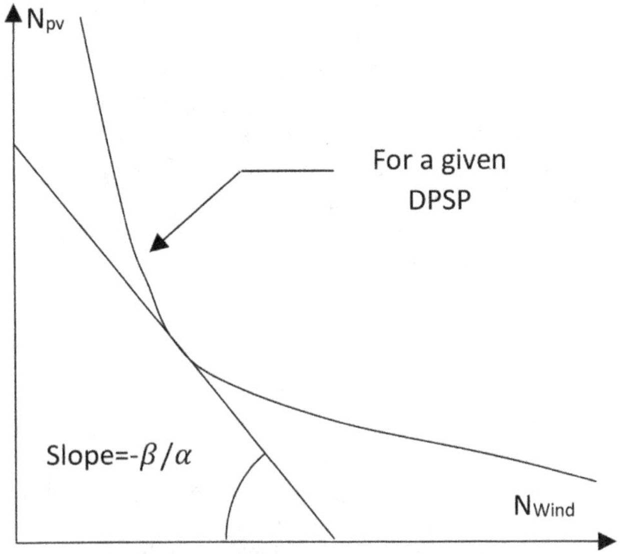

Figure 13.3 *Plot of number of PV modules versus number of batteries for a given DPSP [19]*

of pollutants [10]. However, a reduction of the cost of design concludes an increase in pollutant emissions and vice versa. Hence, these two design objectives end up creating conflicts, which makes the designing with multiple objectives a complicated task. Due to the involvement of a large number of variables that need to be considered and the application of mathematical models, the design process becomes extremely complicated. Conventional optimization methods require higher computational time, while sometimes they are not competent in considering all the characteristics related to the design problem. Several other methodologies have been reported in the literature regarding designing these systems [35, 58, 59]. Mostly, those researches concentrate on searching the configuration and control mechanism that yields the lowest total cost of the system throughout the system lifetime. However, one should also consider the issues related to the environment while designing this type of system.

Most design methods calculate the optimal pollutant emissions while designing the system with minimum costs. Some design methods use HOMER, where it is possible to consider the emission of pollutants by economic evolution and considering them as a part of the cost's objective function. Costs to emissions mapping used in this method decisively affects the results of the design. The technique used by HOMER for the multi-objective designing is recognized as the method of the weights [11]. Another popular multi-objective design task called Multi-Objective Evolutionary Algorithms (MOEAs) has been explored by numerous researchers. Pelet et al. utilized MOEAs for optimizing the system cost and CO_2 emissions for a standalone HRES designed to supply the load demand (thermal and electrical) of three hotels and a town in the Tunisian Sahara [11].

MOEAs were used by Bernal-Agustín *et al.* [36], who presented a multi-objective optimization method considering NPC versus CO_2 emissions for an HRES comprising of solar/wind/diesel with battery storage using MOEAs. Dufo-López and Bernal-Agustín [36] presented a methodology considering triple multi-objective optimization to minimize the total cost during the life cycle of the installation, emission of pollutant (CO_2), and unmet load demand simultaneously. To find the optimum combination of system components and control strategies for the HRES, they used MOEAs and GA together.

F. Hybrid methods

The methodology where two or more optimization methods are combined to optimally size the components of the HRES is called a hybrid method. This technique can be considered the best to solve multi-objective design problems. Two common approaches have been reported in the literature for multi-objective problems. Among them, the first one merges the individual objective functions into a single compound function. The second technique includes the determination of an entire Pareto optimal solution set [60].

Arnette and Zobel discussed a multi-objective linear programming (MOLP) model for optimizing the components of the renewable energy sources and present fossil fuel facilities on a regional basis [61]. A technique for optimal sizing the wind generators, PV panels, and batteries to build an HRES that are optimal in terms of numerous criteria, including cost, reliability, and emissions, has been reported in the literature [51]. A triple multi-objective design technique has been suggested by Dufo-Lopez *et al.* for isolated hybrid systems by minimizing, simultaneously, the total cost throughout the lifetime of the installation, pollutant emissions, and unmet load [62].

G. Computational tools

To evaluate the performance of the HRES, researchers have used various simulation programs/software. Use of the computer simulations and software packages assists in determining the optimum sizes of the components of an HRES by comparing the performance and energy production cost for various system combinations. Among several software tools available for optimization purposes, HOMER, HYBRID2, iHOGA, HYBRIDS, and RETSCREEN are the most popular ones.

HOMER (The Hybrid Optimization Model for Electric Renewables) is the most popular tool for optimizing the system components of HRES designed by the National Renewable Energy Laboratory. HOMER is the most used software tool for optimizing the HRES, both grid-connected and standalone systems [63]. This software can perform multiple analyses at once and help to address a wide range of design queries. HOMER is used for designing the most cost-effective systems, optimum sizing the components of HRES, and economic analysis. It can also be used for the sensitivity analysis of the system under changing conditions like load variation and battery price.

HOMER calculates the system combinations with minimum cost to meet the specific load demand. This is mainly done by three principal steps, including simulation, optimization, and sensitivity analysis.

I. Simulation

HOMER simulates the system operation by calculating the energy balance for each of the 8 760 hours in a year. HOMER can compare both the electric and the thermal load of every hour to the energy that the system can generate in that particular hour. HOMER is also capable of deciding for each hour whether to charge or discharge the batteries for systems. To ensure that the designed system can meet the loads for the entire year, HOMER evaluates the life-cycle cost of the system, accounting for the capital, replacement, operation and maintenance, and interest costs.

II. Optimization

After investigating all feasible system configurations and shorting them by life-cycle cost, HOMER displays the list of the solution configurations. From the presentation, it is easy to select the system with the lowest cost, which would be at the top of the list. One can further scan the list for other feasible solutions.

III. Sensitivity analysis

HOMER is helpful to study the effect of the changes in input conditions on system design. It can successfully conduct sensitivity analysis on all inputs assigning multiple values to the input that needs to be investigated. The optimization process is repeated for each value of the input to examine the effect. So, it gives the user the flexibility to specify the sensitivity variables of their choice HOMER helps them to analyze the results using its graphic capabilities.

HOMER uses hourly simulations to reach the optimum target. The time step simulator uses the hourly load and environmental data inputs for assessing the renewable energy system assessment. The optimum renewable energy system configuration is achieved based on net present cost (NPC) for a specified set of constraint and sensitivity variables.

Researchers have explored HOMER extensively in various RES studies [39, 64–66] and renewable energy system validation tests [62]. HOMER is very straightforward to operate. Thought the simulation process could take time depending upon the number of sensitivity variables used in the process. However, the drawback of this program is that this program does not allow the user to select the appropriate components for the system intuitively. The algorithm and calculations are also not accessible by the user.

Another software used to optimally size the components of the HRES is called HYBRID2, developed by the Renewable Energy Research Laboratory (RERL) of the University of Massachusetts. This software has a very precise simulation process that can define the time interval starting from 10 min to 1 hour. As recommended by the National Renewable Energy Laboratory, the system, once optimized by HOMER, can be further improved using this software [10].

The software for simulation and optimization of Hybrid Standalone Systems of Electric Power Generation based on Renewable Energies, called iHOGA (improved Hybrid Optimization by Genetic Algorithms), was developed by researchers of the University of Zaragoza (Spain) [27].

The software is capable of modeling HRES with AC/DC load and/or Hydrogen and also the consumption of water from a tank or reservoir previously pumped. This software can optimize the sizes of the components of an HRES, including PV generators, wind turbines, hydroelectric turbine, auxiliary generators like diesel or gasoline generators, inverter or inverter-charger and batteries, charger, charge controllers for batteries, as well as hydrogen components like electrolyzer, hydrogen tank, and fuel cell. It can be utilized to optimally size both standalone and grid-connected systems. This software can provide a detailed model of the components and estimate the operation of the designed system precisely. iHOGA is based on GAs, an advanced optimization methodology that enables it to design the optimum system using very low computational times.

Another well-known software called HYBRIDS, created by Solaris Homes, can assess the technical, economic potential of RES for a given configuration and determine the renewable fraction potential. This software also can evaluate the economic viability of the designed system based on NPC. HYBRIDS is a design tool based on a Microsoft Excel spreadsheet that involves the daily-average load and environmental data projected for each month. HYBRIDS cannot simulate more than one system configuration at a time, and it is unable to provide an optimized configuration like HOMER or iHOGA. This software actually demands more details and a higher level of understanding of RES configurations compared to HOMER.

Another Microsoft Excel-based software tool called RETScreen is used to evaluate the energy efficiency and techno-economic viability of the RES. This tool can analyze the energy efficiency of designed HRES considering numerous factors, that is, life-cycle costs, energy production, and greenhouse gas emission reduction [28]

Few more computer tools have been reported in the literature that can design the system components of HRES [42]. Few of them are GAMS (General Algebraic Modelling System) [28], LINDO [33, 67], Opt Quest [67], WDILOG2 [68], Sim Pho Sys (Simulation of Photovoltaic Energy Systems), GRHYSO (Grid-connected Renewable Hybrid Systems Optimization) [69, 70], and H2RES [71].

13.5 Case study

Optimum standalone hybrid renewable energy system design using HOMER for a small community of Portland, Victoria

(P. Ganguly, A. Kalam and A. Zayegh, "Optimum standalone HRES design using HOMER for a small community of Portland, Victoria,"*2017 Australasian Universities Power Engineering Conference (AUPEC)*, Melbourne, VIC, 2017, pp. 1–6, doi: 10.1109/AUPEC.2017.8282486.)

Community energy (CE) is a substantial use of standalone HRES. Community owned and financed renewable energy projects are widespread in European countries and speedily emerging in the USA. However, it is a comparatively new concept in Australia. There are many significant benefits of CE. It maximizes local ownership and participation, creates domestic jobs, and helps to use natural resources efficiently and sustainably. A properly designed CE can meet the energy needs of the community and allow its members to contribute to a cleaner environment.

Numerous factors need to be investigated to design a community-owned standalone HRES. The two most important factors that need to be considered are the reliability of the power supply and the cost of the system. The expected lifetime of these systems is around 25 years. Hence, to design these systems, one needs to consider the expected increase in load demand. This is very crucial to ensure that the designed system would be able to work efficiently for its entire life span.

Based on the available resources, one needs to investigate the most suited renewable energy technology for the targeted location. They also need to consider certain other factors like the total system cost, the excess generated energy, the number and capacity of these sources, storage cost, unmet load, and different load conditions. The optimum sizing of the system components of the HRES for the targeted community is very important to guarantee that the system work can work optimally under various load conditions in terms of power reliability conditions and financial investment.

This work explains a method to optimum sizing of the components of a standalone HRES for a small community of Portland, Victoria, Australia, using the software HOMER. The work aims to find the optimum system configuration for the standalone HRES by minimizing the system cost and fulfilling the load demand of the community with minimum interruption all around the year. The work includes the sensitivity analysis for different values of solar irradiation and hourly wind speed of Portland, Victoria. It takes into consideration the scaled annual average load and a reasonable annual capacity shortage for the targeted community.

13.5.1 Proposed system

The projected standalone HRES is designed to meet the load demand of a small community located in Portland, Australia (38°20′0″ S, 141°36′0″ E). The HRES comprises wind turbines, PV panels as main sources for supplying the load demand of the community and also includes battery bank and converter. The battery bank is

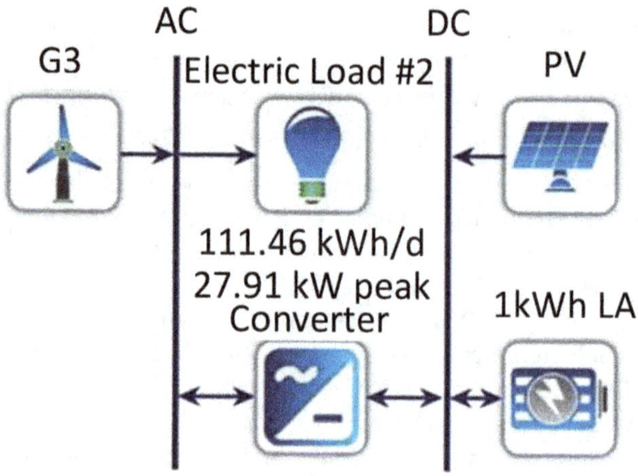

Figure 13.4 *Proposed system configuration in HOMER*

included as the energy storage system to cater to the load demand when there is an electricity production deficiency. It also stores the excess power when the generated power exceeds the load demand.

The planned system configuration is presented in Figure 13.4.

13.5.2 Electrical load

The yearly load demand of the targeted community located at Portland Victoria that consists of 10 households is illustrated in Figure 6.2. With a peak demand of 23.93 kW, the energy demand of the community is 111.46 kWh/day. The load profile of the targeted community is presented in Figure 13.5.

13.5.3 Solar and wind resources

HOMER uses the solar radiation data and wind profile for the targeted location (Portland Victoria (38°20′0″S 141°36′0″E) from the NASA Surface Meteorology

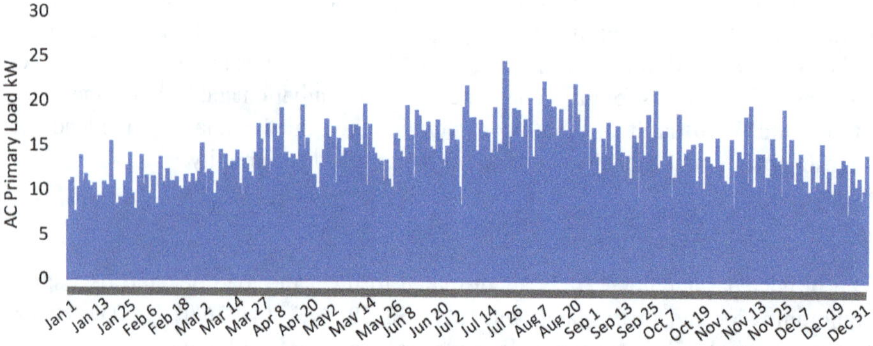

Figure 13.5 *Load profile (yearly) of the targeted community*

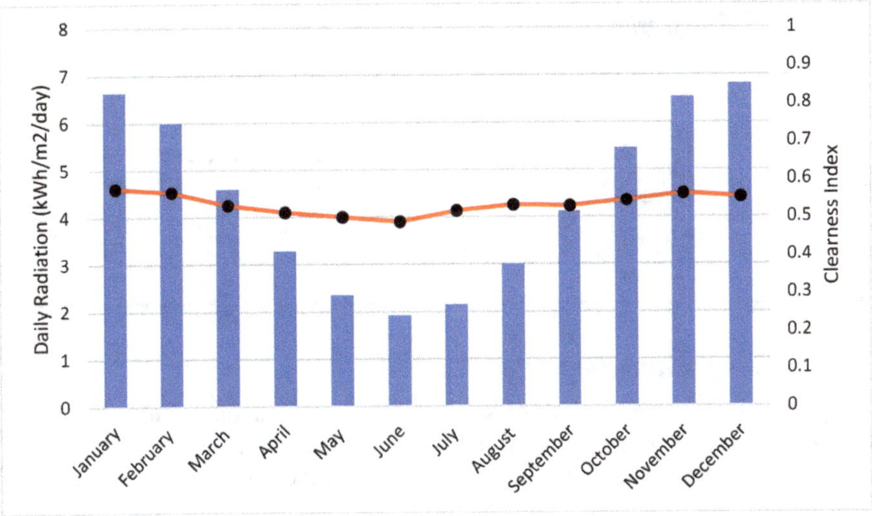

Figure 13.6 *Solar radiation profile of targeted location over a 1-year period (kWh/m²/day)*

and Solar Energy website [72]. The average yearly solar radiation for Portland is 4.43 kWh/m²/day, and the yearly average wind speed of 6.71m/s as per the NASA website. Figure 13.6 shows the solar radiation profile over one year, and Figure 13.7 shows the annual wind profile of the targeted location (as per the NASA website).

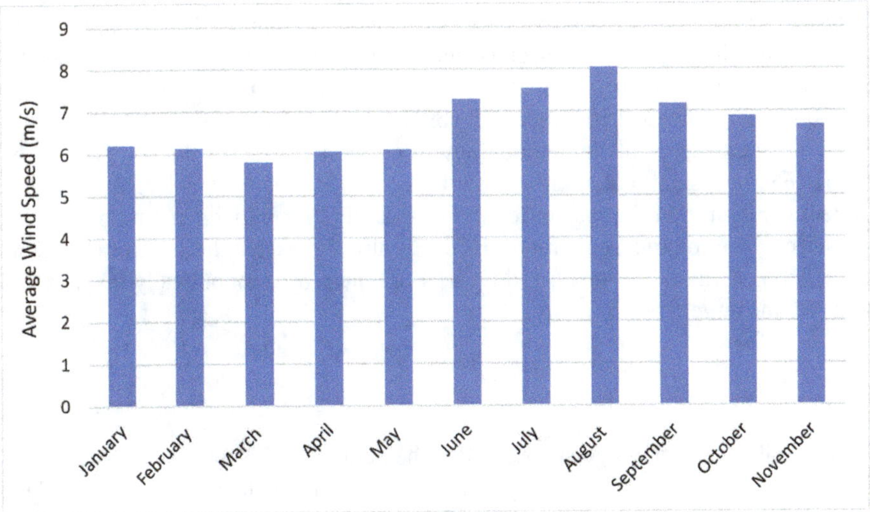

Figure 13.7 *Wind profile of targeted location over a 1-year period (m/hr)*

Table 13.1 The sensitivity variables used and their values

Capacity shortage (%)	Electric load scaled average (kWh/d)	Solar scaled average (kWh/m^2/day)	Wind scaled average (m/s)
0.1	111.457056	4.426666	6.710333
0.2	120	4	5
	130	5	5.8

13.5.4 Results and discussions

The present work considers four sensitivity variables to find out the optimum set of resources under different conditions presented in Table 13.1. HOMER Optimiser™ finds the optimum system configuration utilizing these four sensitivity variables. Two sensitivity test uses two different values of solar and wind average obtained from NASA and BOM. This also considers the increase in yearly average load demand. By running the optimization test for all the different combinations of the sensitivity variables, HOMER obtains the optimum configurations of the system. The results gained from the HOMER optimization test are presented in Table 13.2, where the most expensive ones are the top five rows, and the bottom five rows represent the least expensive systems as per their NPC.

Considering the constraints, the system that was selected as the optimum one for the target community is presented in Table 13.3. The selected system has an NPC of $697 042. The cost of energy (COE) is $1.14. This is higher than the regular grid electricity cost (approximately 26.4740 cents/unit for residential customers [73].

Figure 13.8 shows the cost summary of the system, as per the components. Figure 13.9 represents the possible excess energy production by the system and the capacity shortage for the specified meteorological conditions. From the results obtained, it can be clearly seen that the selected system can perform optimally under measured climatological conditions reliably in supplying load and NPC.

This work represents a methodology of optimum sizing of the components of the HRES for a small community. HOMER could successfully select an optimum system configuration for system reliability and NPC for a given community under the selected metrological conditions. HOMER also shows that the systems consist of PV, wind generators, batteries, and inverter are the most economically viable solutions for the targeted community.

13.6 Conclusion

An incredible growth has been observed in the field of HRES for producing electricity over the last two decades. With the introduction of new technologies in the field of HRES that incorporates solar wind and storage, many challenging problems have emerged, inspiring many research projects. Henceforth, a massive advancement of

Table 13.2 The optimum system configurations obtained by HOMER optimiser™

Capacity shortage (%)	**Electrical load scaled average (kWh/d)**	**Scaled solar AVerage (kWh/ m^2/day)**	**Wind scaled average (m/s)**	**PV (kW)**	**G3**	**1 kWh LA**	**Converter (kW)**	**NPC**
0.1	130	4	5	101	5	566	25	18 578
0.1	130	4.426667	5	64.23	10	537	25	19 274
0.2	130	4	5	87.25	6	539	25	18 064
0.2	120	4	5	104.8	1	623	25	18 700
0.1	120	4.426667	5	104.8	1	605	25	18 201
0.1	111.4571	4.426667	6.710833	44.32	6	320	25	11 564
0.2	120	5	6.710833	43.97	5	338	25	11 686
0.2	111.4571	5	6.710833	61.42	3	319	25	10 586
0.2	111.4571	4.426667	6.710833	50.56	4	332	25	11 212
0.1	111.4571	5	6.710833	45.2	4	348	25	11 602

Table 13.3 Selected system configuration

Capacity shortage (%)	Electrical load scaled average (kWh/d)	Scaled solar average (kWh/m²/day)	Wind scaled average (m/s)	PV (kW)	G3	1 kWh LA	COE ($)	NPC ($)	Operating cost ($)
0.1	130	4	5.8	71.5	8	441	1.14	697 042	15 938

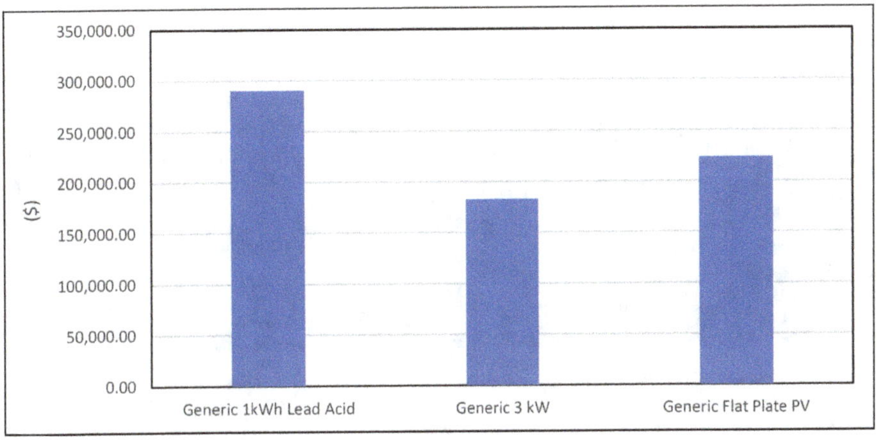

Component	Capital ($)	Replacement ($)	O&M ($)	Fuel ($)	Salvage ($)	Total ($)
Generic 1kWh Lead Acid	132,300	116,878.80	57,010.33	0.00	-15,846.75	290,342.39
Generic 3 kW	144,000	45,908.26	18,615.62	0.00	-25,872.25	182,651.63
Generic Flat Plate PV	214,405	0.00	9,239.12	0.00	0.00	223,644.99
System Converter	300	127.28	0.00	0.00	-23.96	403.33
System	491,005	162,914.34	84,865.09	0.00	-41,742.96	697,042.35

Figure 13.8 The cost summary of the optimum system

technologies has been observed in HRES, and there is a huge prospect of further development.

This chapter briefs the progress of some of these technologies in the field of optimal sizing of the HRES. The most frequently used technologies for optimal sizing of the components of the HRES that evaluate the system power reliability and system cost are explored and reviewed in this chapter. Multiple optimization techniques have been studied and presented, including the probabilistic approach, graphic construction methods, multi-objective design, and iterative technique.

The literature review presented in this study clearly shows that HRES are widely considered as an alternative to conventional energy sources. They are getting increasingly popular in the field of power generation. Most of the systems that have been reviewed in this chapter are comprising of more than one renewable and/or conventional sources like the conventional grid or diesel generators. It has been observed that the hybrid systems that incorporate alternative sources perform better. Hence, there is a present trend in developing multi-objective sizing technologies that can make the HRES reliable, practicable, and environmentally friendly.

This study also indicates that a detailed analysis of renewable energy resources is a crucial first step of the optimum sizing process of an HRES comprising of solar

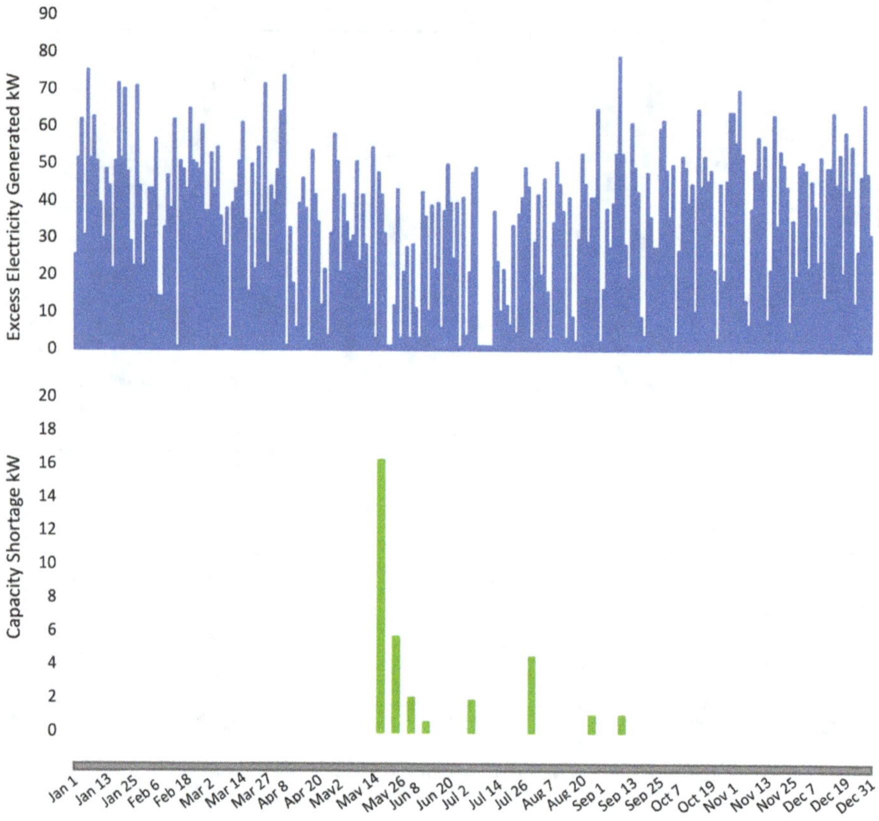

Figure 13.9 *Excess electrical production is presented in the top figure, and the capacity shortage is presented in the bottom figure*

wind and storage. To reduce the operational and establishment cost of the HRES, optimum resource distribution is important.

Renewable energy is an emerging field of research. It can be anticipated that in the coming days more new techniques and algorithms will be developed that might incorporate AI techniques to increase the efficiency of these systems in terms of reliability and cost.

References

[1] Rehman S., El-Amin I.M., Ahmad F., *et al.* 'Feasibility study of hybrid retrofits to an isolated off-grid diesel power plant'. *Renewable and Sustainable Energy Reviews*. 2007;**11**(4):635–53.

[2] Erdinc O., Uzunoglu M. 'Optimum design of hybrid renewable energy systems: overview of different approaches'. *Renewable and Sustainable Energy Reviews*. 2012;**16**(3):1412–25.

[3] Zhou W., Yang H., Fang Z. 'Battery behavior prediction and battery working states analysis of a hybrid solar–wind power generation system'. *Renewable Energy*. 2008;**33**(6):1413–23.

[4] Hepbasli A. 'A key review on exergetic analysis and assessment of renewable energy resources for a sustainable future'. *Renewable and Sustainable Energy Reviews*. 2008;**12**(3):593–661.

[5] 2017. Available from www. worldenergyoutlook.org/weo2015/.

[6] Elhadidy M.A., Shaahid S.M. 'Parametric study of hybrid (wind + solar + diesel) power generating systems'. *Renewable Energy*. 2000;**21**(2):129–39. Elsevier BV.

[7] Khare V., Nema S., Baredar P., *et al*. 'Solar–wind hybrid renewable energy system: a review'. *Renewable and Sustainable Energy Reviews*. 2016;**58**(11):23–33. Elsevier BV.

[8] Nema R.K., Nema S., Agnihotri G., *et al*. 'Design, development and simulation of PC-based scheme for characterisation of solar photovoltaic modules'. *International Journal of Power Electronics*. 2010;**2**(3):304.

[9] Kaundinya D.P., Balachandra P., Ravindranath N.H. 'Grid-connected versus stand-alone energy systems for decentralized power—A review of literature'. *Renewable and Sustainable Energy Reviews*. 2009;**13**(8):2041–50.

[10] Zhou W., Lou C., Li Z., Lu L., Yang H. 'Current status of research on optimum sizing of stand-alone hybrid solar–wind power generation systems'. *Applied Energy*. 2010;**87**(2):380–9.

[11] Yang H., Zhou W., Lu L., Fang Z. 'Optimal sizing method for stand-alone hybrid solar–wind system with LPSP technology by using genetic algorithm'. *Solar Energy*. 2008;**82**(4):354–67.

[12] Merlin J.M., Babu S.R. 'Energy management on grid connected hybrid renewable energy sources using fuzzy logic'. *International Journal of Advanced Research in Electrical, Electronics and Instrumentation Engineering*. 2014;**3**(4):2320–3765.

[13] Panwar N.L., Kaushik S.C., Kothari S. 'Role of renewable energy sources in environmental protection: a review'. *Renewable and Sustainable Energy Reviews*. 2011;**15**(3):1513–24.

[14] Adib R. *Renewables 2018 Global Status Report [online]*. 2018. Available from https://cleanenergysolutions.org/sites/default/files/documents/180605_gsr_2018_presentation_webinar_rad.pdf [Accessed 2 Jan 2021].

[15] Kerr M.J., Cuevas A. 'Generalized analysis of the illumination intensity vs. open-circuit voltage of solar cells'. *Solar Energy*. 2004;**76**(1–3):263–7.

[16] Radziemska E., Klugmann E. 'Thermally affected parameters of the current–voltage characteristics of silicon photocell'. *Energy Conversion and Management*. 2002;**43**(14):1889–900.

[17] Nishioka K., Hatayama T., Uraoka Y., Fuyuki T., Hagihara R., Watanabe M. 'Field-test analysis of PV system output characteristics focusing on module temperature'. *Solar Energy Materials and Solar Cells*. 2003;**75**(3):665–71.

[18] VanOverstraeten R., Mertens R. *Physics, technology and use of photovoltaics*. United States: CRC Press; 1986. p. 288.

[19] Ganguly P., Kalam A., Zayegh A. Optimum sizing of standalone PV/wind power generating system with storage. *International Conference on Research in Education and Science (ICRES)*; Ephesus-Kusadasi/Turkey, May 18--21; 2017.

[20] Zhou W., Yang H., Fang Z. 'A novel model for photovoltaic array performance prediction'. *Applied Energy*. 2007;**84**(12):1187–98.

[21] Mondol J.D., Yohanis Y.G., Smyth M., Norton B. 'Long-term validated simulation of a building integrated photovoltaic system'. *Solar Energy*. 2005;**78**(2):163–76.

[22] Jones A.D., Underwood C.P. 'A modelling method for building-integrated photovoltaic power supply'. *Building Services Engineering Research and Technology*. 2002;**23**(3):167–77. SAGE Publications.

[23] IEA. Renewables 2018, IEA, Paris. 2018. Available from https://www.iea.org/reports/renewables-2018.

[24] Sandeep D., Yajvender P.V., Arthur W. 'Techno-economic analysis of the lithium-ion and lead-acid battery in microgrid systems energy Convers manage'. 2018;**177**:122–42.

[25] Hesse H., Schimpe M., Kucevic D., Jossen A. 'Lithium-ion battery storage for the grid—a review of stationary battery storage system design tailored for applications in modern power grids'. *Energies*. 2017;**10**(12):2107.

[26] Kozaki M., Yamazaki T. 'Remaining battery capacity meter and method for computing remaining capacity'. *U.S. Patent*. 25 Nov. 1997;**5**(691):078.

[27] iHOGA Software. 2017. Available from https://ihoga.unizar.es/en/.

[28] Sinha S., Chandel S.S. 'Review of software tools for hybrid renewable energy systems'. *Renewable and Sustainable Energy Reviews*. 2014;**32**(4–6):192–205.

[29] Yang H., Lu L., Zhou W. 'A novel optimization sizing model for hybrid solar-wind power generation system'. *Solar Energy*. 2007;**81**(1):76–84.

[30] Al-Ashwal A.M., Moghram I.S. 'Proportion assessment of combined PV-wind generating systems'. *Renewable Energy*. 1997;**10**(1):43–51.

[31] Luna-Rubio R., Trejo-Perea M., Vargas-Vázquez D., Ríos-Moreno G.J. 'Optimal sizing of renewable hybrids energy systems: a review of methodologies'. *Solar Energy*. 2012;**86**(4):1077–88.

[32] Lazou A.A., Papatsoris A.D. 'The economics of photovoltaic stand-alone residential households: a case study for various European and Mediterranean locations'. *Solar Energy Materials and Solar Cells*. 2000;**62**(4):411–27.

[33] Akella A.K., Sharma M.P., Saini R.P. 'Optimum utilization of renewable energy sources in a remote area'. *Renewable and Sustainable Energy Reviews*. 2007;**11**(5):894–908.

[34] Luiz C.G.V., Silvio C.A.A. 'Economic analysis of a diesel/photovoltaic hybrid system for decentralized power generation in northern Brazil'. *Energy*. 1998;**23**(4):317–23.

[35] Dufo-López R., Bernal-Agustín J.L. 'Design and control strategies of PV-Diesel systems using genetic algorithms'. *Solar Energy*. 2005;**79**(1):33–46.

[36] Bernal-Agustín J.L., Dufo-López R., Rivas-Ascaso D.M. 'Design of isolated hybrid systems minimizing costs and pollutant emissions'. *Renewable Energy*. 2006;**31**(14):2227–44.

[37] Deshmukh M.K., Deshmukh S.S. 'Modeling of hybrid renewable energy systems'. *Renewable and Sustainable Energy Reviews*. 2008;**12**(1):235–49.

[38] Bucciarelli L.L. 'Estimating loss-of-power probabilities of stand-alone photovoltaic solar energy systems'. *Solar Energy*. 1984;**32**(2):205–9.

[39] Bagul A.D., Salameh Z.M., Borowy B. 'Sizing of a stand-alone hybrid wind-photovoltaic system using a three-event probability density approximation'. *Solar Energy*. 1996;**56**(4):323–35.

[40] Bucciarelli L.L. 'The effect of day-to-day correlation in solar radiation on the probability of loss-of-power in a stand-alone photovoltaic energy system'. *Solar Energy*. 1986;**36**(1):11–14.

[41] Gordon J.M. 'Optimal sizing of stand-alone photovoltaic solar power systems'. *Solar Cells*. 1987;**20**(4):295–313.

[42] Connolly D., Lund H., Mathiesen B.V., Leahy M. 'A review of computer tools for analysing the integration of renewable energy into various energy systems'. *Applied Energy*. 2010;**87**(4):1059–82.

[43] Lal D.K., Dash B.B., Akella A.K. 'Optimization of PV/wind/micro-hydro/diesel hybrid power system in Homer for the study area'. *International Journal on Electrical Engineering and Informatics*. 2011;**3**(3):307.

[44] Munuswamy S., Nakamura K., Katta A. 'Comparing the cost of electricity sourced from a fuel cell-based renewable energy system and the national grid to electrify a rural health centre in India: a case study'. *Renewable Energy*. 2011;**36**(11):2978–83.

[45] Fulzele J.B., Dutt S. 'Optimium planning of hybrid renewable energy system using Homer'. *International Journal of Electrical and Computer Engineering*. 2012;**2**(1):68.

[46] Ajao K.R., Oladosu O.A., Popoola O.T. 'Using Homer power optimization software for cost benefit analysis of hybrid-solar power generation relative to utility cost in Nigeria'. *International Journal of Research and Reviews in Applied Sciences*. 2011;**7**(1):14.

[47] Ashok S. 'Optimised model for community-based hybrid energy system'. *Renewable Energy*. 2007;**32**(7):1155–64.

[48] Mellit A., Kalogirou S.A., Hontoria L., Shaari S. 'Artificial intelligence techniques for sizing photovoltaic systems: a review'. *Renewable and Sustainable Energy Reviews*. 2009;**13**(2):406–19.

[49] Kennedy J. Particle swarm optimization. Encyclopedia of machine learning; Springer US; 2011. pp. 760–6.

[50] Dehghan S. Optimal sizing of a hydrogen-based wind/PV plant considering reliability indices. *Electric Power and Energy Conversion Systems, 2009. EPECS'09*. International Conference on. IEEE; 2009.

[51] Lingfeng W., Singh C. 'Multicriteria design of hybrid power generation systems based on a modified particle Swarm optimization algorithm'. *IEEE transactions on energy conversion*. 2009;**24**(1):163–72.

[52] Upadhyay S., Sharma M.P. 'A review on configurations, control and sizing methodologies of hybrid energy systems'. *Renewable and Sustainable Energy Reviews*. 2014;**38**(6):47–63.

[53] Chauhan A., Saini R.P. 'A review on integrated renewable energy system based power generation for stand-alone applications: Configurations, storage options, sizing methodologies and control'. *Renewable and Sustainable Energy Reviews*. 2014;**38**(2):99–120.

[54] Erdinc O., Uzunoglu M. 'Optimum design of hybrid renewable energy systems: overview of different approaches'. *Renewable and Sustainable Energy Reviews*. 2012;**16**(3):1412–25.

[55] Amer M., Namaane A., M'Sirdi N.K. 'Optimization of hybrid renewable energy systems (HRES) using PSO for cost reduction'. *Energy Procedia*. 2013;**42**(3):318–27.

[56] Al Busaidi A.S., Kazem H.A., Al-Badi A.H., Farooq Khan M. 'A review of optimum sizing of hybrid PV–Wind renewable energy systems in Oman'. *Renewable and Sustainable Energy Reviews*. 2016;**53**:185–93.

[57] Collette Y., Siarry P. *Multi-objective Optimization: Principles and Case Studies*. Springer Science & Business Media; 2013.

[58] Elhadidy M.A. 'Performance evaluation of hybrid (wind/solar/diesel) power systems'. *Renewable Energy*. 2002;**26**(3):401–13.

[59] Muselli M., Notton G., Poggi P., Louche A. 'PV-hybrid power systems sizing incorporating battery storage: An analysis via simulation calculations'. *Renewable Energy*. 2000;**20**(1):1–7.

[60] Luna-Rubio R., Trejo-Perea M., Vargas-Vázquez D., Ríos-Moreno G.J. 'Optimal sizing of renewable hybrids energy systems: a review of methodologies'. *Solar Energy*. 2012;**86**(4):1077–88.

[61] Arnette A., Zobel C.W. 'An optimization model for regional renewable energy development'. *Renewable and Sustainable Energy Reviews*. 2012;**16**(7):4606–15.

[62] Dufo-López R., Bernal-Agustín J.L. 'Multi-objective design of PV–wind–diesel–hydrogen–battery systems'. *Renewable Energy*. 2008;**33**(12):2559–72.

[63] National Renewable Energy Laboratory. *HOMER Getting Started Guide Version 2.1*. NREL; 2005.

[64] Khan M.J., Iqbal M.T. 'Pre-feasibility study of stand-alone hybrid energy systems for applications in Newfoundland'. *Renewable Energy*. 2005;**30**(6):835–54.

[65] Zoulias E.I., Lymberopoulos N. 'Techno-economic analysis of the integration of hydrogen energy technologies in renewable energy-based stand-alone power systems'. *Renewable Energy*. 2007;**32**(4):680–96.

[66] Lambert T. 'Micropower System Modeling with Homer'. *Integration of Alternative Sources of Energy*. Felix A. Farret and M. Godoy Simoes; 2006.

[67] Mousa K., AlZu'bi H., Diabat A. 'Design of a hybrid solar-wind power plant using optimization.'. Engineering Systems Management and Its Applications (ICESMA), 2010 Second International Conference on. IEEE; 2010.

[68] Gupta A., Saini R.P., Sharma M.P. 'Hybrid energy system for remote area-an action plan for cost effective power generation'. Industrial and Information Systems, *2008. ICIIS 2008. IEEE Region 10 and the Third International Conference on*. IEEE; 2008.

[69] Schmitt W. 'Modeling and simulation of photovoltaic hybrid energy systems-optimization of sizing and control'. Photovoltaic Specialists Conference, 2002. Conference Record of the Twenty-Ninth IEEE; IEEE; 2002.

[70] Malinchik S., Roberts A., Fierro S. 'Geo-spatial resource analysis and optimization of investment strategies for renewable energy'. Innovative Technologies for an Efficient and Reliable Electricity Supply (CITRES), 2010 IEEE Conference on. IEEE; 2010.

[71] Sukumar S.R. 'Modeling resource, infrastructure, and policy cost layers for optimizing renewable energy investment and deployment'. Innovative Technologies for an Efficient and Reliable Electricity Supply (CITRES), *2010* IEEE Conference on. IEEE; 2010.

[72] Surface meteorology and Solar Energy. Eosweb.larc.nasa.gov [online]. 2017. Available from http://eosweb.larc.nasa.gov/sse/ [Accessed 27 Jul 2017].

[73] Finance.wa.gov.au. Electricity prices [online]. 2017. Available from https://www.finance.wa.gov.au/cms/Public_Utilities_Office/Homes_and_Communities/Electricity/Electricity_prices.aspx [Accessed 3 Jul 2017].

Part V

Offshore Wind

Chapter 14

Status of offshore wind farms in Europe: the case study of Galicia

Laura Castro-Santos[1] *and Almudena Filgueira-Vizoso*[2]

Maritime wind technology has had a great development in European countries for many years. This significant development is due to installing fixed wind platforms, such as monopiles, jackets, or tripiles in the North Sea. The offshore wind power sector can evolve into main spillways: designing, constructing, and R&D of new platforms at shipyards and installing offshore wind farms in specific locations. For instance, Galicia (a Spanish region located in the North-West of the Iberian Peninsula), traditionally linked to the onshore wind sector, has the potential to take advantage of the opportunities of the offshore wind. The region can establish lines of cooperation between companies, universities, administrations, and other stakeholders to become a world leader in offshore wind power.

14.1 Contextualization

The European Union establishes that by 2020 a contribution of renewable energy to electricity production of 40% is required according to Directive 2009/28/EC on the promotion of the use of energy from renewable sources [1]. Therefore, Spain must increase its electricity production using this type of clean energy.

Galicia is a Spanish region located in the North-West of the Iberian Peninsula. Galicia has the best offshore wind resource in Spain. According to the Galician Energy Balance of 2015 published in June 2017 by the Galician Energy Institute (INEGA), electricity generation with renewable energy sources went from 60% in 2013 and 60.9% in 2014 to 52.6% in 2015 [2]. In addition, within the total of renewable energies, terrestrial wind energy and hydroelectric generation are the ones that cover most of the sector with values in 2015 of 8 482 GWh and 6 502 GWh, which represents 54.3% and 41.7%, respectively [2].

[1]Departamento de Enxeñaría Naval e Industrial, Escola Politécnica Superior, Universidade da Coruña, Esteiro, Ferrol, Spain

[2]Departamento de Química, Escola Politécnica Superior, Universidade da Coruña, Esteiro, Ferrol, Spain

In 2015, Spain generated 36.37% of electricity from renewable sources [2]. Thus, to reach the 40% target set for 2020, it is necessary to understand the main types of renewable energy, both inland and offshore. Many of the onshore sources of energy have their analog at sea (offshore), for instance, onshore wind energy versus offshore wind energy, current energy (onshore and offshore), and geothermal energy (offshore and offshore). However, there are other types of renewable energy that operate only inland (e.g., hydraulic) or exclusively at sea (e.g., wave energy and the energy of salinity gradients).

There are many types of systems to harvest wave energy [3], such as dimmers like the Pelamis [4], point absorbers, oscillating wave converters, or oscillating water columns [5]. The rest of the energies that come from the sea, such as the hydrokinetic energy of the currents (both horizontal and vertical axis), thermal energy, and the energy of the salinity gradients, are much less developed and more oriented to very specific areas of the Earth. In addition, there are also hybrid systems that combine different types of marine renewable energy such as wind energy and wave or current energy on the same platform.

It is essential to point out that 82% of the power of the sea corresponds to the marine wind, 8% to the waves, and 10% to the other energy sources. This is one of the main reasons why offshore wind energy is one of the most technologically developed renewable energy fields, so it will be discussed in this chapter.

Regarding onshore wind, the world's top five markets in 2019 for new installations were China, the US, UK, India, and Spain. Those five markets together made up 70% of the global installation last year. In terms of cumulative installations, the top five markets as the end of 2019 remained unchanged. Those markets are China, the USA, Germany, India, and Spain, which together accounted for 72% of the world's total wind power installation [6] (Figure 14.1). Spain, with 5% of the global onshore wind power installation, lost its position to India comparing to the 2012 ranking, when India was not yet among the top onshore wind power producers [7, 8].

On the other hand, if Spain's contribution to offshore wind is analyzed, it no longer appears in the distribution, neither in the case of fixed offshore wind nor in floating offshore wind. Regarding fixed or non-floating offshore wind, the UK accounts for almost half of the new capacity at 1.7 GW. Germany (1.1 GW), Denmark (374 MW), and Belgium (370 MW) are following the UK. Portugal installed 8 MW of floating offshore wind energy [9]. Currently, Europe has a total of 22 GW of offshore wind energy. The UK and Germany account for three-quarters, while Denmark, Belgium, and the Netherlands share almost all the rest [9].

Comparing the annual onshore and offshore wind energy installations in the European Union (EU) (see Figure 14.2), installing offshore wind farms increases year by year, installing 1.6 GW of offshore wind and 10.9 of onshore wind in the EU in 2016 [10].

It is important to note that, in 2016, most of the offshore wind in Europe were installed in the North Sea (1 504 MW), compared to 56 MW in the Irish Sea and 2 MW that were installed in the Atlantic Ocean [10].

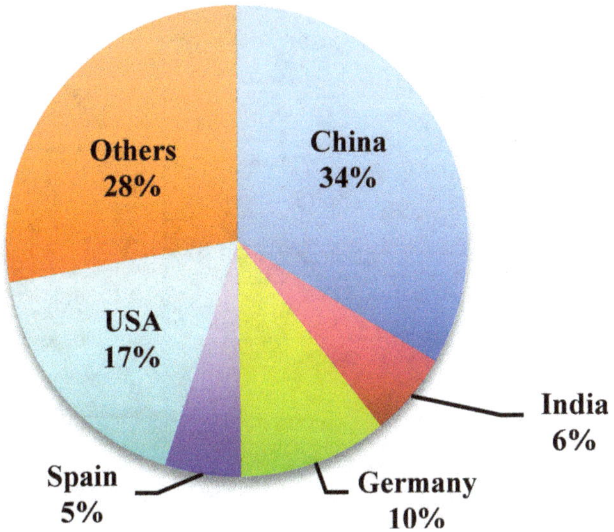

Figure 14.1 *Status of onshore wind power installations in the world in 2015 [7]*

14.2 Offshore wind

Offshore wind energy is currently based on technologies designed for shallow waters up to approximately 50–60 m deep. Of these so-called fixed technologies, 88% are monopiles and 12% jackets [11]. Europe is currently the world leader in installing offshore wind farms, mainly in the North Sea (Figure 14.3).

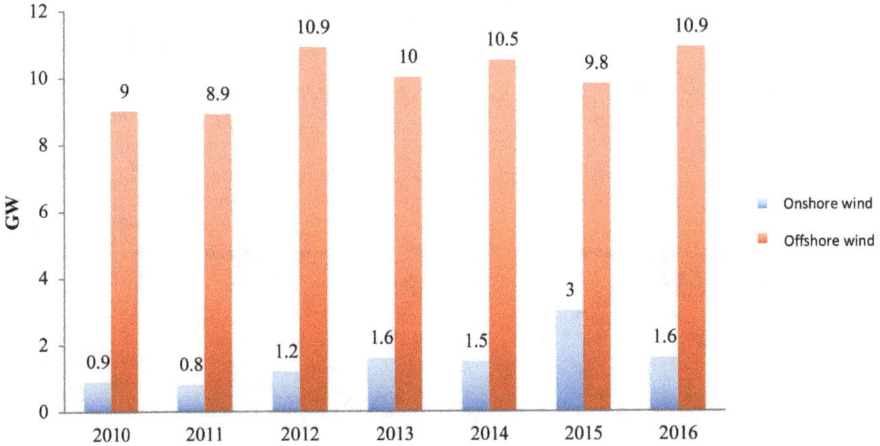

Figure 14.2 *Evolution of annual onshore and offshore wind installations in the EU until 2016 [10]*

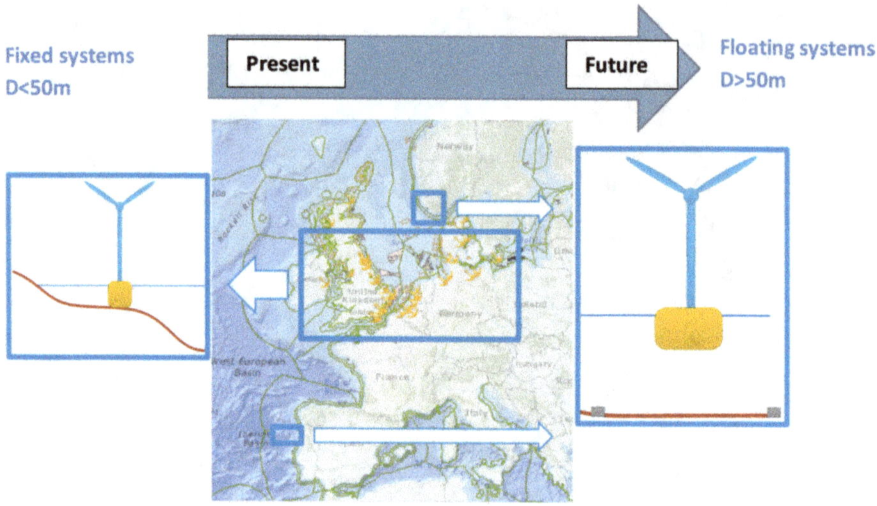

Figure 14.3 *Types of offshore wind systems [12, 13]*

Among the fixed offshore, wind systems are monopiles, jackets, gravity systems, tripiles, or tripods. On the other hand, between floating wind systems [14], the semi-submersible platforms [15], spar systems, and the TLP (Tensioned Leg Platform) [16] represent the majority of the technologies developed. The main difference between fixed and floating systems is that in the latter the platforms are floating and maintain their position using anchors and moorings (Figure 14.4). However, the rest of the components (wind turbine and electrical system) are very similar.

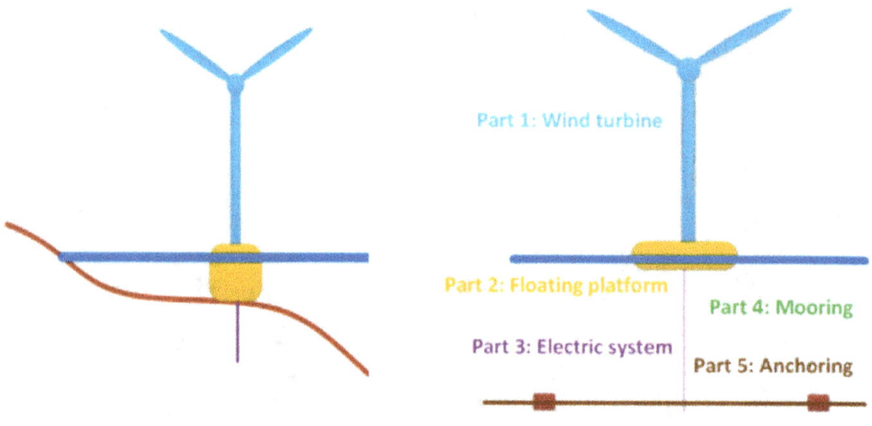

(a) (b)

Figure 14.4 *Fixed (a) and floating (b) offshore wind [12]*

Due to the extensive deep-sea wind potential (over 50–60 m deep), floating wind turbines have great potential to be the core of future offshore wind energy. This is the case of Galicia (Northwest of the Iberian Peninsula), which has very high depths already very close to the coast.

Currently, there are a few floating wind farms. To put this into perspective, with the launch of the new Portuguese floating project, namely, WindFloat Atlantic, Europe's current floating offshore wind energy is only 45 MW, while, according to WINDEUROPE, Europe recorded 3.6 GW new offshore wind energy installation in 2019. The UK accounted for almost half of the recent offshore installation with 1.7 GW. Germany (1.1 GW), Denmark (374 MW), and Belgium (370 MW) followed. Portugal installed 8 MW of floating wind power. The countries of France, the UK, Norway, and Portugal are developing new floating projects. France plans to auction a large-scale floating wind farm in 2021.

In recent years, there have been several offshore prototypes installed. For instance, the Hywind is a spar platform in Norway established by Statoil in 2009 at 200 m depth. This platform carries a 2.3 MW turbine. The WindFloat is a semi-submersible platform installed in 2011 in Aguçadoura (Póvoa de Varzim, Portugal) by Principle Power e Energías de Portugal (EDP) at a depth of 50 m. The platform carries a 2 MW turbine, which was towed from a Portuguese shipyard, where it was built. This platform was decommissioned in the summer of 2016. Other floating platforms have been installed in Japan since the Fukushima disaster.

In 2016, the distance to the coast and the depth of European offshore wind farms increased. Most of these parks have depths between 30 and 45 m and distances to the coast between 40 and 100 km [11]. There is a strong need for extending offshore wind to great depths in the coming years. Technological developments in the field of offshore wind come from the hand of floating wind.

14.3 Present and future of offshore wind in Galicia

The presence of wind energy in Galicia is in the hands of the Navantia Fene shipyard, located in A Coruña. The Temporary Business Union (UTE) Navantia-Windar built 29 fixed jacket platforms for the Wikinger offshore wind farm (Baltic Sea at Deutschland GmbH) in this shipyard. Figure 14.5 shows the construction of the jackets in the Ría de Ferrol shipyard. The Wikinger wind farm comprises 705 MW offshore wind turbines, each with jacket-type structures and an electrical substation, built by Navantia Puerto Real (Cádiz), which connects the park to the German electricity grid. Bladt Industries in Denmark built the other 41 jackets.

The same UTE constructed five spar-type platforms in the same shipyard to install in the first floating offshore wind farm in Scotland in 2017, which was awarded to a Statoil ASA subsidiary company called Hywind (Scotland) Ltd. This project represents the first commercial series of floating offshore wind in the world. The spar platform consists of a steel cylinder made up of several internally reinforced ring sections. In its widest part, it reaches a diameter of 14.4 m. It has two watertight compartments inside; the bottom one has bulk ballast to achieve the stability of the

Figure 14.5 Jackets being built in Navantia Fene [Source: own]

platform [17]. In Figure 14.6, a Hywind platform can be seen at the Navantia Fene shipyard.

Therefore, the future of offshore wind in Galicia comes from the hand of shipbuilding, which must diversify the sectors in which it traditionally worked (shipbuilding, military sector, repair, etc.) in order to adapt to an industry emerging from offshore wind, which promises a great development in the coming years. However, it is important to highlight that the future of offshore wind in Galicia may not only focus on exporting platforms but also on the possibility of installing parks on our coasts. In this sense, two important aspects must be analyzed: the economic and legislative-environmental aspects. From the economic point of view, the most influential parameters in the internal rate of return (IRR) of an investment in a floating offshore wind farm are (Figure 14.7): the wind scale parameter (cw), that is, the wind generated at a certain location; the electric rate (PkWh); the number of wind turbines (NWT), that is, the size of the park; and the cost of capital (r).

Figure 14.6 Spar built in Navantia Fene

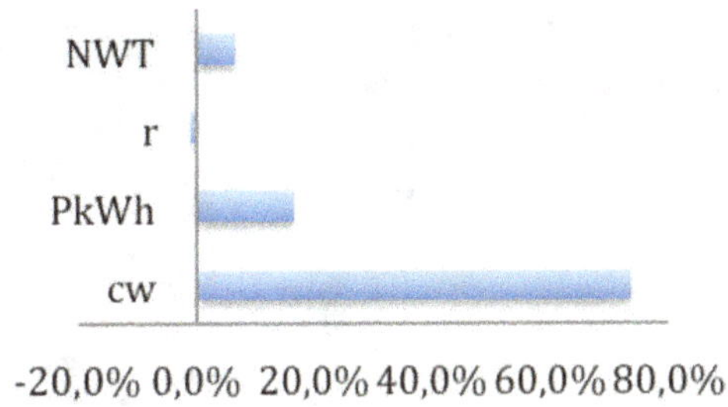

	cw	PkWh	r	NWT
IRR influence	73,8%	16,9%	-0,8%	6,9%

Figure 14.7 *Most influential parameters for the economic viability of a floating offshore wind farm [18]*

As can be seen, one of the most influential parameters is the electricity rate, which in the case of Spain, varies continuously. This variation causes significant changes in the IRR of an investment in a floating offshore wind farm. This means that the economic viability of the park is compromised by the continuous variations in the value of the fee, as can be seen in Figure 14.8. This figure represents the IRR maps created for the coast of Galicia based on the different electricity rates. It is observed that the lower the rate, the feasibility results are worse. This rate instability makes investors prefer to invest in other countries where the value of the electricity rate is assured and, therefore, the economic viability of the installation.

On the other hand, there is the legislative-environmental factor, since there may be areas where the economic viability is excellent, but the park cannot be installed legislatively or environmentally. Some examples are:

- "Decree 19/2011, of February 10, which definitively approves the guidelines for spatial planning" [20] says that:

  (...)the possible implementation of marine energy production facilities of any kind will be subject to the following criteria and in line with the agreed parliamentary agreement:
  (a) The distance, arrangement, and characteristics of the generators will be conditioned by the landscape, ecological, cultural, and productive fishing-shellfish values of the stretch of the studied coast.
  (b) The evacuation lines of the energy produced will be carried on the seabed without modifying or affecting the continental shelf or the bottoms of the estuaries.
  (c) The entry into the mainland will take place underground, with no other topographic alterations or earthworks than those essential for the burial of the lines. (...) [20].
- The *"Real Decree 1028/2007, of July 29, which establishes the administrative procedure for processing applications for authorization of electricity*

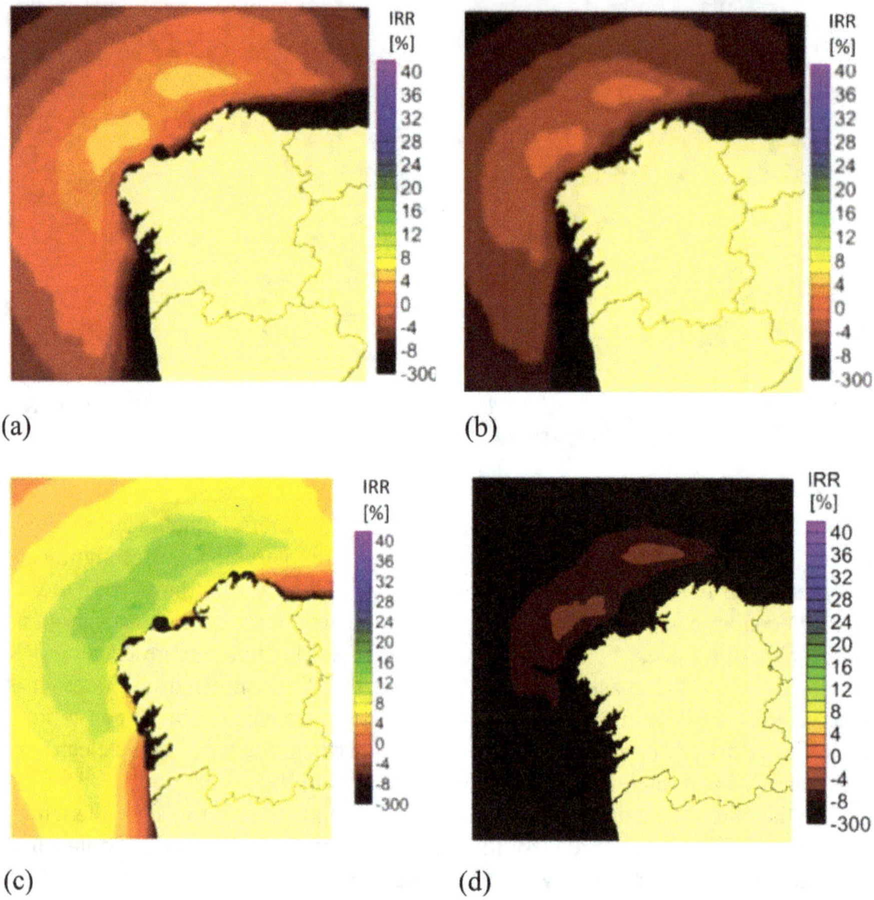

Figure 14.8 Maps of the IRR of an offshore wind farm with the electricity rate of 2012 (a), 2013 (b), 2014 (c), and 2015 (d) [12, 19]

generation facilities in the territorial sea" [21] its purpose is "to regulate the procedures and determine the conditions and criteria for obtaining the necessary administrative authorizations and concessions for the construction and expansion of the electricity generation facilities that are physically located in the territorial sea" [21].

- The "*Strategic Environmental Study of the Spanish coast for the installation of offshore wind farms"* [22] its purpose is "to determine the areas of the public or maritime terrestrial domain that, for environmental purposes, meet the favorable conditions for the installation of offshore wind farms" [22]. This allowed defining the exclusion zones and suitable zones with environmental conditions.

In this sense, using tools based on geographic information systems (SIX) that allow analyzing different restrictions in a specific location is interesting. For

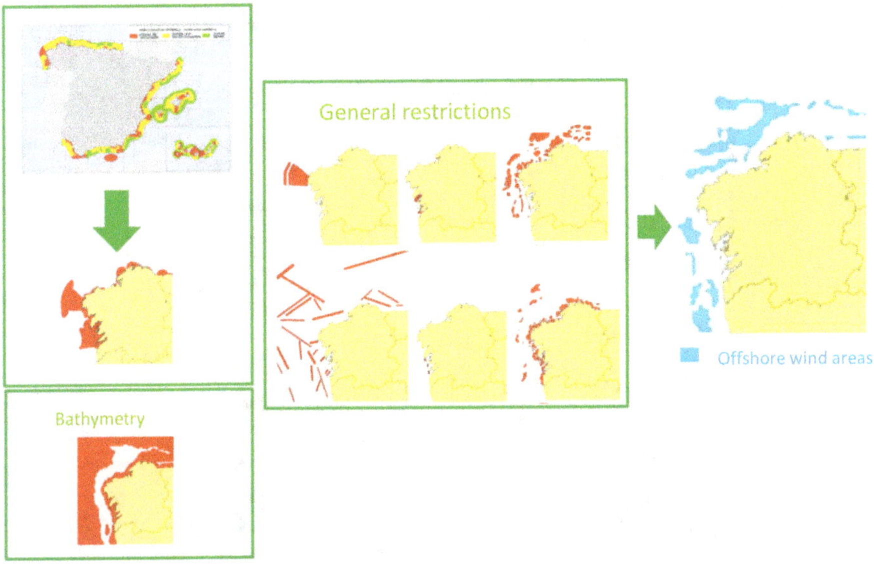

Figure 14.9 *Legislative-environmental restrictions in Galicia for the installation of floating offshore wind farms [23]*

example, it can be seen how restrictions were introduced (in red on the maps) for navigation areas, legislative restrictions, schools of fish, fishing grounds, seismic failures, environmentally protected areas (e.g., the Atlantic Islands National Park), rocky areas, etc. for the analysis of the installation of offshore wind farms in Galicia. The result is blue areas of the map, where the parks could be installed (Figure 14.9).

After the joint consideration of the economic factors and the legislative-environmental factors, a map is obtained that combines both, so it allows knowing which are the most economically viable areas taking into account the areas where this floating technology cannot be installed. Figure 14.10 shows that Galicia can be divided into three areas (G1, G2, and G3), of which G1 has the highest viability.

14.4 Conclusions

Offshore wind technology has been developed in many European countries for many years, including fixed platforms, such as monopiles, jackets, and tripiles, among others, in the North Sea.

Galicia, traditionally linked to the onshore wind sector, should take advantage of its offshore wind potential. This can happen on two different fronts: (1) design, construction, and R&D in new platforms in the Galician shipyards; and (2) installation of offshore wind farms in its territory, by taking into account all the stakeholders involved (sailors, shellfish workers, and administration, universities) and respecting the environment.

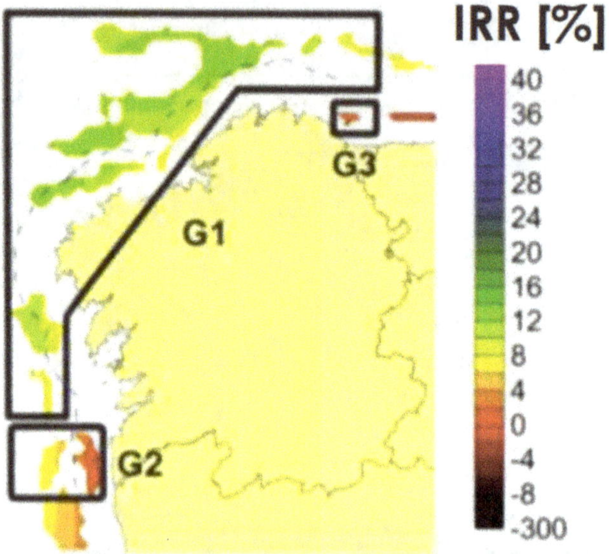

Figure 14.10 *IRR of offshore wind farms in Galicia considering a series of legislative-environmental restrictions and a rate of 190 €/MWh where there would be economically viable areas in Galicia*

It is important to note that the continental shelf in Galicia is quite deep, so the platforms to be used would be mostly floating. This technology is still being developed commercially, so there is a wide range of research and development possibilities in this sector. Thus, universities, companies, administrations, and other interest groups must go hand in hand in this new technological challenge offered by the sea.

References

[1] Diario Oficial de la Unión Europea. *DIRECTIVA 2009/28/CE DEL PARLAMENTO EUROPEO Y DEL CONSEJO de 23 de abril de 2009 relativa al fomento del uso de energía procedente de fuentes renovables y por la que se modifican y se derogan las Directivas 2001/77/CE y 2003/30/CE*; 2009.

[2] Instituto Enerxético de Galicia (INEGA). *Balance enerxético de Galicia 2015*; 2017.

[3] Uihlein A., Magagna D. 'Wave and tidal current energy – a review of the current state of research beyond technology'. *Renewable and Sustainable Energy Reviews*. 2016;**58**(152):1070–81.

[4] Dalton G.J., Alcorn R., Lewis T. 'Case study feasibility analysis of the pelamis wave energy convertor in Ireland, Portugal and North America'. *Renewable Energy*. 2010;**35**(2):443–55.

[5] Toyota K., Nagata S., Imai Y., Setoguchi T. *Research for evaluating performance of OWC-type Wave Energy Converter "Backward Bent Duct Buoy"*. EWTEC; 2009. pp. 901–13.

[6] Global Wind Energy Council (GWEC). *Global Wind Report 2019*. Belgium; 2020.

[7] GWEC. *Global Wind Report 2015. Annual Market Update*; 2016.

[8] Global Wind Energy Council (GWEC). *Global Wind Statistics 2010*; 2012.

[9] Wind Europe. Wind energy in Europe in 2018 n.d. 2018. Available from www.ewea.org/ [Accessed 31 Jul 2019].

[10] Wind Europe. The European offshore wind industry – key trends and statistics. 2016. Available from https://windeurope.org/about-wind/statistics/offshore/european-offshore-wind-industry-key-trends-and-statistics-2016/ [Accessed 19 May 2021].

[11] Wind Europe. *Wind in Power. 2016 European Statistics*; 2017.

[12] Castro-Santos L. *Methodology related to the development of the economic evaluation of floating offshore wind farms in terms of the analysis of the cost of their life-cycle phases*. Universidade da Coruña; 2013.

[13] Castro-Santos L., Prado G., Diaz-Casas V. 'Methodology to study the life cycle cost of floating offshore wind farms'. *10th Deep Sea Wind R&D Conference*. Trondheim, Norway; 2013. pp. 179–86.

[14] Jonkman J.M., Matha D. 'Dynamics of offshore floating wind turbines-analysis of three concepts'. *Wind Energy*. 2011;**14**(4):557–69.

[15] ECN, MARIN, Lagerwey the Windmaster, TNO, TUD, MSC. *Study to feasibility of boundary conditions for floating offshore wind turbines*. Delft Netherlands; 2002.

[16] Sclavounos P.D., Lee S., DiPietro J. 'Floating offshore wind turbines: tension leg platform and taught leg buoy concepts supporting 3–5 MW wind turbines'. *European Wind Energy Conference, Warsaw, Poland*. 2010:1–7.

[17] Navantia. *Páxina web Navantia*; 2017.

[18] Castro-Santos L., Diaz-Casas V. 'Sensitivity analysis of floating offshore wind farms'. *Energy Conversion and Management*. 2015;**101**(1):271–7.

[19] Castro-Santos L., Diaz-Casas V. 'Perspectivas de futuro de los parques eólicos marinos flotantes'. *Dínamo Técnica*. 2014;**14**:11–13.

[20] Boletín Oficial del Estado (B.O.E.). *Decreto 19/2011, de 10 de febrero, por el que se aprueban definitivamente las directrices de ordenación del territorio*; 2011.

[21] Boletín Oficial del Estado (B.O.E.). *Real Decreto 1028/2007, de 20 de julio, por el que se establece el procedimiento administrativo para la tramitación de las solicitudes de autorización de instalaciones de generación eléctrica en el mar territorial*; 2007.

[22] Secretaría General de Energía y de la Secretaría General del Mar. *Estudio estratégico ambiental del litoral español para la instalación de parques eólicos marinos*; 2009.

[23] Castro-Santos L., Diaz-Casas V. 'Economic influence of location in floating offshore wind farms'. *Ocean Engineering*. 2015;**107**(6):13–22.

Index

www.ingramcontent.com/pod-product-compliance
Lightning Source LLC
LaVergne TN
LVHW011837200926
848484LV00002B/368

* 9 7 8 1 8 3 9 5 3 0 9 9 9 *